Statistics in Engineering

CHAPMAN & HALL STATISTICS TEXTBOOK SERIES

Editors:

Dr Chris Chatfield
Reader in Statistics
School of Mathematical Sciences
University of Bath, UK

Professor Jim V. Zidek
Department of Statistics
University of British Columbia, Canada

OTHER TITLES IN THE SERIES INCLUDE

Practical Statistics for Medical Research
D. G. Altman

Interpreting Data
A. J. B. Anderson

Statistical Methods for SPC and TQM
D. Bissell

Statistics in Research and Development
Second edition
R. Caulcutt

The Analysis of Time Series
Fourth edition
C. Chatfield

Problem Solving – A Statistician's Guide
C. Chatfield

Statistics for Technology
Third edition
C. Chatfield

Introduction to Multivariate Analysis
C. Chatfield and A. J. Collins

Modelling Binary Data
D. Collett

Modelling Survival Data in Medical Research
D. Collett

Applied Statistics
D. R. Cox and E. J. Snell

Statistical Analysis of Reliability Data
M. J. Crowder, A. C. Kimber, T. J. Sweeting and R. L. Smith

An Introduction to Generalized Linear Models
A.J. Dobson

Multivariate Analysis of Variance and Repeated Measures
D. J. Hand and C. C. Taylor

The Theory of Linear Models
B. Jorgensen

Statistical Theory
Fourth edition
B. Lindgren

Essential Statistics
Second edition
D. G. Rees

Decision Analysis: A Bayesian Approach
J. Q. Smith

Applied Nonparametric Statistical Methods
Second edition
P. Sprent

Elementary Applications of Probability Theory
H. C. Tuckwell

Statistical Process Control: Theory and practice
Third edition
G. B. Wetherill and D. W. Brown

Full information on the complete range of Chapman & Hall statistics books is available from the publishers.

Statistics in Engineering

A practical approach

Andrew V. Metcalfe

Department of Engineering Mathematics,
University of Newcastle,
Newcastle upon Tyne, UK

CHAPMAN & HALL

London · Glasgow · Weinheim · New York · Tokyo · Melbourne · Madras

Published by Chapman & Hall, 2–6 Boundary Row, London SE1 8HN, UK

Chapman & Hall, 2–6 Boundary Row, London SE1 8HN, UK

Blackie Academic & Professional, Wester Cleddens Road, Bishopbriggs, Glasgow G64 2NZ, UK

Chapman & Hall GmbH, Pappelallee 3, 69469 Weinheim, Germany

Chapman & Hall USA, One Penn Plaza, 41st Floor, New York NY 10119, USA

Chapman & Hall Japan, ITP-Japan, Kyowa Building, 3F, 2-2-1 Hirakawacho, Chiyoda-ku, Tokyo 102, Japan

Chapman & Hall Australia, Thomas Nelson Australia, 102 Dodds Street, South Melbourne, Victoria 3205, Australia

Chapman & Hall India, R. Seshadri, 32 Second Main Road, CIT East, Madras 600 035, India

First edition 1994

© 1994 Andrew V. Metcalfe

Typeset in 10/12 Times by Thomson Press (India) Ltd, New Delhi
Printed in Great Britain at Clays Ltd, Bungay, Suffolk

ISBN 0 412 49220 2

A catalogue record for this book is available from the British Library

Library of Congress Catalog Card Number: 94-70979

⊚ Printed on permanent acid-free text paper, manufactured in accordance with ANSI/NISO Z39.48-1992 and ANSI/NISO Z39.48-1984 (Permanence of Paper).

Contents

Preface

Two common features of many modern engineering projects are a need for realistic modelling of random phenomena and a requirement that this be achieved within tight budgets and deadlines. This book is written for engineering students and professional engineers, and also for students of mathematics, statistics, and operations research who may be members of teams working on such projects.

It has been my intention to motivate readers to use the methods I describe by demonstrating they are an essential part of practical problem-solving and decision-making. Examples range from oil rigs to designs of prosthetic heart valves. There is an emphasis on the valuable technique of multiple regression, which involves describing the variation of one 'dependent' variable in terms of several 'explanatory' variables. I have used this instead of algebraically equivalent but more formal models whenever possible. The advantages are simplicity and the availability of multiple regression routines in popular spreadsheet software. The more formal models have advantages for complex experimental designs but these are well covered in more advanced texts. Many industrial experiments are based on relatively simple 'factorial designs' augmented with some extra runs when necessary. These are naturally described in terms of multiple regression, and the experimental design package DEX, written for the statistical novice, adopts this approach. I am grateful for permission to reference DEX and quote examples from the users' guide. There are many rather more specialist statistical packages, and I have chosen to refer to MINITAB at various points in the text (MINITAB is a registered trademark of Minitab Inc.). It is particularly easy to use, widely available, and I find its multiple regression routine one of the best I have used for standard applications. I am grateful to Minitab Inc. for permission to reference MINITAB throughout the book, and for the assistance provided under Minitab's Author Assistance Program.

I have assumed that the reader is familiar with elementary algebra, and has had some acquaintance with simple calculus. Some of the appendices and exercises are slightly more mathematical than the main text, but these can be passed over by those without the time or inclination to follow them through. A list of notation is given in Appendix B1. A suggested short course, for a reader in a hurry, is given at the end of the problem-solving guide in Appendix B4.

I am grateful to many people for their contributions, especially my industrial contacts and the Master of Science graduates from our Department who have so clearly shown the value of statistical methods during their placements with companies. I have tried to acknowledge their work explicitly, but requests for commercial confidentiality have sometimes prevented this. I wish to thank Alan Jeffrey for encouraging me to submit the proposal for this book; Nicki Dennis of Chapman & Hall for her enthusiasm about it; and Tony Greenfield for many helpful comments during the first draft, several practical industrial examples, a case study, and his overall support for the project. Richard Leigh made further useful suggestions during his thorough final editing. I have greatly appreciated the editorial and production services provided by Lynne Maddock and other staff at Chapman & Hall. However, I am responsible for any errors and obscurities. It is also a pleasure to thank Mandy Knox for cheerfully typing the manuscript with great efficiency.

Andrew Metcalfe
Department of Engineering Mathematics
University of Newcastle upon Tyne

1
Why understand 'statistics'?

In 1940, four months after the Tacoma Narrows Bridge was·opened, a mild gale set up resonant vibrations along its half-mile centre span. It collapsed within a few hours. It was a slender bridge and, although some benign longitudinal oscillations had been allowed for in the design, the torsional vibrations that destroyed it were quite unexpected. The statistical techniques which give an average frequency composition of non-deterministic disturbances, such as wind gusts, have been an essential part of the investigations into what went wrong, now believed to be a von Karman vortex effect, and how to avoid similar disasters.

All physical processes are subject to some random disturbances, and statistical methods help engineers to allow for these. Many modern control systems, their applications include flexible robot arms and space telescopes, are designed to take account of sensor measurement errors and random disturbances. Manufacturers need to design products which can easily be made within their specification, and which will be reliable in a changing operating environment, if they are to survive in today's competitive markets. Electronic engineers design receivers which can recover a signal from background noise, and so on.

Statistics is an essential part of many engineering projects, and several cases will be looked at in detail throughout the book. The first includes aspects of the design of drilling rigs for oil and gas fields in the North Sea. In shallow waters the structures are relatively stiff, and simple calculations based on records of wave heights and directions may be adequate. In deep water the average frequency composition of wave forces becomes important and more sophisticated methods are used. The statistical contributions discussed are the estimation of the distribution of lifetimes of joints, the design of an experiment to investigate the effect of different welding and anti-corrosion techniques on these lifetimes, and a method for estimating the response of rigs in heavy seas.

A second case study, which I refer to in several places, is a typical asset management plan (AMP) for water supply and sewerage systems. Such plans are prepared by water companies in the UK to justify their pricing policies to the government regulatory body. The objectives of an AMP are to assess the performance and condition of the systems, and to estimate the investment required over a 20-year period to satisfy defined standards and levels of service. From a statistical point of view it is necessary to identify the population, draw random samples when appropriate, identify the work to be done, cost this work, estimate the total cost and quantify the uncertainty associated with it. This will include uncertainty over individual schemes, uncertainty over average unit costs and sampling error. The AMP has to be cross-certified by independent consultants. In particular, the methods for obtaining the estimate of uncertainty in the total cost have to be justified. If this uncertainty is excessive, companies will be penalized.

Another case is built around a typical chemical engineering process. A range of feasible operating conditions can usually be established from physical principles. For a process which involves oxidation in a kiln the variables under operator control can include the burner setting, composition of the meal, feed rates of the meal, fan speeds and rotation speed. The best combination of variables within their feasible ranges will have to be established by experiment. Changing one variable at a time can be misleading because the effect of any one variable is likely to depend on the values of the others. For example, a fast feed rate may be sustainable with a high burner setting, fast rotation and high fan speeds, but result in partial oxidation, and hence waste product, at other settings. Methods for analysing any available records to obtain preliminary information before designing an experiment are described and their limitations discussed. The reason for using a systematically designed experiment, or sequence of experiments, is to obtain reliable information about the effects of variables for a minimum amount of effort and expense.

A fourth case study is an application of a rather more 'subjective' approach to the analysis of data. One of the projects I am now involved with is the assessment of the effect of replacing priority controlled junctions (without traffic lights) by mini-roundabouts. A randomized trial would involve selecting pairs of junctions, randomly choosing one from each pair to be changed to a mini-roundabout, and comparing the results from the 'treated' and 'untreated' junctions. However, experienced traffic engineers may be sufficiently confident that a road safety measure will reduce accidents to implement it straight away at suitable junctions. In such cases evaluations rely on before and after studies. Control areas may be constructed, but there is controversy over how best to do this, and no strong empirical evidence exists in favour of any particular strategy. Although there is a legal requirement to record all accidents involving injury, detailed analyses of the data relating to particular road safety measures are limited. However, the Country Surveyors' Society Standing Advisory Group on Accident Reduction (SAGAR) does aim to collate all studies of selected safety measures. We combine results from all studies of mini-roundabouts which have been noted by SAGAR.

The 'cases' emphasize the importance of statistical methods to engineers involved in planning, and make use of several techniques. Many much shorter engineering examples are also given, to illustrate particular statistical procedures and to convey an impression of the variety of applications. Several examples are taken from a bicycle factory. Is a 'quick look' an effective visual test for run-out (buckle) in wheels? Is there a difference between crown races from two suppliers? Other examples include: whether or not to accept a large delivery of concrete paving blocks; the monitoring of a machine that produces domestic gas burners from steel plates; the smoothness of the paint finish on ship hulls; the electroplating of pistons for hydraulic machinery; the effectiveness of fuel additives; flood prediction; and a rather strange application involving piles of rocks!

2

Probability in engineering decisions

2.1 Introduction

A port authority intends building a container park to improve its cargo handling facilities. The park will be constructed from approximately 1 million small interlocking concrete blocks called pavers (commonly used for garage forecourts). Both the crushing strength of the pavers and their cement content are important. The strength is, to some extent, associated with the cement content, but the latter is important in its own right as it affects the resistance of the blocks to frost damage. The authority would like some reassurance that blocks delivered by the manufacturer satisfy the specifications for strength and cement content before laying them. Testing all the blocks is out of the question, because the tests are destructive and cement content analyses are long and expensive, so the authority will only test a sample. However, a suitable procedure for selecting this sample must be devised. It would not be prudent for the authority to base decisions on some 'typical' blocks provided by the manufacturer's representative! Neither would the manufacturer be happy with a sample of blocks chosen by the authority. The authority's engineer might choose blocks which appear to have more aggregate in their mix, or blocks which are paler in colour.

Both parties should be willing to agree to the principle of selecting a random sample of blocks. Conceptually, this could be done by allocating them an identifying number, putting chips marked with corresponding numbers into a drum, mixing them thoroughly, drawing out a number of chips equal to the chosen sample size and locating the corresponding blocks. In practice the drum and chips would be replaced by a table of random numbers, and an example of this procedure is given later. When the selected blocks have been found their crushing strengths and cement will be measured.

Now suppose the authority has received the first delivery of 100 000 pavers. A sample of 24 was chosen, using random numbers, and a laboratory has returned the results in Table 2.1. We can look for any relationship between strengths and cement contents by plotting the 24 pairs (Fig. 2.1). We can summarize the strengths by giving their average (add them up and divide by 24 to obtain $54\,\mathrm{N\,mm^{-2}}$), and their range (the difference between the largest

Table 2.1 Compressive   strengths
and cement contents of 24 pavers

Compressive strength $(N\,mm^{-2})$	Percentage dry weight (%)
38.4	16.6
75.8	16.6
40.0	15.3
38.0	17.1
60.3	20.7
70.0	20.8
63.6	17.2
59.8	16.6
71.0	20.9
70.6	20.0
48.8	15.3
29.3	16.6
40.5	15.3
48.8	15.3
35.8	16.8
56.8	17.3
37.2	16.0
48.8	16.3
52.0	15.5
50.1	15.7
74.6	17.1
73.6	18.1
43.8	18.8
60.4	16.6

and the smallest, which equals $46\,N\,mm^{-2}$). The sample average is used as an estimate of the average strengths for the entire population of 100 000 blocks. The range gives us an indication of the variation of strengths. The randomization makes the sample 'fair', inasmuch as we have no reason to suspect that the sample average will overestimate the mean rather than underestimate it, or that these data will be more or less spread out than in other samples of the same size. A summary of the cement contents is that their average and range are 17.2% and 5.6%, respectively (assuming a dry concrete density of $2250\,kg\,m^{-3}$). In addition to making the selection of the sample 'fair', randomization is the basis for producing measures of precision of the estimates of population quantities. These measures involve the notion of probability, which is the subject of this chapter.

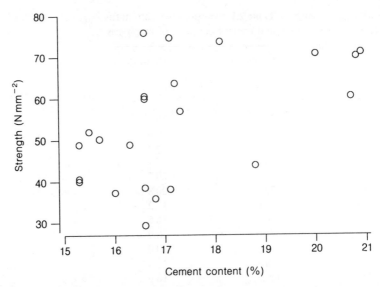

Fig. 2.1 Compressive strength $(N\,mm^{-2})$ against cement content (%) for 24 pavers.

Unfortunately, being fair may not be a compelling reason for the manufacturer to accept the results from a sample, if they indicate that a delivery fails to meet the specification. Courts in the UK are not yet used to considering statistical evidence and the glib – but, as you will read later, unfounded – defence that only a small proportion of the blocks has been tested could sound plausible. It is far preferable to agree sampling procedures in advance (British Standards often include them) and include them in contracts. Specifications for quantities that are measured on some continuous scale, such as strength and cement content, should take account of variability and specify both upper and lower limits, or, if more appropriate, some lower (or upper) value that a stated proportion should exceed.

Acceptance sampling is valuable when single or occasional purchases are likely to be made from a supplier, but is not always appropriate in manufacturing industry if a company consistently uses a few reliable suppliers. A certain car manufacturer has a long-term contract with a tyre manufacturer to supply the tyres for its vehicles. The tyre manufacturer checks all tyres for serious faults, and monitors samples during production for minor discrepancies. A buyer from the car manufacturing company has visited the tyre manufacturer, seen all the systems in operation, and decided they are sufficiently reliable for acceptance sampling to be an unnecessary expense. Another example is provided by a local manufacturer of air filters. Many parts, such as plastic endcaps, are regularly bought from other companies. The filter manufacturer reduces the levels of acceptance sampling as good working relationships are built up with suppliers, eventually inspecting only one per batch to ensure that the exact items ordered have been delivered. The filter manufacturer has established record systems which enable the origin of source components to be traced, should problems arise later in the manufacturing process. Aiming for sound manufacturing processes, rather than removing defective items by inspection, reduces scrap levels and saves time and costs. Statistical methods play a major part in achieving this, and have many other engineering applications.

2.2 Defining probability

Most people have some notion about probability from games of chance, gambling and everyday language. Statements such as 'The oil company will probably strike oil in this sector of the North Sea' and 'The chance of both pumps for the cooling water failing at the same time is low' need no further explanation. Some numerical measure of probability is needed to make more precise statements, and a natural scale goes from 0, representing impossibility, to 1, representing certainty. Most events in life will have fractional probabilities, which are often expressed as percentages, associated with them. Weather forecasts in the UK, now sometimes include the probability of rain, for example, 'A 20% chance of rain in the Glasgow area' or 'An even chance of rain at Wimbledon', and sometimes more detailed statements such as 'The probability of thunderstorms in the North East region is between 40% and 60%'. Why do you think they give a range instead of a single figure of, say, 50%? Similar probabilities have been given in the USA for some time.

There are three ways of arriving at probabilities for events, all of which have their applications, and a few definitions are now needed. The first is that of an experiment, which is any operation whose outcome cannot be predicted with certainty. A set of possible outcomes of the experiment, defined so that exactly one must occur, is known as a sample space. It is not necessarily unique. With some experiments we can appeal to symmetry and claim that all the outcomes are equally likely to occur. An event is any collection of outcomes from the sample space, and it is said to occur if one of its elements is the result of the experiment. The probability of an event A occurring is defined as:

$$\Pr(A) = \frac{\text{Number of outcomes which result in } A}{\text{Total number of equally likely possible outcomes}}$$

You can see that the probability of A occurring lies between 0 and 1: if it is impossible it will have probability 0; and if it is certain it will have probability 1.

Example 2.1

One way of constructing a random number table is by spinning the ten-sided regular polygon shown in Fig. 2.2 and noting the digits corresponding to the

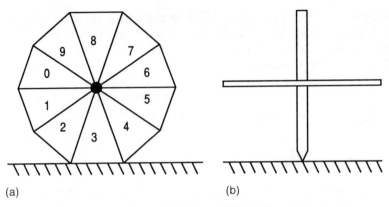

Fig. 2.2 Decagon spinner: (a) front elevation at rest; (b) spinning.

edges that come to rest on the table. If the device really is symmetrical, and the torques it is given are slightly different, the ten digits from 0 to 9 can be considered equally likely to occur at any spin. The sample space is

$$\{0, 1, 2, 3, 4, 5, 6, 7, 8, 9\}$$

The event 'spinning a seven' is

$$\{7\}$$

and the probability of this event, which is conveniently written Pr(7), is 1/10, or 0.1.

The event 'spinning an odd number' is

$$\{1, 3, 5, 7, 9\}$$

and

$$\text{Pr(odd number)} = 5/10$$

A probability such as:

Pr(sum of the digits obtained with two spins equals 4)

is a little more complicated. A possible sample space for the sum of the two digits would be:

$$\{0, 1, 2, \ldots, 18\}$$

but the outcomes are not equally likely and the definition of probability given above is not applicable. A sample space of 100 equally likely pairs of digits is shown in Fig. 2.3. The pairs

$$\{(0, 4), (1, 3), (2, 2), (3, 1), (4, 0)\}$$

correspond to the event that the total of the two numbers is 4, and the required

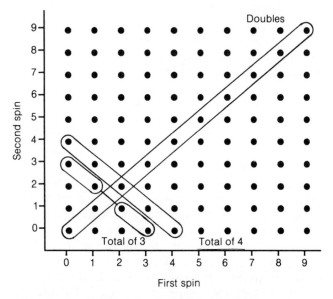

Fig. 2.3 A sample space for two spins of decagon spinner.

probability is 5/100, or 0.05. Similar arguments show that:

$$Pr(doubles) = 10/100 = 0.10$$

and

$$Pr(total\ of\ 3) = 4/100 = 0.04$$

It is reasonable to expect that probabilities such as

Pr((total of 3) or (doubles))

can be expressed in terms of the constituent events. From Fig. 2.3 it can be seen that because there are 14 events which either add to 3 or are doubles

Pr((total of 3) or (doubles)) = 14/100 = 0.14 = Pr(total of 3) + Pr(doubles)

But, in contrast to this,

Pr((total of 4) or (doubles)) = 14/100 = 0.14

which is less than the sum of probabilities of the two constituent events. The explanation is that the sample point (2, 2) is both a double and has a total of 4. If Pr(total of 4) is added to Pr(doubles), the sample point (2, 2) is counted twice. To compensate for this 'double counting' it needs to be removed once. It follows that

Pr((total of 4) or (doubles)) = Pr(total of 4) + Pr(doubles)
$$- Pr((total\ of\ 4)\ and\ (doubles))$$

It should be noted that the point (2, 2) is still counted once, and that in probability calculations 'or' is conventionally taken to include both events. This calculation is an example of the addition rule of probability. It is quite general, and equally applicable to Pr((total of 3) or (doubles)), but in this case Pr((total of 3) and (doubles)) is zero because it is impossible to obtain a double which adds to 3.

While mechanical apparatuses can be used to construct random number tables, electronic devices are much more convenient. ERNIE, the original computer that drew winning premium bond numbers, is a well-known example. The letters actually stand for 'electronic random number indicator equipment'. Modern computers, and scientific calculators, usually provide software-generated pseudo-random numbers. These are calculated from certain types of mathematical recurrence relations with either a chosen starting number, in which case the sequence can be repeated, or with date and clock time used at the start. While sequences of such numbers are in fact determined by the starting number and the rule, quite long sequences cannot be distinguished from genuine random number sequences by empirical tests. An example of a simple pseudo-random number generator is given in Exercise 2.16. A table of random digits is given in Appendix E (Table E.1). The format of this table is designed purely for user convenience, and there is no other significance to it. The table should be worked through in a systematic manner – down columns is probably most convenient – which has been decided before a starting point is selected. The starting point can be selected by tossing coins, throwing dice or some other procedure involving chance. A typical application follows.

Example 2.2

A manager in a water company intends surveying a random sample of 8 from the company's 38 water towers as part of the asset management plan. She identifies each tower by a number from 01 to 38. She decides to work down the columns in Table E.1. There are 10 sets of five digit columns. She will start with the top two-digit pair of the set corresponding to the remainder after division by 10 of the page number in a dictionary opened at an arbitrary point; a remainder of 0 will correspond to the twentieth set. Tower number m will have two random numbers in the table associated with it, m and $40 + m$, so every tower has the same probability, 8/38, of being in the sample.

She opened the dictionary at page 652, so she starts with the left-hand pair of digits at the top of the second set. The sequence of two digit numbers is:

73   tower 33 selected
44   tower 4 selected
03   tower 3 selected
79   ignored as no corresponding tower

03   ignored as tower 3 already in sample
94   ignored as no corresponding tower
85   ignored as no corresponding tower
79   ignored as no corresponding tower

and so on down to

31   tower 31 selected

The sample is

{tower 33, tower 4, tower 3, tower 11, tower 23, tower 29, tower 2, tower 31}

This is an example of **simple random sampling without replacement**, and, throughout this book, you may assume that a 'random sample' has been drawn in this way or by some approximation to it.

The equally likely definition of probability was developed in the context of gambling games by Gerolamo Cardano (born 1501), and such famous mathematicians as Galileo, Pascal and Fermat in the sixteenth century (David, 1955). In practical applications it is often not possible to construct a sample space of equally likely outcomes. A more general approach is to define the probability of an event A as the proportion of occurrences of A in a very long series of experiments. If the number of experiments is N, the ratio

(number of times A occurs)$/N$

is known as the **relative frequency** of occurrence of A, and is expected to become closer to $\Pr(A)$ as N increases. Insurance companies make risk calculations for commonly occurring accidents on the basis of relative frequencies, often pooling their claims records to do so. It does not, however, form a sound basis for unusual risks such as the launching of communications satellites.

Example 2.3

From extensive past records a marketing manager in a car manufacturing company knows that 5% of cars sold will have bodywork faults (B) and 3% will have mechanical faults (M) which need correcting under warranty agreements. These figures include the 1% of cars with both types of fault. The manager wishes to know the probability that a customer will be sold a car which has to be returned under warranty. That is,

Pr((bodywork fault) or (mechanical fault))

The answer is 7%, taking care to avoid counting the 1% of cars with both faults twice. A useful way to see this is to draw a **Venn diagram**, in which relative frequencies are represented by areas (Fig. 2.4). The addition rule of probability applies to relative frequencies in the same way as to probabilities defined in

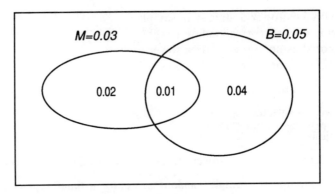

Fig. 2.4 Venn diagram for car faults.

terms of equally likely outcomes. For this example,

$$\Pr(B \text{ or } M) = \Pr(B) + \Pr(M) - \Pr(B \text{ and } M)$$
$$= 0.05 + 0.03 - 0.01$$
$$= 0.07$$

In some situations an experiment cannot be repeated, even conceptually, and a relative frequency approach is not appropriate. In such cases events can have **subjective probabilities** assigned to them. To give an example, water authorities in the United Kingdom have recently been asked to estimate their operating costs for the next 20 years. Some work can be identified as essential, but there are projects that engineers assess as having a 'fifty–fifty' chance of being carried out. Treating these as having a probability of 0.5 of being carried out is preferable to ignoring them or treating them as certain. Other applications of subjective probabilities include decision-making and possible 'scenarios' in computer simulations.

Example 2.4

A small oil company has options on concessions to drill in, and extract any oil over a one-year period from, three locations in the North Sea which will be referred to as A, B and C. The cost of any of the concessions is 20% of revenue. The cost of drilling depends on the location and is £450 million, £300 million and £500 million at A, B and C, respectively. If the drill strikes oil the revenue over the year would be £2000 million. The company can raise capital for only one venture. A geologist thinks it would be 'fair' to bet £3 against £7 that the drill would strike oil at location A. That is, he imagines that if oil were to be found he would gain £7, and if not he would lose £3. This is equivalent to assigning a probability p to the drill striking oil, where

$$7p + (-3)(1 - p) = 0$$

$(1 - p)$ being the probability of not finding oil. The solution to this equation is a value of p equal to 0.3. In a similar manner, the geologist assigns probabilities of 0.3 and 0.4 to the drill striking oil at B and C, respectively. If everyone concerned was as familiar with probabilities as with betting he could have assigned probabilities directly.

A manager in the company now works out the expected profit at each of the locations. The expected profit at location A, in millions, is the 80% of the revenue the company keeps multiplied by the probability that it strikes oil less the cost of drilling:

$$(80\% \text{ of } 2000)(0.3) - 450 = 30$$

In fact the company will either lose £450 million or gain £1150 million if it drills at A, and 'expected' in the statistical sense does not mean most likely. The 'expected profit' can be thought of as the average profit that would result from a very large number of such deals, a proportion of 0.3 resulting in an oil strike. The expected profits at locations B and C are £180 million and £140 million, respectively. The manager recommends that the company should take up the option to drill at location B.

2.3 The addition rule of probability

For two events A and B,

$$\Pr(A \text{ or } B) = \Pr(A) + \Pr(B) - \Pr(A \text{ and } B)$$

The following remarks are important.

 (i) $\Pr(A \text{ or } B)$ includes the probability that both occur.
 (ii) If A and B cannot occur at the same time they are said to be **mutually exclusive** and $\Pr(A \text{ and } B) = 0$. If A and B are mutually exclusive the addition rule simplifies to

$$\Pr(A \text{ or } B) = \Pr(A) + \Pr(B)$$

(iii) It is sometimes easier to calculate the probability that A does not occur than the probability that A does occur. The two probabilities are related by the formula,

$$\Pr(A \text{ occurs}) = 1 - \Pr(A \text{ does not occur}).$$

This follows from the addition rule, by writing 'not A' for the event that A does not occur, and replacing B by 'not A'. This gives

$$\Pr(A \text{ or not } A) = \Pr(A) + \Pr(\text{not } A) - \Pr(A \text{ and not } A).$$

But the event (A and not A) is impossible and the event (A or not A) is certain, so

$$1 = \Pr(A) + \Pr(\text{not } A)$$

and

$$\Pr(A) = 1 - \Pr(\text{not } A)$$

In fact this last result has already been used in Example 2.4, for the probability of not striking oil. Another example follows.

Example 2.5

A store will sell jeans as 'seconds' if they do not conform to standard sizes or if the dye has taken badly. The buyer knows from past experience that 5% of a certain manufacturer's deliveries do not conform to size and 4% are badly dyed. These figures include 0.2% with both defects. What is the probability that a randomly selected pair of jeans is of top quality?

$$\Pr(\text{pair classified as top quality}) = 1 - \Pr(\text{pair classified as seconds})$$
$$= 1 - (0.05 + 0.04 - 0.002) = 0.912$$

Addition rules for three, or more, events can be deduced algebraically from the rule for two events. In practice, it is often easier to use Venn diagrams for three events, and this advice is followed below.

Example 2.6

A new block of offices has recently been built near the centre of London. The building has been entered for design awards under three categories, appearance (A), energy efficiency (E) and working environment (W). An independent architect made a first assessment of the probability that it would win the various awards as 0.4, 0.3 and 0.3, respectively, and added that it would not win more than one award. On reflection, she decided that it might win any two awards with probabilities of 0.1 and all three awards with a probability of 0.01. Why did she modify the original assessments, and what is the modified probability that the building wins at least one award?

If it is assumed that the building cannot win more than one award, the events are mutually exclusive, and

$$\Pr(\text{wins an award}) = \Pr(A) + \Pr(E) + \Pr(W)$$
$$= 0.40 + 0.30 + 0.30 = 1$$

The architect revised her assessments because she did not think the building was certain to win an award. This is a typical hazard of assigning subjective probabilities and it is necessary to check that they make sense when considered together! With the revised assessments, the probability that it wins at least one award can be seen to be 0.71 (Fig. 2.5). When putting the numbers in Venn diagrams you should start with the intersection of all three areas. Also note

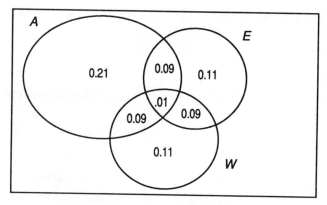

Fig. 2.5 Venn diagram for design awards.

that the probabilities of winning any two awards, *E* and *W* for example, include the possibility of winning all three, in the same way as the probability of winning *A* includes the possibilities of winning two or three awards including *A*.

2.4 Conditional probability

The fact that a particular situation already prevails may influence the occurrence of another event, so assessment of the probability of the occurrence of an event must take this into account. A factory manufactures outboard motors, and it is known from extensive past records, of *N* motors, that 40% needed carburettor adjustments (*C*), 30% needed timing adjustments (*T*) and 21% needed both adjustments. This is conveniently shown on a Venn diagram in Fig. 2.6(a). The probability that a motor selected at random requires a carburettor adjustment is 0.4. Now think about the probability that a randomly selected motor requires

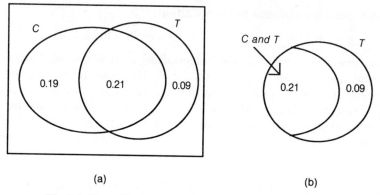

(a)   (b)

Fig. 2.6 Venn diagrams for outboard motor adjustments.

a carburettor adjustment, given that you already know it requires a timing adjustment. This **conditional probability** is written as follows, where the vertical line is read as 'given that' or 'conditional on':

$$\Pr(C|T)$$

This probability is the proportion of those motors needing timing adjustments which also require carburettor adjustments:

$$\Pr(C|T) = \frac{\text{frequency of carb and timing adjustments}}{\text{frequency of timing adjustments}}$$

$$= \frac{\text{frequency of carb and timing adjustments}/N}{\text{frequency of timing adjustments}/N}$$

$$= \frac{\Pr(C \text{ and } T)}{\Pr(T)} = \frac{0.21}{0.30} = 0.70$$

In Fig. 2.6(a) this is represented by the proportion of T which is covered by C, that is 0.21 divided by 0.30. This is emphasized in Fig. 2.6(b).

In general, the probability of B conditional on A is defined by

$$\Pr(B|A) = \Pr(A \text{ and } B)/\Pr(A)$$

and justified by the arguments given above. The definition of conditional probability can be rearranged to give the **multiplicative rule of probability**:

$$\Pr(A \text{ and } B) = \Pr(A)\Pr(B|A)$$

and also, by similar reasoning,

$$= \Pr(B)\Pr(A|B)$$

The events A and B are **independent** if and only if

$$\Pr(B|A) = \Pr(B)$$

This condition is equivalent to the conditions

$$\Pr(A|B) = \Pr(A) \quad \text{and} \quad \Pr(A \text{ and } B) = \Pr(A)\Pr(B)$$

A statement that A and B are 'independent' must not be confused with a statement that they are 'mutually exclusive'. The latter is asserting that the probability of A and B equals 0. The carburettor adjustments are not independent of the timing adjustments. Motors which need timing adjustments are more likely to need carburettor adjustments than the others.

Example 2.7

A case of 12 bottles of wine includes three that contain less than the declared volume of 700 ml. If two are chosen at random, what is the probability that

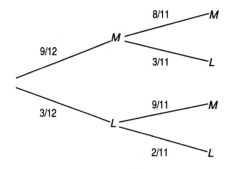

Fig. 2.7 A sample space for sampling from the case of wine bottles.

one contains more, and one less, than 700 ml? The sample space is shown in Fig. 2.7. It is easier to follow the argument if the two bottles are considered to be chosen consecutively rather than simultaneously, but this does not affect the probability of the event. The probability that the first bottle chosen exceeds 700 ml is 9/12. There would then be 11 remaining bottles, 3 containing less that 700 ml, so the probability that the second is underfilled, given that the first was overfilled, is 3/11. Therefore the probability of choosing an overfilled bottle followed by an underfilled bottle is $(9/12) \times (3/11)$ which is approximately 0.205. Similarly, the probability of choosing an underfilled bottle followed by an overfilled bottle is $(3/12) \times (9/11)$. The two sequences are mutually exclusive and together make up the event of one overfilled and one underfilled bottle. The probability of this event is therefore 0.409.

The multiplicative rule can be extended to more than two events. For example:

$$\Pr(A \text{ and } B \text{ and } C) = \Pr(A \text{ and } B) \Pr(C|A \text{ and } B)$$
$$= \Pr(A) \Pr(B|A) \Pr(C|A \text{ and } B)$$

It is sometimes easiest to handle such probabilities using a tree diagram to represent the sample space.

Example 2.8

In a batch of 90 transistor radios 10 are defective. Find the probability of exactly one defective in a random sample of 3.

In Fig. 2.8 *GGD*, for example, represents the outcome that the first chosen is good, the second is good and the third is defective. The probability that the first drawn is good is 80/90. The probability that the second drawn is good, given that the first is good, is 79/89. The probability that the third drawn is defective given that the first two are good is 10/88. The probability of the outcome *GGD* is therefore $(80/90) \times (79/89) \times (10/88)$ which equals 0.0897. The outcomes *GGD*, *GDG*, *DGG* are mutually exclusive and together make up the event of obtaining

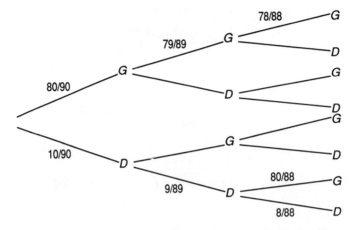

Fig. 2.8 A sample space for sampling from the batch of radios.

exactly one defective in a sample of 3. The probabilities of *GDG* and *DGG* also equal 0.0897 so the probability of exactly one defective is 0.269.

Example 2.9

Establishing communications satellites in space is a hazardous business. There is a 10% chance that the rocket launch will not be successful. If L is the probability of success, we can denote the probability of failure, 'not L', by $\bar{L}$. $\bar{L}$ is the complement of L. Even when the launch goes well, there is a 20% chance that the satellite will not be positioned correctly ($\bar{P}$). If it is not positioned correctly there is a 50% chance of correcting this using small rocket motors on the satellite itself (R). There is also a 30% chance that the solar-powered batteries do not survive a year ($\bar{S}$). A telecommunications engineer works out the probability that a satellite which is due for launch will be handling messages

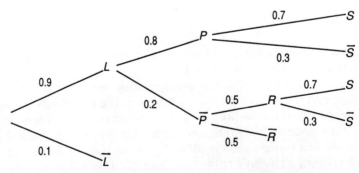

Fig. 2.9 A sample space for the telecommunications satellite.

in one year's time (Fig. 2.9). This is

$$(0.9 \times 0.8 \times 0.7) + (0.9 \times 0.2 \times 0.5 \times 0.7) = 0.567$$

The following calculation is relatively simple because events are assumed to be independent.

Example 2.10

A storm water sewer system has been designed so that the yearly maximum discharge will cause flooding, on average, once every 10 years. If it is assumed that such years occur randomly and independently, the probability of flooding within the next 5 years is given by:

$1 - $ Pr(no flooding in years 1 and 2 and 3 and 4 and 5)

$= 1 - $ Pr(no flood year 1) Pr(no flood year 2)... Pr(no flood year 5)

$= 1 - (1 - 0.1)^5 = 0.41$

A simple reliability problem follows.

Example 2.11

A space probe has a sensing system which can stop the launch if the pressure in the fuel manifold of the engine falls below a critical level. The system consists of four pressure switches which should open if there is a critical pressure drop. They are linked into an electrical connection between terminals A and B as shown in Fig. 2.10. If current cannot flow from A to B the launch is stopped. The switches fail to open when they should with probability q and such failures are independent. The probability that the launch will not be stopped if the pressure drops below the critical level is:

Pr((S1 and S2 fail to open) or (S3 and S4 fail to open))

$= $ Pr(S1 and S2 fail to open) $+ $ Pr(S3 and S4 fail to open)

$\quad - $ Pr((S1 and S2 fail to open) and (S3 and S4 fail to open))

$= q^2 + q^2 - q^2 q^2 = 2q^2 - q^4$

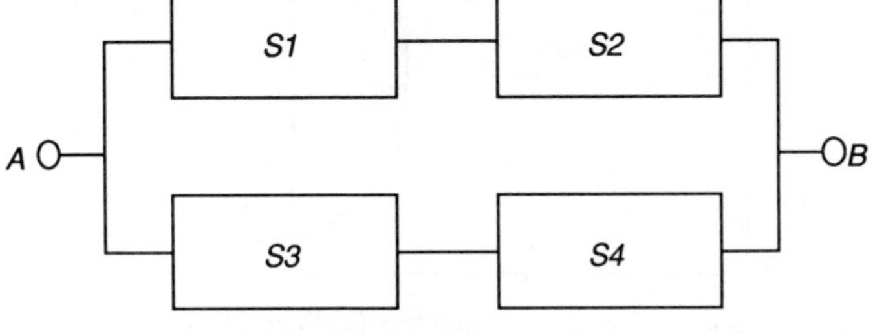

Fig. 2.10 Circuit for fuel manifold safety device.

Pressure switches can also open when they should not, and the analysis of this situation is one of the exercises at the end of the chapter.

The last example in this section is an application of a theorem which was described explicitly in an article by the Reverend Thomas Bayes (1763). **Bayes' theorem** has far-reaching consequences and is the basis for a system of inference which we will use in the case study of the mini-roundabouts.

Example 2.12

A chemist has developed a screening test for a serious disease which is curable if diagnosed in its early stages. The test is not infallible and even if a person has the disease there is a 1% chance that the test will be negative. Also if a person does not have the disease there is a 5% chance that the test will be positive. A proportion p of a certain population has the disease. What is the probability that a person selected at random has the disease if the test is positive?
   The sample space is shown in Fig. 2.11(a), where the probability that a person has the disease is denoted by D, and that that person has not by $\bar{D}$, and where the probability that the test is positive is denoted by T, and that it is negative

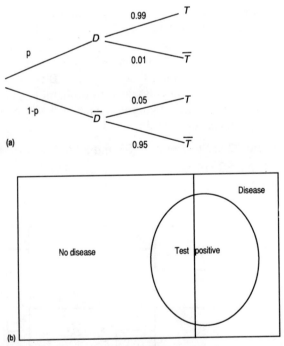

Fig. 2.11 (a) Sample space for the screening test. (b) Pr(test positive) is the sum of Pr(test positive and disease) and Pr(test positive and no disease).

by $\bar{T}$. The required probability is

$$\Pr(D|T) = \Pr(D \text{ and } T)/\Pr(T)$$

The numerator is

$$\Pr(D \text{ and } T) = \Pr(D)\Pr(T|D) = 0.99p$$

The denominator is

$$\Pr(T) = \Pr(D \text{ and } T) + \Pr(\bar{D} \text{ and } T)$$

since the two events D and $\bar{D}$ are mutually exclusive and exhaustive. That is, exactly one of them must be the case (Fig. 2.11(b)). So

$$\Pr(T) = 0.99p + 0.05(1 - p)$$

Hence the required probability is $0.99p/(0.99p + 0.05(1 - p))$. If p happens to be 1 in 1000, the conditional probability of having the disease if the test is positive is less than 2 in 100. Follow-up tests are carried out quickly and the unnecessary worry imposed on those wrongly diagnosed by the screening test is outweighed by the benefits of early treatment for people with the disease.

2.5 Arrangements and choices

A car manufacturer is running a competition in which contestants are asked to rank eight safety features in the same order as a panel of experts. The number of possible arrangements of the eight features, is

$$8 \times 7 \times 6 \times 5 \times 4 \times 3 \times 2 \times 1 = 40\,320$$

because the first feature can be any one of eight, the second feature can be any of the remaining seven, and so on. Figure 2.12 attempts to show this without listing all 40 320 possibilities! In general, the number of arrangements, also called **permutations**, of r objects is written as $r!$, read as 'r factorial', and defined by

$$r! = r \times (r - 1) \times (r - 2) \times \cdots \times 2 \times 1$$

A company has 12 nominees for four seats on the board. Each shareholder is sent a form containing all 12 names and may vote for any four nominees by marking an order of preference using the numbers 1 to 4. The form can be correctly completed in

$$12 \times 11 \times 10 \times 9 = 11\,880$$

different ways; because the 1 can be placed by any of the 12 names, the 2 by any of the remaining 11 names, the 3 by any of the remaining 10 names and the 4 by any of the remaining 9 names. In general, the number of arrangements

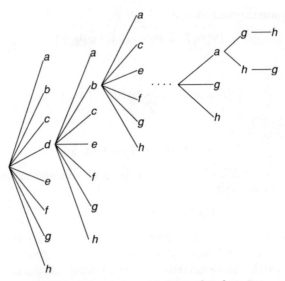

Fig. 2.12 Arrangements of eight safety features.

of r objects from n, which I shall denote by $_nP_r$, is

$$_nP_r = n(n-1)(n-2)\cdots(n-r+1) = n!/(n-r)!$$

where 0! is conventionally taken as 1. This conventional value, which is perhaps surprising at first sight, is a natural consequence of the definition of a more general factorial function known as the **gamma function**. You are asked to investigate this in one of the exercises.

In many applications the number of choices of r objects from n is required. This is related to the number of arrangements because each choice can be arranged in $r!$ different ways. Writing the number of choices, also called **combinations**, of r objects from n as $_nC_r$, the relationship can be written

$$[_nC_r]r! = {_nP_r}$$

and so

$$_nC_r = n!/((n-r)!r!)$$

I have chosen to use the notation $_nC_r$ for the number of choices because it most nearly corresponds to that used on calculator keys.

Example 2.13

A garage is offering a choice of any three from five extras free with a certain car bought at list price. The number of choices is

$$5!/(2!3!) = (5 \times 4)/(2 \times 1) = 10$$

Notice that the 3! was divided into 5! to leave 5×4.

Example 2.14 Simple random sampling without replacement

If a sample is chosen by allocating to members of the population numbers from 1 to N, putting chips with these numbers on into a drum, mixing, and drawing out n, then every possible sample of size n has the same probability of selection. There are $_NC_n$ possible samples. The number of samples that include a particular member of the population is the number of ways of choosing $n-1$ members from the remaining $N-1$, which is $_{N-1}C_{n-1}$. The probability of selecting any particular member is, therefore,

$$_{N-1}C_{n-1}/_NC_n$$

which easily simplifies to n/N. In simple random sampling every member of the population has the same probability of selection, but other random sampling schemes also have this property and it is not a sufficient definition of simple random sampling. Some more complicated sampling schemes will be introduced in the context of asset management plans.

2.6 Decision trees

Decision trees are a useful aid to decision-making which use the concept of expected value introduced in Example 2.4. They also ensure that people think about the steps involved in the production process! The following fictitious case study is loosely based on an example given by Moore (1972).

The engineering manager of Macaw Motors Limited has been told by the sales director that if Macaw produces a prototype DC direct drive torque motor and submits it to a manufacturer of robots, Crow Cybernetics, Macaw may get an order for 1000. The order will be placed on 31 March with Macaw, or some other company, for delivery on 30 September. Crow is prepared to pay £800 per motor.

The engineering manager has the following additional information.

1. The cost of producing a prototype motor, which will have to be made using machined parts, is £48 000. This cost must be borne by Macaw whether or not it is given the order.
2. The cost of producing the order, using machined parts, will be £80 000 for tooling and a further £560 per motor.
3. The senior production engineer thinks it may be possible to make the motors to specification using plastic injection mouldings for some components. The cost of the moulds for the injection moulding machine would be £40 000 but the remaining tooling cost would be reduced by £16 000 and the marginal production cost per motor would be reduced to £480.
4. If the plastic moulded parts are not successful there will be time to revert to using machined parts but the full cost of tooling will also be incurred.

5. The production engineer estimates the probability that the moulded parts will be successful as 0.5.
6. The sales director estimates the probability of obtaining the order as 0.4.

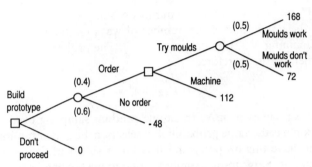

Fig. 2.13 Decision tree for Macaw Motors.

Our task is to advise the engineering manager. The first step is to draw a tree diagram (Fig. 2.13), distinguishing between decisions and chance occurrences by squares and circles, respectively. We now work back from the right-hand side to decision points, ruling out the more expensive options. The profit, ignoring interest for the moment, of meeting an order by using machined parts is calculated below (in units of £1000).

INCOME
    revenue from 1000 motors   800
OUTGOINGS
    prototype   48
    tooling   80
    production   560
                             ‾‾‾‾‾
                            688

Hence the profit would be the difference between 800 and 688, which is 112. The profit of meeting an assumed order if plastic injection mouldings are tried is more complicated because we do not know if they will work. If they do, the profit would be increased to 168 but, if they are not satisfactory, the profit would be reduced to 72. There is a 50% chance that they work, so the expected profit is

$$0.5 \times 168 + 0.5 \times 72 = 120$$

This exceeds 112 so the better decision, using an expected profit criterion, would be to try injection mouldings. We can now calculate the expected profit if Macaw competes for the order. There is a 40% chance that it will make an expected profit of 120 and a 60% chance that it will produce a prototype but is not given the order, a loss of 48. The overall expected value of trying for the order is therefore

$$0.4 \times 120 + 0.6 \times (-48) = 19.2$$

However, we have ignored interest on money between March and September. Let us now suppose Macaw borrows money to meet the order at an interest rate of 20% per annum. The outgoings should be increased by approximately 10%, reducing the expected profit of 120 to 52. The overall expected value is now

$$0.4 \times 52 + 0.6 \times (-48) = -8$$

The decision is clearly sensitive to the exact interest calculations and the assumed probabilities. Macaw may have some other business opportunities that are more promising, in which case this possibility would not be pursued. However, it would be worth the risk if Macaw has few orders and could stand the potential loss.

2.7 Summary

Population

The population is the collection of all the objects which we wish to be represented in some investigation.

Randomization

Imagine assigning a number to every member of the population and putting chips with the same numbers into a large drum. Now suppose that every chip is equally likely to be drawn out of the drum. Then a draw of n chips will identify a simple random sample of size n. In practice, random number tables are used instead of any mechanical apparatus. The advantages of randomization are that it can be seen to be fair and enables us to estimate the precision of estimates of population quantities from the sample.

Probability rules

$$\Pr(A \text{ or } B) = \Pr(A) + \Pr(B) - \Pr(A \text{ and } B)$$
$$\Pr(A) = 1 - \Pr(\bar{A})$$
$$\Pr(B|A) = \Pr(A \text{ and } B)/\Pr(A)$$
$$\Pr(A \text{ and } B) = \Pr(A)\Pr(B|A) = \Pr(B)\Pr(A|B)$$

Events A and B are mutually exclusive if $\Pr(A \text{ and } B)$ equals zero and independent if $\Pr(A \text{ and } B)$ equals the product of $\Pr(A)$ with $\Pr(B)$.

Bayes' theorem

$$\Pr(A|B) = \frac{\Pr(B|A)\Pr(A)}{\Pr(B|A)\Pr(A) + \Pr(B|\bar{A})\Pr(\bar{A})}$$

In general suppose we have k mutually exclusive and exhaustive events $\{A_1, \ldots, A_k\}$, and some observed event B.

$$\Pr(A_i|B) = \frac{\Pr(B|A_i)\Pr(A_i)}{\sum_{j=1}^{k} \Pr(B|A_j)\Pr(A_j)}$$

Arrangements and choices

There are

$$_nP_r = n(n-1)\cdots(n-r+1)$$

ways of arranging r objects from n.

There are

$$_nC_r = {_nP_r}/r!$$

choices of r objects from n. ($_nC_r$ is also commonly written ^nC_r or $\binom{n}{r}$. I have chosen to use the first because it most nearly corresponds to that used on calculator keys.)

Exercises

2.1   A student has built an electronic fruit machine as a final year project. It has two 'reels' which can display one of raspberry (R), strawberry (S), apple (A), pear (P) or orange (O). The reels 'spin' independently, and each of the five fruits is equally likely to appear.

   (i) A turn on the machine is the outcome (pair of fruits) when both reels are spun. Draw a sample space consisting of equally likely points. Hence deduce the probability of obtaining both reels showing summer fruits (R and S are summer fruits).

   (ii) The machine has now been programmed to spin the reels again, without displaying the result, if both give the same fruit. Construct a sample space of equally likely points, and write down the probability of both reels showing summer fruits.

   (iii) Explain why a sample space {0 summer fruits, 1 summer fruit, 2 summer fruits} is not very useful.

2.2   Two regular perfectly balanced decagons are spun (come to rest indicating 0 or 1··· or 9). Find the probabilities of the events described, if the spins are probabilistically independent:

(i) the total of the two numbers is a multiple of 3;
(ii) the total of the two numbers is a multiple of 4;
(iii) the total of the two numbers is a multiple of 3 or 4;
(iv) the total of the two numbers is a multiple of one of 3 or 4, but not both.

2.3   Five faulty television tubes are mixed up with 12 good ones. Find the probabilities of the events described if selection is random:

(i) one selected tube is good;
(ii) two selected tubes are good;
(iii) two selected tubes are bad;
(iv) out of two selected tubes one is good and one is bad;
(v) at least one out of two selected tubes is good.

2.4   A machine displays numbers in binary form by the use of lights. The probability of an incorrect digit is 0.01 and errors in digits occur independently of one another. What is the probability of

(i) a 2-digit number being incorrect?
(ii) a 3-digit number being incorrect?
(iii) a n-digit number being incorrect?

2.5   A garage owner estimates, from extensive past records, that 70% of cars submitted for a Ministry of Transport test need lamp adjustments and 60% need brake adjustments or repairs. 50% of the cars need both adjustments.

(i) What is the probability that a car selected at random needs at least one adjustment?
(ii) What is the conditional probability of a car requiring a lamp adjustment given that a brake adjustment was necessary?
(iii) What is the conditional probability of a car requiring a brake adjustment given that a lamp adjustment was necessary?

2.6   The failure of pumps A or B, and C will stop the water supply system of Fig. 2.14 functioning. What is the probability that the supply is maintained, if the pumps A, B and C have independent probabilities of failure a, b and c?

2.7   Refer to Example 2.11. Suppose the switches open when they should not with probability p. What is the probability that the launch is stopped unnecessarily? If the probability of a critical pressure drop is c, what is the probability that the sensing system operates correctly? (After Bajpai et al., 1968.)

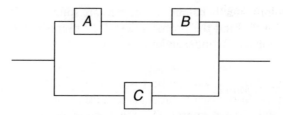

Fig. 2.14 System of three pumps.

2.8 Suppose that the probability of failure of an aircraft engine in flight is q and that an aircraft is considered to make a succesful flight if at least half of its engines do not fail. For what values of q is a two-engined aircraft to be preferred to a four-engined one? Assume that engine failures are independent.

2.9 In a twin-engined plane the unconditional probability an engine fails at some time during a flight is q. However, engine failures are not independent:

Pr(right engine fails at some time during flight|

left engine fails at some time during the flight) $= a$

similarly for the left engine failing given that the right fails at some time. The probability that one engine fails when the other does not is given by

Pr(right fails|left engine does not fail) $= b$

and similarly for left fails given right does not. Find the probabilities of 0, 1 and 2 engines failing in terms of a, b and q. Then use the symmetry of the situation to eliminate b.

2.10 A razor blade manufacturer purchases steel strip from three suppliers A, B and C. On average 70% are bought from A, 20% from B, and 10% from C. The probability that a roll of steel strip from A causes the line to stop because of width or thickness variation is 0.001. Corresponding probabilities for rolls from B and C are 0.002 and 0.005, respectively. If an unmarked roll causes a stoppage, what are the probabilities it was from A, B and C?

2.11 The 100 000 pavers described at the beginning of this chapter were delivered as 100 palettes of 1000 pavers. The 1000 pavers in a palette are stacked 10 by 10 by 10. Set up a procedure for assigning a number to each paver, so that the 24 selected pavers can be found relatively easily.

2.12 An oil company manufactures its own offshore drilling platforms. An ultrasonic device will detect a crack in a weld with a probability of 0.95. It will also wrongly signal cracks in 10% of good welds. It costs about £5 per weld to test it in this way. From past experience about 0.5% of new welds contain slight cracks. A thorough X-ray analysis of welds is completely accurate but costs £50 per weld. An undetected crack may lead to a visual failure of the joint when the platform is in service, and remedial work will then cost £500. The probability of a slight undetected crack causing the joint to fail in this way is 40%. Assume this is the only significant mechanism for joint failure and compare the expected costs of the following policies.

(i) Test all joints with the ultrasonic device and follow up signals of cracks with the X-ray analysis. Repair of a cracked weld on land will cost £50, but assume the repaired weld will then be good.
(ii) X-ray all joints and repair those with cracks.
(iii) No inspection.

2.13 In how many different ways can you:

(i) link together seven different components into a line;
(ii) link seven components into a line if two are now identical;
(iii) seat seven people at a round table;
(iv) thread seven different round semi-precious stones on to a necklace if the position of the knot is ignored?

2.14 A manufacturer makes radios with three wavebands. In how many ways can three from seven non-overlapping wavebands be chosen?

2.15 Explain why

(i) ${}_NC_n = {}_NC_{N-n}$
(ii) ${}_NC_n = {}_{N-1}C_{n-1} + {}_{N-1}C_n$

Demonstrate that these results are identities by applying the definitions.

2.16 **Pseudo-random number generator.** The following example illustrates a general method but should not be relied on for serious work. Many statistical packages and mathematical packages, MINITAB and NAG for example, provide thoroughly tested generators, and Press et al. (1992) discuss the principles. The recurrence relationship

$$I_{j+1} = 106I_j + 1283 \quad \text{(modulo 6075)}$$

where the instruction (modulo 6075), means let I_{j+1} equal the remainder after dividing $(106I_j + 1283)$ by 6075, will generate a sequence of integers $\{I_j\}$ between 0 and 6074. These integers are then divided by 6075 to give a sequence of decimals,

$$r_j = I_j/6075$$

between 0 and 1. The sequence $\{r_j\}$ is often useful in its own right (Example 4.13), but a sequence of random digits can be obtained by multiplying r_j by 10 and truncating, that is dropping the fractional part.

(i) Generate a sequence of ten I_j starting from

$$I_0 = 3142$$

(ii) Use your ten I_j to construct a sequence of 10 digits.
(iii) Explain why it would not be satisfactory to use all the digits in the I_j as a random sequence.
(iv) Would it be valid to round the r_j multiplied by 10 to the nearest integer instead of truncating?

Press *et al.* (1992) recommend a random shuffle of the sequence by this sort of algorithm before using it.

2.17 The gamma function is defined by

$$\Gamma(\alpha) = \int_0^\infty t^{\alpha-1} e^{-t} dt \qquad \text{for } \alpha > 0$$

(i) Use integration by parts to show that

$$\Gamma(\alpha) = (\alpha - 1)!$$

Hence deduce a value for 0!

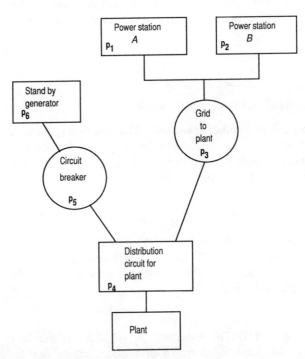

Fig. 2.15 Schematic of power supply to plant.

The gamma function is a generalization of the factorial function for integers.

2.18 A pharmaceutical company manufactures vaccines at a particular site. The processes are crucially dependent on an electricity supply for heating and control. The vaccines have to be stored in refrigerators, and these, too, are powered by electricity. The grid can be supplied by either of two power stations, and the site has its own standby generator.

The site manager intends making a simple, interim, **probabilistic risk assessment** from the schematic of the supply shown in Fig. 2.15. The p_i in the figure are the probabilities the components fail in any year.

(a) (i) Write down the probability that the site has an electricity power cut in any year in terms of the p_i, stating any assumptions you need to make.

   (ii) What features of the problem have so far been ignored?

(b) Write a computer program that will analyse the situation if the cost of a power cut is C_1, and the probabilities of failures in the standby generator link can all be reduced by a factor of θ_k for an investment of C_k in new plant. Test your program with plausible p_i, C_1, C_k, θ_k values.

3

Justifying engineering decisions

Although probability theory allows us to describe random variation and investigate its effects, we still have to justify the probabilities used. These are based on our experience, records of which are known as data.

3.1 Presenting data

3.1.1 Population and sample

Sometimes data are collected from every member of a population, and this is often known as a **census**. The national census in the United Kingdom, which takes place every ten years, aims to record some basic information about every member of the population. All legal imports and exports of certain goods are monitored by the Customs and Excise, and published figures are based on this information. In the United Kingdom all traffic accidents which result in injury or fatality are recorded, in principle, and the data published. In practice, some accidents causing injury are not recorded, and this illustrates the general point that there will still be errors in census data. Many other government statistics are based on samples, and relatively accurate data from random samples can be much more reliable than dubious data from the population. Furthermore, it is not only governments that carry out censuses of populations: companies usually keep records of customers and their orders, educational institutions keep records of all their students, and so on.

In any investigation it is essential to know whether data have come from the entire population or from a sample, and, if the latter, what the population is. This is not always straightforward, and it often depends on the investigator's point of view. For instance, civil engineers might consider traffic accidents during the past year as a random sample in time from all future years if present conditions were to persist, whereas a police administrator accounting for resources used during the past year might, in this context, view them as the population. The population of 'all future years if present conditions persist' is an example of an **imaginary infinite population**. Such populations are commonly assumed in engineering and scientific applications.

Example 3.1

A pharmaceutical company uses benzene as a reagent for making several of its products. It buys a large consignment of 1000 drums from a chemical engineering company. The batch of 1000 drums is a population from the pharmaceutical company's point of view, and a sample will be tested for purity.

The chemical company has its own quality assurance systems. As part of these an inspector monitors the output from the benzene process at approximately, but not exactly, half-hour intervals. At each inspection the last 20 drums from the production line are put to one side. They are considered a random sample in time from the imaginary infinite population of drums that would be produced if the process continues on its present settings indefinitely. He then draws a simple random sample of 3 from these 20 drums. The three drums are opened and four 100 ml volumes are taken at different depths, using a specially designed sampler, from each one. The 100 ml volumes are analysed for purity, and if the results are not satisfactory the process will be stopped and corrected. The pharmaceutical company decides to make regular purchases of benzene from this chemical company, and, being impressed by the chemical company's quality assurance systems, decides it need no longer sample every batch of benzene delivered. The saving in inspection costs is one of the advantages of using regular suppliers rather than always taking the cheapest quotation.

3.1.2 Types of variable

Some variables are naturally measured on a numerical scale and can conveniently be classified as **discrete** or **continuous**. Discrete variables are usually 'counts' and therefore non-negative whole numbers.

Example 3.2

A typical very large-scale integrated circuit chip has thousands of contact windows. A window is a hole, with an approximate diameter of 3.5 μm, etched through an oxide layer, which is about 2 μm deep, by photolithography. A factory produces wafers containing 400 chips. Each wafer consists of four sectors designated north, south, east, and west. There is a test pattern of 20 holes located in each sector. Some windows do not pass through all the oxide layer and are said to be 'closed'. Closed windows result in loss of contacts, and the sector is scrapped if any closed windows are found in the test pattern. An inspector records the number of closed windows in each test pattern, a discrete variable which can take any value in the range 0, 1, 2, 3, ..., 20.

Other examples of discrete variables include the number of ships arriving at a port per day, the number of paintwork flaws on the bodywork of finished cars, and the number of bacteria in litre samples of water. In these situations the discrete variables can take any value from 0, 1, 2, ... and there are no specific upper limits.

Continuous variables can take any value between certain limits which need not be specified. Examples include: length, weight, temperature and time.

Example 3.3

Antimony oxide is a flame retardant used in the manufacture of paint and paper, among other applications. A company sells its top grade, for use when colour considerations are critical, in 25 kg bags. The bags are filled and weighed automatically. The weighing machine has a digital display which has a resolution of 1 g. A mass of 25 218 g can be interpreted as a mass between 25 217.50 and 25 218.49 ... and it is both reasonable and convenient to treat such measurements as observations of a continuous variable.

There are many other variables which provide a verbal, rather than numerical, description. These will be referred to as **categorical** throughout this book. Examples are the make of a car tyre (such as Dunlop, Michelin, Pirelli and Uniroyal), and the material from which water pipes are made (such as concrete, iron and polypropylene). Some categorical variables can be sensibly ordered, and this blurs the distinction between discrete and categorical variables. A manufacturer grades a whitish powder, added to paints and plastics as a flame retardant, as 'white', 'off-white', and 'grey'. However, it may not be reasonable to assume some equal increment discrete scale.

Example 3.4

Table 3.1, from *Social Trends 17* (Central Statistical Office (CSO), 1987) gives estimates of water quality in waterways in England and Wales during 1985. The classification of quality is an ordered categorical variable. The classification of waterways is a categorical variable which cannot be sensibly ordered. The majority of waterways are freshwater rivers, and later volumes of *Social Trends* give a breakdown by Regional Water Authority and exclude estuaries. It is nevertheless possible to compare the percentages of canals and freshwater rivers in each quality classification. They hardly change between 1985 and 1990.

3.1.3 Diagrams

Most people find diagrams far more interesting than the tables of numbers on which they are based. People reading reports can assimilate information presented in the form of diagrams quickly, a crucial advantage when they can only devote a short amount of time to such reports. It is therefore not surprising that a considerable amount of effort is made to produce entertaining and informative diagrams. There is also plenty of scope for them to be subtly, and sometimes blatantly, misleading. The most common device is to draw heights of two-dimensional representations of three-dimensional objects in proportion to the variable being promoted. The champagne glasses in Fig. 3.1 appeared in

Table 3.1 Waterway quality in England and Wales

	Canals		Freshwater rivers		Estuaries		
	1980[1]	1985	1980[1]	1985		1980[1]	1985
Quality classification[2]							
Class 1A—Good	233	198	13 599	13 271	Class A—Good	1 865	1 860
Class 1B—Good	762	757	13 458	13 229	Class B—Fair	619	652
Class 2—Fair	974	1 271	7 697	8 463	Class C—Poor	130	129
Class 3—Poor	382	240	2 882	3 315	Class D—Bad	114	89
Class 4—Bad	25	31	615	619			
Total[3]	2 377	2 498	38 250	38 896		2 734	2 730

[1]In the 1980 survey two new classification schemes, one a more objective use-related scheme for freshwater rivers and canals and the other a rather subjective scheme for estuaries and saline rivers, were used together with the former classification scheme which applied to all waters in the survey.
[2]The classifications are described fully in Department of the Environment, *River Quality in England and Wales, 1985* (HMSO).
[3]Changes in the total lengths surveyed in different years are due to remeasurement and the inclusion of different stretches.

Source: Department of the Environment

Fig. 3.1 Champagne glasses (based on an investment advertisement).

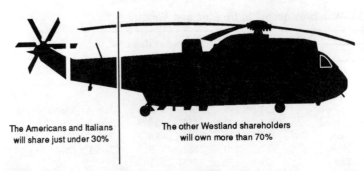

The Americans and Italians
will share just under 30%

The other Westland shareholders
will own more than 70%

So who'll be at the controls?

Fig. 3.2 Helicopter from Westland advertisement.

an advertisement in a recent issue of the *Bulletin of the Association of University Teachers*. The value of the investment has increased by a factor of 1.83 over 3 years but the largest glass is taller than the smallest by a factor of 2.4 and holds nearly 14 times as much champagne. The area of the leading 70% by length of the helicopter in Fig. 3.2 is rather more than 70% of the total area. You are left to judge whether or not this is misleading.

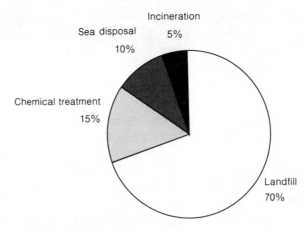

(Sea disposal includes sea incineration)

Source: Her Majesty's Inspectorate of Pollution

Fig. 3.3 Special waste: by disposal route.

Slicing Profit to Good Use...

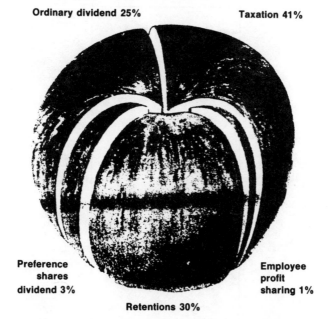

Ordinary dividend 25%

Taxation 41%

**Preference
shares
dividend 3%**

**Employee
profit
sharing 1%**

Retentions 30%

One apple = £1828 million pre-tax profit.
British Telecom's turnover in 1986 was £8387 million. Its
operating profit was £2095 million and its pre-tax profit was
£1828 million. The apple diagram shows the percentages
taken by the various slices given to dividends, employee profit
sharing, tax, and the amount kept by British Telecom for
investing in its business to improve service to customers.

Fig. 3.4 Apple from BT annual report of 1988 (courtesy of BT).

Pie charts are a good way of presenting data and Fig. 3.3 (CSO, 1992) shows
the disposal routes for 'special waste' arising in the UK in 1989–90. Special
waste is controlled waste which consists of, or is contaminated by, substances
which are dangerous to life. It is estimated that there were between 2.0 and 2.5
million tonnes of special waste during this period! Does this sound less formidable if it is described as less than 0.4% of total waste? The apple in Fig. 3.4,
taken from the 1986 BT *Report to Customers*, is a nice variation. The winner
of a recent 'misleading statistic competition' submitted an empty can of lager
which is illustrated in the scale drawing in Fig. 3.5. You are left to think about
why it is misleading!

3.1.4 Line chart for discrete data

The data in Table 3.2 are the numbers of asbestos-type fibres (of length greater
than 5 μm, width less than 3 μm and length-to-width ratio greater than 3) found

Fig. 3.5 Lager can (based on product promotion).

Table 3.2 Asbestos-type fibres in 1 litre samples of air

Observed number of particles	Number of occasions	Relative frequencies
0	34	0.2378
1	46	0.3217
2	38	0.2657
3	19	0.1329
4	4	0.0280
5	2	0.0140
6 or more	0	0.0000
Total	143	1.000

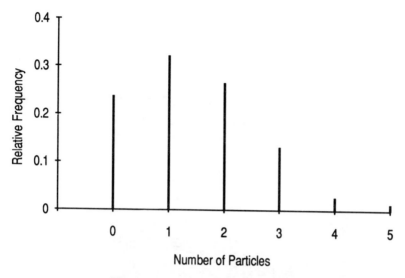

Fig. 3.6 Line chart for particle data.

in 1 litre samples of air, centred 1.5 m above ground level, at various times and locations in a factory workshop. The workshop was once used for the manufacture of brake linings but now houses several microprocessor controlled lathes. These data are represented by Fig. 3.6. Notice that the lengths of the lines are the **relative frequencies**, that is, for m equal to 0, 1, ..., 5,

$$\frac{\text{number of occasions on which } m \text{ particles seen}}{\text{number of volumes of air looked at}}$$

The data are thought of as a random sample 'in time' of fibres in this workshop. This density of asbestos-type fibres would probably not be classed as a health hazard under current UK occupational exposure control limits.

3.1.5 Bar chart, histogram and cumulative frequency polygon for continuous data

Wave data are essential for investigating the fatigue life of offshore structures. For gas drilling rigs in shallow water the height of waves is the most relevant variable. In deep water the predominant frequencies are also vital information, because natural frequencies of the structure can be excited. The data in Table 3.3 are wave heights at a location in the North Sea over one year.

The range of observed heights has been divided into 13 equal intervals, called **class intervals**. The choice of end points has been made to avoid any ambiguity over which interval a wave height (measured to the nearest 0.1 m) belongs to. The first interval is assumed to include all waves with actual heights between 0 m and 0.949 ... m, the second all waves between 0.950 m and 1.949 ... m, and

Table 3.3 Number of individual wave, classified by height and direction, in one year at a location in the North Sea

Wave height (metres)	N	NE	E	SE	S	SW	W	NW	Total
0.0–0.9	453 036	327 419	403 390	571 102	947 294	1 074 776	804 533	439 132	5 020 682
1.0–1.9	178 960	109 887	113 367	111 698	174 120	291 679	268 823	179 909	1 428 443
2.0–2.9	45 305	23 131	25 478	17 797	22 798	42 112	57 887	51 699	286 207
3.0–3.9	12 718	5 851	6 193	3 105	3 024	5 636	12 837	15 288	64 652
4.0–4.9	3 939	1 798	1 543	535	416	804	3 056	4 772	16 863
5.0–5.9	1 319	593	384	91	63	123	749	1 556	4 878
6.0–6.9	473	193	93	17	12	19	183	520	1 510
7.0–7.9	175	60	21	3	2	3	45	173	482
8.0–8.9	65	17	5	0	0	0	10	58	155
9.0–9.9	24	5	1	0	0	0	4	18	52
10.0–10.9	9	1	0	0	0	0	0	6	16
11.0–11.9	3	0	0	0	0	0	0	2	5
12.0–12.9	1	0	0	0	0	0	0	0	1
Total	696 027	468 955	550 475	704 348	1 147 729	1 415 152	1 148 127	693 133	6 823 946

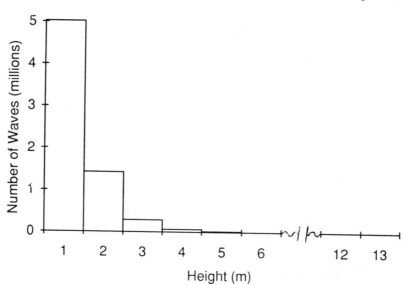

Fig. 3.7 Bar chart for wave heights.

so on. It follows that the width of all the intervals can be taken as 1 (this is not exact, but near enough, for the first interval). A reasonable alternative would be to define intervals as 0–1, 1–2, etc., and to allocate heights on boundaries equally between the intervals, but I have reproduced the table I was given. In Fig. 3.7 the height of each bar equals the total number of waves in the corresponding class interval.

It is sometimes useful to use class intervals of different lengths, and hence provide more detail where there are sufficient data to justify this. If this is done it is no longer appropriate to make the heights of bars equal to the frequencies. Some data on ship hull roughness are used to illustrate this point. A large proportion of the total resistance for large slow-moving merchant ships is due to friction from the hull. A poor paint finish will result in a waste of fuel; even a moderate degree of underwater hull roughness will lead to increases of 20% in fuel bills. A relatively simple measure of roughness is given by measurements of lowest trough to highest peak over 50 mm transects. These can be measured with a hull roughness analyser, which is a hand-held instrument with a stylus diameter of about 1.6 mm (any smaller diameter would damage the paint finish). The data in Table 3.4 are measurements from 550 such transects of a ship's hull. The wetted area was treated as a rectangular grid of 100 squares of equal area, and five readings were taken at random within each square. It is not sensible to make the height of bars equal frequencies because, for example, the 63 readings in the third class interval, which has a length of 10, would correspond to a higher bar than the 42 readings in the fourth interval which only has a length of 5. Taking the heights of bars as frequency per unit length of interval

Table 3.4 Lowest trough to highest peak in 50 mm transects of a ship's hull

Trough-to-peak height (μm)	Frequency	Cumulative frequency
47.5–57.4	7	7
57.5–67.4	24	31
67.5–77.4	63	94
77.5–82.4	42	136
82.5–87.4	57	193
87.5–92.4	85	278
92.5–97.4	78	356
97.5–107.4	70	426
107.5–117.4	46	472
117.5–127.4	26	498
127.5–137.4	14	512
137.5–207.4	25	537
207.5–277.4	13	550

will give a fair visual representation, since 63/10 is less than 42/5. It is also convenient to scale heights so that the areas of bars equal the proportions of the data in the corresponding class intervals, and the total area equals 1. That is, the heights of the bars are equal to

$$\frac{\text{relative frequency}}{\text{length of class interval}}$$

which can be described as **relative frequency density**. Remember that the relative frequency of observations in any interval is the frequency divided by the total number of observations, and is just the proportion of the data in that interval. A bar chart which has been scaled to have a total area of 1 will be referred to as a **histogram** throughout this book. An alternative way of representing the data is to plot the proportion (often expressed as a percentage) of data less than a given trough-to-peak value against that value. This is known as a **cumulative frequency polygon**. The histogram and cumulative frequency polygon for the roughness data are drawn in Fig. 3.8. A typical calculation for the former is the height of the highest bar, corresponding to the sixth interval:

$$\frac{85/550}{5\,\mu m} = 0.031\,\mu m^{-1}$$

Notice that end points of intervals are appropriate when drawing the cumulative frequency polygon: for example we can say 31 trough-to-peak values are less than 67.45, and all the values are less than 277.45.

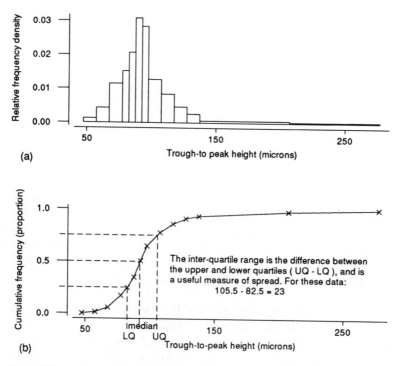

Fig. 3.8 (a) Histogram for hull roughness data. (b) Cumulative frequency polygon for hull roughness data.

3.2 Summarizing data

It was convenient to summarize the strengths, and cement contents, of the 24 pavers with which we began Chapter 2, by the average and the range. With larger data sets it is even harder to see the main features from the list, and some summary measures are essential. While the average is the most common measure of the 'centre' of a data set, it is often useful to consider other possibilities as well, or instead. The range is a simple measure of variability within a sample, but it is difficult to identify the corresponding quantity in the population because the range will tend to increase as the sample size increases. Other summaries of variability are usually preferred, except in quality control when samples of the same size are repeatedly taken.

3.2.1 Measures of centrality

The usual 'average' is obtained by adding up the data and dividing by their number. This is defined as the arithmetic mean, usually abbreviated to the **mean**. Formally, if data $\{x_1, \ldots, x_n\}$ are a sample of size n from some population, the sample mean ($\bar{x}$) is defined by

$$\bar{x} = \sum_{i=1}^{n} x_i/n.$$

If the population is finite, of size N, the corresponding data can be represented by $\{x_1, \ldots, x_N\}$. The population mean is denoted by the Greek letter 'mu' (μ) and defined by

$$\mu = \sum_{i=1}^{N} x_i/N.$$

The sample mean is an **unbiased estimate** of the population mean. That is, if you imagine taking very many random samples of size n and calculating an $\bar{x}$ for each one, the average of all these $\bar{x}$ would equal μ. In practice, you will only have a single $\bar{x}$ but the larger the sample size the closer you can expect it to be to μ. A definition of μ for an infinite population will be given in the next chapter, but imagining N tending to infinity is adequate for the moment. Other types of mean are more appropriate in some situations and examples are given in the exercises.

Another measure of centrality is the 'middle' value when the data are arranged in ascending order. This is called the **median**. If there are an odd number of data the median is the datum in the middle, and if there are an even number it is the mean of the two data in the middle. It can conveniently be found from the cumulative frequency polygon by reading across from 0.5 (or 50%). The median of the roughness data (Fig. 3.8(b)) is approximately 92. Unlike the mean, the median is not sensitive to extreme observations, so a significant difference between the two is informative. A few anomalously large values will result in the mean being larger than the median.

Example 3.5

The data in Table 3.5 (Adamson, 1989) are the annual maximum floodpeak inflows to the Hardap Dam in Namibia, covering the period from October 1962 to September 1987. The range of these data is from 30 to 6100. The median and mean are 412 and 859, respectively, and such a large difference is a warning that the data are rather strange.

A MINITAB **stem-and-leaf plot** and a **boxplot** are shown in Fig. 3.9. Both clearly display the unusual features of the data. The stem-and-leaf plot is almost self-explanatory, if it is compared with the original data put into ascending order. There are four data less than 100, three between 100 and 199, two between 200 and 299, and so on. The left-hand column needs rather more explanation. The bracketed figure in the left-hand column gives the number of observations on the same line as the median. The others are the number of observations on their line or beyond (moving away from the median). Turning to the boxplot, the + in the box represents the median. The lower limit of the box is the lower

Table 3.5 Annual maximum floodpeak inflows to Hardap Dam
(Namibia): catchment area 12 600 km^2

Year	Inflow (m^3 s^{-1})	Year	Inflow (m^3 s^{-1})
1962–3	1864	1975–6	1506
1963–4	44	1976–7	1508
1964–5	46	1977–8	236
1965–6	364	1978–9	635
1966–7	911	1979–0	230
1967–8	83	1980–1	125
1968–9	477	1981–2	131
1969–0	457	1982–3	30
1970–1	782	1983–4	765
1971–2	6100	1984–5	408
1972–3	197	1985–6	347
1973–4	3259	1986–7	412
1974–5	554		

quartile (LQ), the value below which 25% of the data lie. It is calculated by ordering the n data from smallest to largest and taking the observation at position $(n + 1)/4$, interpolating when this is not an integer. The upper limit of the box is the upper quartile (UQ), which is the observation at position $3(n + 1)/4$. The upper and lower quartile of the Hardap Dam data are 164 and 846, respectively. With larger data sets they can be calculated from the cumulative frequency polygon by reading across from 25% and 75%.

It follows that the box represents the middle half of the data. Inner and outer **fences** are defined as:

inner fence $LQ - 1.5 (UQ - LQ)$ and $UQ + 1.5 (UQ - LQ)$

outer fence $LQ - 3.0 (UQ - LQ)$ and $UQ + 3.0 (UQ - LQ)$.

Dotted lines are drawn to the most extreme observations that are still inside the inner fences. Values between the fences are shown by * and values beyond the outer fences by 0.

A third measure of centrality is the **modal value** or **mode**. This is the most commonly occurring value when the data are restricted to specific values. If continuous data are grouped the **modal class interval** is the interval with the highest frequency density, and the construction shown in Fig. 3.10 can be used to obtain a single value for the mode from a histogram with equal length class intervals either side of the modal class. The definitions of the median and mode given here are applicable to both samples and finite populations. Definitions for infinite populations will be deferred until the next chapter.

MTB > Describe Cl.

	N	MEAN	MEDIAN	TRMEAN	STDEV	SEMEAN
peakinfl	25	859	412	667	1313	263

	MIN	MAX	Q1	Q3
peakinfl	30	6100	164	846

MTB > DotPlot 'peakinf'.

```
   . . . :
   : : : : : : . . .        :      .                    .
   : : : : : : . . .
+ - - - - - - - + - - - - - - + - - - - - - + - - - - - - + - - - - - + - - - - -
0         1200        2400        3600        4800        6000  peakinfl
```

MTB > Stem-and -Leaf 'peakinfl'.

Stem-and-leaf of peakinfl N=25
Leaf Unit=100

```
     (13)   0   0001112223444
      12    0   5566889
       5    1
       5    1   559
       2    2
       2    2
       2    3   3
       1    3
       1    4
       1    4
       1    5
       1    5
       1    6   1
```

MTB > Gstd.
MTB > BoxPlot Cl.

```
      - - - - -
 - - |+     |- - - - -     *              0                      0
      - - - -.
+ - - - - - - + - - - - - - + - - - - - - + - - - - - - + - - - - - - + - - - -
0         1200        2400        3600        4800        6000
                                                          peakinfl
```

Fig. 3.9 Dot plot, stem-and-leaf plot and boxplot of Hardap Dam data.

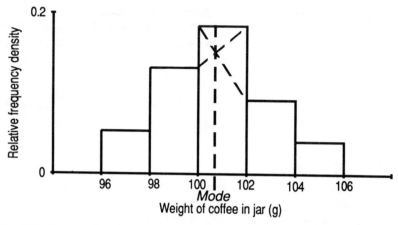

Fig. 3.10 Construction of mode from histogram with equal length class intervals.

Example 3.6

A manufacturer of men's shoes makes them in sizes 4, 5, 6, 7, 8, 9, 10 and 11. Sales over the past year, in hundreds of pairs, have been 24, 38, 70, 250, 640, 580, 200, and 32 for the eight sizes, respectively. The modal size is 8 and this information would be useful if the manufacturer intended introducing some half-sizes. There would probably be most demand for sizes $7\frac{1}{2}$ and $8\frac{1}{2}$. The mean shoe size would not be useful.

Example 3.7

The mode of the ship hull roughness data is 91 μm.

In situations such as the following it is necessary to use a **weighted mean**.

Example 3.8

A water company operates in three divisions A, B and C. Each division has estimated the cost of bringing the levels of service up to a new standard. The costs are expressed as pounds per year per property over a five-year improvement period.

	Division A	Division B	Division C
Estimated cost (£/year per property)	10	14	20
Number of properties (thousands)	300	200	100

48 Justifying engineering decisions

The estimated total cost is £7.8 million per year, which is equivalent to an average of £13 per year per property. This is a weighted mean of the mean costs for the three divisions with weights proportional to the number of properties in the division. In general, the quantity

$$\sum x_i w_i / \sum w_i$$

is a weighted mean of the data $\{x_1, \ldots, x_n\}$. (Notice that the limits of the summation can be left off the sigma sign when they are clear from the context.)

3.2.2 Measures of spread

The data in Table 3.6 are the resistances of random samples of six 1 kΩ resistors from stocks of two types, F and M. The mean of the type F resistors is 1002 and the mean of the type M resistors is 1000. The specification for a resistor in a TV circuit is 1 kΩ ± 1%. Would you recommend the manufacturer use type F or type M? The type M resistors are clearly too variable and the manufacturer should use type F. In fact F and M are BS1852 resistance tolerance codes for ±1% and ±20%, respectively.

With a sample size as small as 6 the spread is apparent from the numbers, but for larger data sets some summary measure is essential. The simplest measure is the difference between the largest datum and the smallest datum, which is known as the **range**. One of its disadvantages is that it is difficult to make comparisons between samples of different sizes.

Another possibility is the average of the absolute deviations from the mean. Formally, if $\{x_1, \ldots, x_n\}$ is a data set, its **mean absolute deviation** is

$$\sum |(x_i - \bar{x})| / n$$

While this is sometimes used, it is awkward to handle mathematically.

The usual measure is the average of the squared deviations from the mean, which is called the **variance**. The units of variance are the square of the units of the measurements. The **standard deviation** is the positive square root of the variance and therefore has the same units as the measurements. If $\{x_1, \ldots, x_N\}$

Table 3.6 Resistances of resistors (ohms)

Type F	Type M
1002	1018
998	1060
1003	1053
1000	930
1004	964
1005	975

is a population the variance (σ^2) is defined by

$$\sigma^2 = \sum_{i=1}^{N} (x_i - \mu)^2/N$$

The population standard deviation (σ) is the square root of this. Now suppose we have a random sample $\{x_1, \ldots, x_n\}$ from this population. The sample estimate of the population standard deviation is given by

$$s^2 = \sum_{i=1}^{n} (x_i - \bar{x})^2/(n-1)$$

It is usual to use the divisor $n-1$ because s^2 is then an unbiased estimator of σ^2. If you imagine millions of independent random samples of n being drawn from the population the average of the corresponding s^2 would be σ^2. However, we do not demand that all our estimators are exactly unbiased. The average value of s would not be exactly σ but the difference, technically the bias, is negligible compared with the variability from sample to sample. Notice that the population standard deviation is denoted by the Greek lower-case 'sigma' and the Roman equivalent 's' is used for the sample estimate. The variance calculated from the sample, s^2, is an estimate of the population variance with $n-1$ **degrees of freedom**. The term 'degrees of freedom' reflects the fact that if $n-1$ of the deviations from the mean are known they are all known, because they add to zero. Formally

$$\sum(x_i - \bar{x}) = \sum x_i - n\bar{x} = \sum x_i - n(\sum x_i/n) = 0$$

The denominator of $n-1$ in the definition implies that s^2 would be undefined (0/0) in a sample of size 1. This emphasizes that a sample of one provides no information about variability. In large samples there is little difference between variances calculated using $n-1$ and n. MINITAB calculates standard deviations with a divisor $n-1$. Typing

LET K1 = STDEV(C1)

in the **session window** stores s in K1.

Example 3.9

The mean resistance of the six type F resistors is $1002\,\Omega$. The estimate of population variance is

$$[(1002 - 1002)^2 + \cdots + (1005 - 1002)^2]/(6-1)$$
$$= (0 + 16 + 1 + 4 + 4 + 9)/5 = 6.8$$

The standard deviation is therefore $2.61\,\Omega$. The standard deviation of the six type M resistors is $52.1\,\Omega$.

The standard deviation is easier to interpret than the variance. If data have

Table 3.7 Crushing strengths of concrete cubes

Class interval (N mm^{-2})	Frequency
42.0–49.9	3
50.0–51.9	0
52.0–53.9	2
54.0–55.9	7
56.0–57.9	28
58.0–59.9	27
60.0–61.9	39
62.0–63.9	30
64.0–65.9	24
66.0–67.9	13
68.0–69.9	7

a histogram which is approximately 'bell-shaped', about two-thirds lie within one standard deviation of the mean and about 95% lie within two standard deviations of the mean.

Example 3.10

The grouped data in Table 3.7 are the crushing strengths of 180 concrete cubes. The mean and standard deviation, calculated from the original data (Table D.1 in Appendix D), are 61.10 and 3.96. The histogram is shown in Fig. 3.11. Despite the fact that it is not particularly close to a bell shape, almost exactly two-thirds of the data are within 3.96 of 61.10. Five values (3%) are more than two standard deviations below the mean and none is more than two standard deviations above it, consequences of the asymmetry in the tails.

Although the standard deviation is not so easily understood as the range, it is becoming more widely used. The information in Table 3.8 is taken from the instructions for use of the hearing protectors I was given on a recent visit to a factory. The apparent reduction in attenuation at 500 Hz might be due to chance, and it would be interesting to know the sample size.

When variables are restricted to positive values (weights, volumes, etc.) it may be more appropriate to measure the spread in relative terms. The **coefficient of variation** is the ratio of the standard deviation to the mean and is often expressed as a percentage.

$$\text{sample coefficient of variation} = (s/\bar{x}) \times 100\%$$

The sample coefficients of variation for the type F and type M resistors are 0.26% and 5.2%, respectively. These can be compared with the British Standard

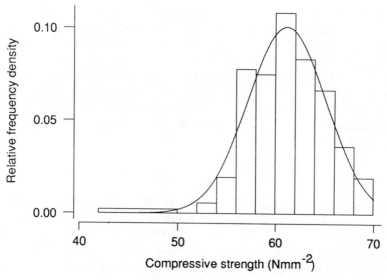

Fig. 3.11 Compressive strengths of concrete cubes.

Table 3.8 Attenuation of EAR Semi-Aural Hearing Protectors (USA Tested to ANSI S 12.6 – 1984 NRR 16)

Frequency (Hz)	125	250	500	1000	2000	3150	4000	6300	8000
Mean attenuation (dB)	19.3	19.5	17.7	20.5	29.8	37.2	37.2	38.6	39.0
Standard deviation (dB)	2.9	2.9	2.7	3.1	3.2	2.7	2.7	2.9	4.4
Assumed protection (dB)	16.4	16.6	15.0	17.4	26.6	34.5	34.5	35.7	34.6

tolerance codes, which are also expressed as percentages. It is very unlikely that the resistance of any resistor will be more than four standard deviations from the mean of all such resistors, so it seems that the stocks are within tolerances. However, the means and standard deviations were calculated from sample sizes of only 6, and as resistance is easily measured it would be sensible to increase the sample sizes. We could then make a more confident assessment.

3.2.3 Statistics and parameters

A statistic is a numerical quantity which is calculated from a sample. The corresponding population characteristic is known as a **parameter** of the popu-

52 Justifying engineering decisions

lation. The statistic is used as an estimate of the corresponding unknown parameter. The more common statistics are distinguished from the corresponding population quantities by using Roman letters for the former and Greek letters for the latter: $\bar{x}$ and μ, s^2 and σ^2, for example. However, we soon run out of convenient letters and it is another statistical convention to distinguish a statistic from the population quantity by placing a chevron (usually read as 'hat') over the statistic. As an example, '$\widehat{cv}$' is a fairly obvious shorthand for the sample coefficient of variation, which is our estimate of the coefficient of variation in the population (cv).

3.2.4 Grouped data

Paper clips are commonly sold in boxes of 100, with 'contents approximate' written somewhere on the box. The results in Table 3.9 were obtained by counting the contents of a random sample of 25 boxes. When calculating the mean it would be rather a waste of time to count 99 three times, 100 five times and so on! The obvious thing to do is to multiply the number by the frequency. That is,

$$\bar{x} = \sum x_k f_k / n$$

where

$$n = \sum f_k$$

and the summation is over the eight values of x_k. Similar considerations lead to

$$s^2 = \sum (x_k - \bar{x})^2 f_k / (n - 1)$$

Table 3.9 Contents of 25 boxes of paper clips

Number of clips in box	Number of boxes containing x_k clips
x_k	f_k
96	1
98	1
99	3
100	5
101	8
102	4
103	2
104	1

For the paper clips,

$$\bar{x} = 100.7 \quad \text{and} \quad s = 1.70$$

It is quite common to have access to grouped data only, available in a report perhaps, in which case a similar situation arises. Approximate values of statistics can be calculated by assuming that all the data in an interval are at its mid-point. The same formulae apply with x_k equal to the mid-point of an interval and f_k equal to the number of data in it. For the ship hull roughness data the mid-points of the intervals are 44.95, 64.95, 74.95, 82.45, ..., 97.45, 104.95, ..., 134.95, 169.95, 249.95. The approximate values of $\bar{x}$ and s are 99 and 30, respectively. Recall that the median was read off the cumulative frequency polygon as 92, which is less than the mean. This is a consequence of the histogram having a longer tail on the right. The mean is at the balance point of the histogram, whereas the areas on either side of the median are equal. In such cases the mode will usually be less than both the median and mean.

3.2.5 MINITAB calculations

The DESCRIBE command gives the number of data, mean, median, trimmed mean, standard deviation, standard error of the mean, minimum, maximum, and the quartiles of the data in the column. The latest, Release 9, runs under Windows. If the data are in column 1 (C1) you can type

DESCRIBE C1

in the session window, exactly as for earlier releases. But many people prefer to use the mouse, as follows:

Stat ▶ Basic Statistics ▶ Descriptive Statistics ▶
    Variables: C1

Furthermore N, MEAN, MEDIAN, STDEV, MIN and MAX can be used as functions with the column as argument. For example, typing

LET K1 = MAX(C1) − MIN(C1)

in the session window will give the range of the data. If you prefer to use the mouse:

Calc ▶ Mathematical Expressions ▶
    Variable: K1
    Expression: MAX(C1) − MIN(C1)

To save space in the rest of the book I shall only give calculations in the session window form. The effect of the DESCRIBE command on the Hardap Dam data is shown in the upper half of Fig. 3.9. The trimmed mean is a 5% trimmed mean. MINITAB removes the smallest 5% and the largest 5% of values and averages the rest. The standard error of the mean is explained in Chapter 5.

You will have to type in the following formulae to deal with grouped data. Suppose the mid-points and frequencies are in columns 1 and 2, respectively. Then the mean is calculated from

LET K1 = SUM(C1 * C2)/SUM(C2)

and the standard deviation is calculated from

LET K2 = SUM((C1 − K1) * (C1 − K1) * C2)

LET K2 = SQRT (K2/(SUM(C2) − 1))

You can combine these two commands into one longer command if you are sufficiently careful with brackets! It may be worth storing these commands in a file for repeated use. The effects of the display commands HISTOGRAM, DOTPLOT and BOXPLOT can be seen in Figs 3.12 and 3.9.

3.3 Fatigue damage

A finite element model for a static analysis of an offshore structure design resulted in stress against wave height relationships for each element at each node. Separate relationships were give for four wave headings N–S, NE–SW, E–W, and SE–NW. We use a simple approximation for fatigue damage with constant amplitude stress cycling, which is known as **Miner's rule**, although Miner's work extended well beyond it. The wave data in Table 3.3 are grouped, so let f_k represent the number of waves in an interval which has a mid-point wave height corresponding to a stress x_k. Then the total damage D is calculated as

$$D = \sum x_k^3 f_k / A$$

where A is a constant which depends on the geometry. The joint is assumed to fail when D equals 1. The stress against wave height relationship for a particular joint is given in Table 3.10, and the value of A is 8.75×10^{12}. The total damage is 0.139 per year and the corresponding average lifetime is 7.2 years. Although this estimate is based on a simple approximation and a single year of wave data, it is useful for design purposes. When the calculations are repeated for all the other joints, it is possible to assess the safety of the structure and locate its weakest points. Teams of divers monitor the joints on operational structures and carry out remedial work where necessary. Furthermore, sufficient redundancy is built into the structure for it to remain safe even if a joint were to fail. However, it is possible to check the accuracy of predictions by comparing with observations of crack propagation.

3.4 Shape of distributions

While the location and spread of a set of data are the most important features, two other aspects which may be seen in histograms of large data sets are

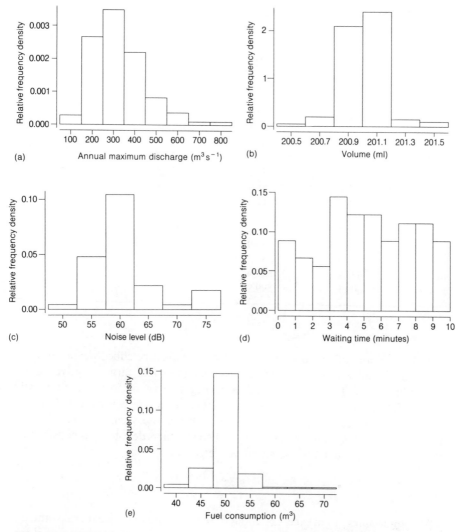

Fig. 3.12 MINITAB histograms for: (a) Thames annual floods (m^{3}s^{-1}); (b) contents of shampoo bottles (ml); (c) metro noise readings (dB); (d) waiting time for train (min); and (e) fuel consumption (m^3).

asymmetry and a lack of data in the flanks because of a high peak or heavy tails. This detail becomes particularly important when we attempt to predict the extreme values taken by variables. Applications include river flood levels, high tides, and tensile failure of certain materials. Another consequence is that some general results for approximate bell-shaped distributions, such as the fact that 95% of the data fall within two standard deviations of the mean, are no longer applicable.

Table 3.10 Stress corresponding to wave height at a joint on an offshore structure

Wave height (m)	Stress wave heading N–S (Pa)	Stress wave heading NE–SW (Pa)	Stress wave heading E–W (Pa)	Stress wave heading SE–NW (Pa)
0.475	26.3	13.4	14.8	25.2
1.450	78.8	40.1	44.4	75.5
2.450	121.2	67.6	75.7	128.4
3.450	167.1	134.4	100.0	178.3
4.450	211.9	174.9	124.8	226.4
5.450	255.6	188.9	150.2	272.5
6.450	305.9	211.6	175.7	319.0
7.450	362.9	243.0	201.5	365.9
8.450	423.9	294.4	239.3	416.7
9.450	489.1	365.9	289.2	471.6
10.450	554.3	437.4	339.1	526.5
11.450	624.3	515.4	392.9	589.2
12.450	699.1	599.8	450.5	659.5

3.4.1 Skewness

A useful measure of asymmetry is based on the average cubed deviation from the mean. Values near zero indicate symmetry, while large positive values indicate a long tail to the right and large negative values indicate a long tail to the left (see Exercise 3.9). The average cubed deviation will have dimensions of the measurements cubed and it is difficult to assess what is 'large', so a non-dimensional measure of symmetry, called **skewness** denoted $\hat{\gamma}$, is constructed by dividing by the cube of the standard deviation.

$$\hat{\gamma} = \frac{\sum (x_i - \bar{x})^3/(n-1)}{s^3}$$

Absolute values of $\hat{\gamma}$ in excess of about 0.5 correspond to noticeable asymmetry in a histogram and absolute values in excess of 2 are rare. The skewness of the crushing strength of the 180 concrete cubes (Fig. 3.11) is -0.60. The following data sets are also given in Appendix D: annual maximum discharges of the River Thames at Teddington for 109 years (water years for the period 1883–1991) (Table D.2); volumes of shampoo in 100 plastic containers with nominal contents of 200 ml (Table D.3); peak noise levels at 46 sites near the Tyne and Wear Metro (Table D.4); waiting times for 90 passengers waiting for an 'every ten minute' tube train service without reference to a timetable (Table D.5); and fuel used by a ferry for 141 passages between the same two ports (Table D.6). Their MINITAB histograms are shown in Fig. 3.12 and their skewnesses are 1.08, 0.33, 1.43, -0.14 and 2.36, respectively.

3.4.2 *Kurtosis*

The extent of the tails of a distribution, relative to the standard deviation, is measured by a non-dimensional quantity, based on the average fourth power of deviations from the mean, known as kurtosis (denoted $\hat{\kappa}$):

$$\hat{\kappa} = \frac{\sum(x_i - \bar{x})^4/(n-1)}{s^4}$$

Values greater than 3 usually indicate relatively extensive tails and a high peak compared with a typical bell-shaped distribution. A flat distribution would have a kurtosis nearer 2. The values of $\hat{\kappa}$ for the concrete cubes, annual maximum discharges, shampoo bottles, Metro noise readings, waiting times for the tube, and fuel consumption records for the ferry are 4.5, 4.9, 6.3, 4.5, 2.0 and 15.9, respectively. A possible mechanism for the high kurtosis of the volumes of shampoo is that there are several filling nozzles, one of which is much more erratic than the others. The extraordinarily high skewness and kurtosis of the fuel consumption volumes is due to two outlying observations. The temptation to dismiss such data as probable recording errors must be avoided! I was able to check the original records, taken from the ship's log, and found there had been gale force headwinds on those passages. If these data were ignored when calculating the mean, average fuel costs would probably be underestimated. They are even more important when it comes to deciding minimum fuel requirements.

Hydrologists calculate skewness and kurtosis for flood records at many sites to help decide on appropriate theoretical distributions for predicting floods (section 4.2.5). In a manufacturing context, high kurtosis may indicate some contaminating distribution which is more variable than the predominant distribution. This may warrant investigation. Furthermore, range charts used in statistical quality control (section 5.7) are sensitive to deviation in kurtosis from 3.

Calculating $\hat{\gamma}$ and $\hat{\kappa}$ is easy in MINITAB. The skewness of data in column 1 can be calculated from

```
LET K5 = SUM((C1 − MEAN (C1)) **3)/(N(C1) − 1)
LET K5 = K5/(STDEV(C5)) **3
```

If you are careful with the brackets it could be written as one line. It may be worth storing as a file.

3.5 Summary

Population and sample

- Define the population carefully – assume it is of size N, where N may be infinite.

- Draw a random sample of size n.
- Estimate population characteristics from the sample.

	Sample estimate of population parameter	Population parameter
Mean	$\bar{x} = \sum x_i / n$	$\mu = \sum x_j / N$
Variance	$s^2 = \sum (x_i - \bar{x})^2 / (n - 1)$	$\sigma^2 = \sum (x_j - \mu)^2 / N$
Standard deviation	$s = \sqrt{s^2}$	$\sigma = \sqrt{\sigma^2}$
Skewness	$\hat{\gamma} = [\sum (x_i - \bar{x})^3 / (n - 1)] / s^3$	$\gamma = [\sum (x_j - \mu)^3 / N] / \sigma^3$
Kurtosis	$\hat{\kappa} = [\sum (x_i - \bar{x})^4 / (n - 1)] / s^4$	$\kappa = [\sum (x_j - \mu)^4 / N] / \sigma^4$

Nearly all of this book is about estimating unknown characteristics of a population from a sample, so the standard deviation calculated from the sample will always be s unless stated otherwise. Although the population parameters are properly defined, they cannot be calculated unless we know the values of the x_j for the entire population.

Histogram

- Label axes clearly.
- The vertical axis of the histogram is relative frequency density (relative frequency/width of class interval).
- Provided all class intervals have the same width, a bar chart, frequency plotted vertically, gives a fair visual impression, and the histogram is the same shape scaled to have an area of 1.
- Use mid-points of class intervals when calculating approximate values of $\bar{x}$ and s.

Cumulative frequency polygon

- Label axes clearly.
- The vertical axis is the proportion, or percentage, of the data less than a particular value of the variable (x) which is plotted horizontally.
- Cumulative frequencies are less than the right-hand end points of intervals.

MINITAB

The DESCRIBE, BOXPLOT, HISTOGRAM, DOTPLOT and STEM-AND-LEAF commands give you a good algebraic and visual summary of the data in column 1. Using the mouse:

Stat ▶ Basic Statistics ▶ Descriptive Statistics ▶
   Variables: C1
Graph ▶ Character Graphs ▶ Boxplot...
Graph ▶ Histogram...
Graph ▶ Character Graphs ▶ Dotplot...
Stat ▶ EDA ▶ Stem-and-leaf...

If you have an earlier release, just type the command name followed by C1.

Exercises

3.1 The following data are carbon contents (%) of coal. Calculate the mean, variances with divisors n and $n-1$, standard deviations with divisors n and $n-1$, median and range.

   87 86 85 87 86 87 86 81 77 85
   86 84 83 83 82 84 83 79 82 73

3.2 The yield points for 120 sample lengths of a given cable, measured in newtons per square millimetre to the nearest integer, can be grouped as follows:

Yield point ($N\,mm^{-2}$)	50–79	80–89	90–99	100–109	110–119	120–129	130–149	150–179	180–239
Frequency	3	6	13	25	24	21	18	7	3

   (i) Draw a histogram, ensuring that the total area is 1.
   (ii) Draw a cumulative frequency polygon.
   (iii) Calculate the approximate mean and standard deviation (s) of the data.
   (iv) Calculate the approximate median, lower quartile, upper quartile, and interquartile range.

3.3 If you travel 60 km at 30 km h^{-1}, 60 km at 60 km h^{-1} and 60 km at 90 km h^{-1}, what is your 'average' speed?

3.4 You invest £1000 for 3 years with interest rates of 10%, 20% and 15%, compounded yearly, applied for the first, second and third year, respectively. What single rate compounded over 3 years would leave you with the same amount of money? Would your answer change if the three interest rates had been 20%, 15% and 10% during the first, second and third year, respectively?

3.5 (i) A company manufactures bedroom furniture, and the length of the door for a particular unit is specified as 2 m. Measurements of the lengths of 100 doors are made in metres $\{x_i\}$. The mean and standard deviation of these measurements are 1.997 m and 0.002 m, respectively. Now suppose that the deviations from 2 m were measured in millimetres $\{y_i\}$. Write down the mean, variance, and the standard deviation of the y_i in millimetres.

(ii) More generally, if we have n data $\{x_i\}$ and they are scaled (or coded) to give,

$$y_i = ax_i + b$$

express $\bar{y}$, s_y^2, and s_y in terms of $\bar{x}$, s_x^2 and s_x. I suggest you start:

$$\bar{y} = \sum y_i/n = \sum (ax_i + b)/n$$

3.6 The following method of calculating quartiles has the advantage that it leads to the original observations or averages of two of them. For the lower quartile:

- Work out $n/4$.
- If $n/4$ is not an integer move up to the next integer and take the corresponding observation.
- If $n/4$ is an integer take the average of the $n/4$ and $(n/4 + 1)$th observations.

For the upper quartile start by working out $3n/4$ and move up or average the $3n/4$ and $(3n/4 + 1)$th observations.

(i) Try this on the Hardap Dam data.
(ii) Compare with the method described in the text $((n + 1)/4)$ and that of taking the median of the observations to the left and right of the median of the original data.
(iii) Would any of these methods generalize to calculating a lower and upper 10% percentile?

3.7 Prove the 'hand-computation' formula

$$\sum (x_i - \bar{x})^2 = \sum x_i^2 - \left(\sum x_i\right)^2 \bigg/ n$$

I suggest you start with

$$\sum (x_i - \bar{x})^2 = \sum (x_i - \bar{x})(x_i - \bar{x})$$
$$= \sum (x_i - \bar{x})x_i - \sum (x_i - \bar{x})\bar{x}$$

and note that $\bar{x}$ is constant with respect to the summations.

3.8 'Hand-computation' formulae such as the right-hand side of the identity

$$\sum_{i=1}^{n} (x_i - \bar{x})^2 = \sum_{i=1}^{n} x_i^2 - \left(\sum_{i=1}^{n} x_i\right)^2 \bigg/ n$$

can be very sensitive to rounding errors and they should never be programmed on a computer.

(a) Calculate s for the following set of numbers.

   96.351   96.021   96.138   96.251   96.283

by:

- (i) using the left-hand side of the above identity;
- (ii) using the right-hand side of the above identity and working to the full accuracy of a hand-held calculator;
- (iii) using the right-hand side of the above identity and rounding all interim results to 5 significant figures.

(b) What 'obvious' constant could be subtracted from each datum to reduce rounding errors when using the hand-computation formula with these data?

(c) Calculate the standard deviation of the following five estimates (in metres per second) of the speed of light:

   299 792 458.351, 299 792 458.021, 299 792 458.138,
   299 792 458.251, 299 792 458.283

by using the standard deviation key on your calculator. (The deviations from 299 792 458 which was the value accepted by the 17th General Congress on Weights and Measures (1983) are fictitious. See the article by Greenfield and Siday (1980) for a more detailed discussion of rounding errors.)

3.9 A manufacturer supplies paint in cans with declared contents of 1 litre. The following data are deviations from 1 litre in units of 5 ml for a sample of 31 cans.

Deviation ($\times 5$ ml)	Number of cans
-1	10
0	15
1	3
2	2
3	1

- (i) Draw a line chart for these data.
- (ii) Calculate the mean, standard deviation (s), skewness and kurtosis.

3.10 The following series is the minimum monthly flow ($m^3 s^{-1}$) in each of the 20 years 1957 to 1976 at Bywell on the River Tyne.

21   36   4   16   21   21   23   11   46   10
25   12   9   16   10   6   11   12   17   3

(i) Draw a boxplot of the data.
(ii) Calculate the mean, standard deviation (s), and skewness.
(iii) Take the logarithms of the data, repeat (i) and (ii) and compare the results.

3.11 Look up the article by Dalal *et al.* (1989). Comment on the difference between Figures 1(a) and 1(b).

4

Modelling variability

4.1 Discrete probability distributions

A cycle manufacturer builds wheels using a semi-automatic lacing machine and a fully automated wheel truing robot. A random sample of 20 wheels is taken from each shift, and measurements of spoke tension, run-out (buckle), and 'hop' (eccentricity) are made. As a result of these tests wheels are classified as 'top', 'second', or 'poor' quality. All assembled bicycles are inspected before dispatch, and any poor-quality wheels should be detected. They can usually be trued by hand. The second-quality wheels are adequate for touring bicycles, but the manufacturer knows from past experience that when the process is running well 80% of wheels are of top quality, and only a few of the remainder are of poor quality.

4.1.1 Binomial distribution

The number of second- and poor-quality wheels in a sample is a discrete variable, and its probability distribution gives probabilities that it will take values in its range $(0, 1, \ldots, 20)$. We can express these probabilities in terms of process parameters, if we are prepared to make some assumptions. These expressions allow us to investigate the consequences of various rules for stopping the process and adjusting the machines. Suppose that the process is producing a proportion (p) of non-top-quality wheels. The random selection of the 20 wheels justifies an assumption that the probability that any one is not of top quality is p, independently of the quality of any of the other wheels in the sample. The number of non-top-quality wheels will have a binomial distribution. This distribution was described by James Bernoulli (1654–1705) in his posthumous work *Ars Conjectandi* (1713).

Example 4.1

A process for assembling bicycle wheels produces a proportion p of 'defective' (non-top-quality) wheels. A random sample of three wheels is taken. Let the variable X be the number of defective wheels in the sample. The sample space is shown in the tree diagram of Fig. 4.1. The probability that X takes the values

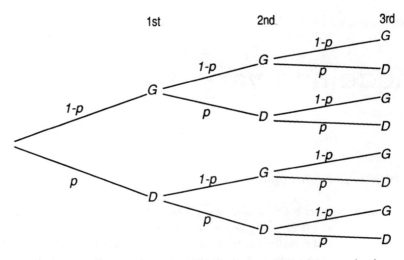

Fig. 4.1 Sample space for three wheels randomly selected from production.

0, 1, 2 and 3 can be summarized by the formula

$$\Pr(X = x) = {}_3C_x p^x (1 - p)^{3-x} \qquad \text{for } x = 0, 1, 2, 3$$

The distinction between X and x is that the former is the name of the variable and the latter represents a particular value. The formula can easily be verified from the diagram, but we can generalize it if we note that ${}_3C_2$ is the number of ways of labelling two of the three wheels 'defective' and, less specifically, ${}_3C_x$ is the number of ways of labelling x of the three wheels 'defective'.

The following general description has wide applications. Suppose we can describe a situation as a sequence of trials, each of which has two possible outcomes commonly referred to as 'success' or 'failure'. If the probability of a success at each trial is constant then the number of successes has a binomial distribution. We can summarize this formally as follows:

1. There is a fixed number of trials (n).
2. There are two possible outcomes for each trial ('success' or 'failure').
3. There is a constant probability of success (p). This implies that the outcomes of trials are independent.

Let the variable X be the number of successes. Then $\Pr(X = x)$, which is usually abbreviated to $P(x)$, is given by

$$P(x) = {}_nC_x p^x (1 - p)^{n-x} \qquad \text{for } x = 0, 1, \ldots, n$$

The proof follows from Example 4.1 because ${}_nC_x$ is the number of ways of labelling x of the outcomes of the n trials 'defective'. For any probability distribution we must have

$$\sum_{\substack{\text{Possible} \\ \text{values} \\ \text{of } x}} P(x) = 1$$

If you are familiar with the binomial expansion you can see that the binomial expansion of

$$(p + (1 - p))^n$$

is the sum of the terms of the binomial distribution. Since $(p + (1 - p))$ is just 1 we have verified the general result for the binomial distribution:

$$\sum_{x=0}^{n} {}_nC_x p^x (1 - p)^{n-x} = 1$$

The binomial distribution has two parameters, n and p, and there is a convenient

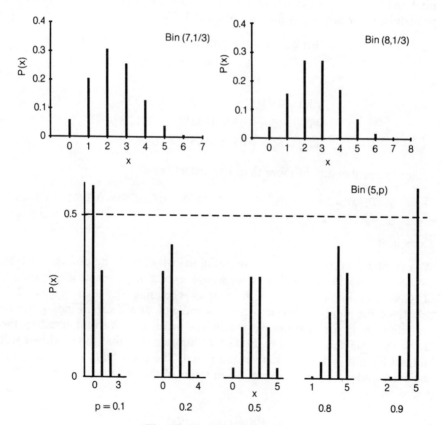

Fig. 4.2 Binomial distributions.

shorthand for saying that X has a binomial distribution with n trials and probability of success p:

$X \sim \text{Bin}(n, p)$

The $\sim$ is read as 'distributed as'. Line charts for binomial probabilities with a few choices of n and p are shown in Fig. 4.2.

Example 4.2

Five per cent of a certain manufacturer's fireworks are defective (do not ignite). Find the probability that more than two in a random sample of 12 are defective.
   Let X be the number of defective fireworks. Then

$X \sim \text{Bin}(12, 0.05)$

More than two defectives includes 3, 4,... and so on up to 12. Therefore finding the probability directly would involve calculating ten individual probabilities from the formula. We can save time if we use the fact that the sum of the probabilities for all 13 possible outcomes is 1. Then

$$\begin{aligned} \Pr(X > 2) &= 1 - \Pr(X \leqslant 2) \\ &= 1 - (P(0) + P(1) + P(2)) \end{aligned}$$

Now

$$\begin{aligned} P(0) &= 0.95^{12} & &= 0.540\,36 \\ P(1) &= 12 \times 0.05 \times 0.95^{11} & &= 0.341\,28 \\ P(2) &= 66 \times 0.05^2 \times 0.95^{10} &&= 0.098\,79 \end{aligned}$$

Hence the probability of more than two defectives is

$$\Pr(X > 2) = 1 - (0.540\,36 + 0.341\,28 + 0.098\,79) = 0.019\,57$$

Example 4.3

We can now investigate rules for stopping the wheel building process. At best, 20% of wheels will be defective so we would 'expect' four defectives per sample of 20. We also know that even if the process is running with an average of 20% defectives, the number of defectives in a sample of 20 will vary. Finding five or six defectives in a single sample would not necessarily warrant resetting the lacing machine and robot. Let us start with a rule (R1) that the machines will be reset if eight or more defectives are found. The probability of resetting the machines under R1 can be expressed in terms of the underlying proportion of defectives (p).

$$\Pr(\text{reset}) = \sum_{8}^{20} P(x)$$

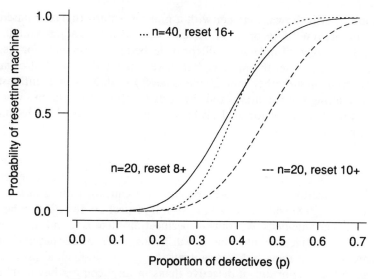

Fig. 4.3 Reset rules for wheel building robots.

where

$$P(x) = {}_{20}C_x p^x (1 - p)^{20 - x}$$

which is plotted against p in Fig. 4.3. The calculations are easily programmed. If you use MINITAB the command and subcommand

```
CDF   7   K9;
   BINOMIAL   20   K1.
```

will put

$$P(0) + P(1) + \cdots + P(7)$$

for a binomial distribution with $n = 20$ and p equal to K1, into K9. Alternatively use the mouse:

Calc ▶ Probability Distribution ▶ Binomial ▶

    Cumulative probability
    Number of trials: 20
    Probability of success: K1
    Input constant: 7
    Optional storage: K9

This has the advantage that you can clearly see all the other facilities from the menus. A similar plot for another possible rule (R2), reset if ten or more defectives are found, is also shown, together with a plot for a third rule (R3) based on a larger sample, reset if 16 or more defectives found in a random sample of 40 wheels. The larger sample size will provide better information for deciding whether or not to reset the process, but there is the additional cost of making measurements on 40 rather than 20 wheels every shift. Measurement costs, the costs of resetting the machines, and the costs of truing poor-quality wheels by hand at the final inspection will all have to be considered when deciding on the best policy.

Example 4.4

A filter manufacturer (M) has ordered chromed endcaps from a new supplier (S). Batches of 1000 will be delivered. M will inspect 30 and accept the batch if no more than one defective (out-of-specification) item is found. If more than one item is defective the entire batch will be inspected. S will pay the costs of this inspection and replace defective items.

Let p be the proportion of defective items in an incoming batch. Then the probability that the first item inspected is defective is p, but subsequent probabilities will vary slightly depending on previous results. For example, if p is 0.05 the probability the first item is defective is 50/1000. The probability the second item is defective is either 50/999 or 49/999 depending on whether the first item was good or defective, and so on. However, the binomial distribution is a very good approximation for reasonably large batch sizes (see Exercise 4.4).

The **average outgoing quality** (AOQ) is the average proportion, over many batches, of defectives leaving the inspection process, and depends on the incoming proportion of defectives (p). The following expression gives the probability of accepting the batch

$$\text{Pr(accept batch)} = (1 - p)^{30} + 30p(1 - p)^{29}$$

Three example calculations are

$$
\begin{aligned}
p &= 0.001 & \text{Pr(accept)} &= 0.9996 \\
p &= 0.01 & \text{Pr(accept)} &= 0.9639 \\
p &= 0.1 & \text{Pr(accept)} &= 0.1837
\end{aligned}
$$

The calculation of the AOQ is based on two assumptions, both of which are questionable. The first is that 100% inspection will find all the defective items. The second is that all the defectives found are replaced with good items. Then, at the end of the process of inspecting the entire batch, which should be taken to include the replacement of the defective items found, the outgoing batch will contain no defectives and

$$AOQ = \text{Pr(accept batch)} \times p + \text{Pr(inspect entire batch)} \times 0$$

p	AOQ (%)
0.001	0.1
0.01	0.96
0.02	1.76
0.03	2.32
0.04	2.64
0.05	2.77
0.06	2.73
0.07	2.59
0.10	1.84

AOQL is approximately 2.8%

The maximum value of the average outgoing quality is the **average outgoing quality limit** (AOQL). In this case the AOQL is about 2.8% (Fig. 4.4), but it is generally accepted that 100% inspection is not completely reliable (see, for example, Hill, 1962), in which case the AOQ may not decrease as p increases beyond 0.05. Similar sampling plans are covered in detail in British Standards BS6000–BS6002.

The AOQL is a worst possible scenario, and as it is in both the manufacturer's and suppliers' interests to keep p very low some manufacturers will help trusted suppliers achieve this.

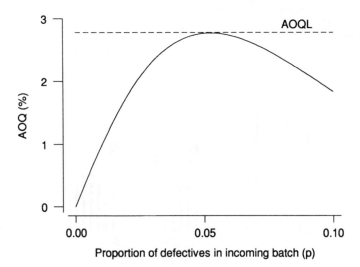

Fig. 4.4 Average outgoing quality for inspection of chromed endcaps.

4.1.2 Poisson distribution

A chemical process produces a white powder which may contain unwanted specks. These are usually a consequence of impurities in raw materials, but could be rust particles from the steel pipework through which the powder passes. A few specks per kilogram are acceptable, but this is a feature of the process which needs monitoring. The Poisson distribution, published by Siméon Denis Poisson in 1837, describes the variability of the number of specks per kilogram if they occur at random. It can be used as a basis for deciding whether or not production is proceeding satisfactorily.

The general formulation is based on the idea of a **Poisson process**. In a Poisson process events occur in some continuum (such as time, volume, area and length) subject to the following assumptions. For the moment we will refer to the continuum as 'time' (t).

1. Events occur at random and independently. This is a strong assumption, and one consequence is that if you have just seen an event this has no effect on your chance of seeing another.
2. Two or more events cannot occur simultaneously.
3. The events themselves occupy a negligible amount of time.
4. Events occur at an average rate of λ per unit of time, and λ will be assumed constant for the derivation of the formula.

Let X be the number of events in a length of time t. Then the probability distribution of X is given by the formula

$$P(x) = \frac{e^{-\lambda t}(\lambda t)^x}{x!} \qquad \text{for } x = 0, 1, 2, \dots$$

Although it is defined for x ranging from 0 to infinity the probabilities are

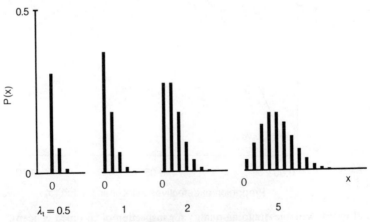

Fig. 4.5 Poisson distributions.

negligible for x much larger than λt and add to 1, as they must:

$$\sum_0^\infty P(x) = e^{-\lambda t} \sum_{x=0}^\infty \frac{(\lambda t)^x}{x!} = e^{-\lambda t} e^{\lambda t} = 1$$

A simple derivation of the formula is given in Appendix A1. The probabilities depend on a single parameter only, which is the product of λ and t. Line charts for a few choices of λt are shown in Fig. 4.5.

Example 4.5

It has been decided that an average rate of 25 specks per kilogram of powder is just acceptable. The process will be monitored by checking 100 gram samples at approximate hourly intervals. The probabilities of observing $0, 1, \ldots, 8, 9$ or more specks, if the process is running at an acceptable level, can be calculated as follows. The rate λ is 25 specks per kilogram and the mass of continuum we require probabilities for is 0.1 kg. Therefore $\lambda t = 2.5$.

$$
\begin{aligned}
P(0) &= e^{-2.5} & &= 0.0821 \\
P(1) &= (2.5)e^{-2.5} & = 2.5 \times P(0) & = 0.2052 \\
P(2) &= (2.5)^2 e^{-2.5}/2! = 2.5 \times P(1)/2 & &= 0.2565 \\
P(3) &= (2.5)^3 e^{-2.5}/3! = 2.5 \times P(2)/3 & &= 0.2138 \\
P(4) &= 2.5 \times P(3)/4 & &= 0.1336 \\
P(5) &= 0.0668 \\
P(6) &= 0.0278 \\
P(7) &= 0.0099 \\
P(8) &= 0.0031 \\
\Pr(X \geqslant 9) &= 1 - (P(0) + \cdots + P(8)) = 0.0011
\end{aligned}
$$

You can check these against MINITAB with the command and subcommand

    PDF;
        POISSON 2.5.

As usual you can use the mouse:

    Calc ▶ Set Patterned Data ▶
        Store result in column: C8
        Start at: 0
        End at: 8
        Increment: 1
    Calc ▶ Probability Distribution ▶ Poisson ▶
        • Probability

Mean: 2.5
Input column: C8

The number of unwanted specks could suddenly increase because of a change in raw materials, and a possible solution would be to use a different batch. In contrast to this, a problem with rust in pipework would lead to a slow increase in unwanted specks, and the only solution would be to replace it. Even short halts to the process are expensive, but the company cannot afford to lose its good reputation by supplying a poor-quality product. If nine or more specks are detected in a particular sample their chemical content will be analysed so that their origin can be traced. The probability of doing this chemical analysis if the process is running with an acceptable average level of specks is about 0.001. If a problem with raw materials is indicated, 100 g samples will be looked at every 10 minutes. If a high level of specks is maintained the raw material will be changed. It would be unwise to take this action immediately after recording a large number of specks in one sample! Apart from the fact that the unusual does sometimes occur – for example, nine or more specks will be noted in one in a thousand samples on average – specks may tend to occur in clusters, because the impurities in the raw material are larger particles which are broken down into specks. The distribution of the number of specks would then be better modelled by the negative binomial distribution (see Exercise 4.22). The variance of this distribution is greater than that of the Poisson distribution with the same mean, and this increases the probability of seeing large numbers of specks. If records are kept of the numbers of specks in 100 g samples it will be possible to compare relative frequencies with the hypothetical probabilities. If the chemical analysis shows some specks are small particles of rust the trend will be monitored, and if the problem persists pipework will have to be replaced.

The Poisson distribution is often taken as a simple model for the number of road accidents at junctions, and used to assess the effects of road safety measures. It is plausible, provided accidents are counted rather than the number of people or cars involved. The accident rate may vary systematically throughout a year, so λt is replaced by the integral of the time-varying rate over the length of time t (if rates are being compared over full calendar years this is the yearly accident rate).

There are many other discrete probability distributions, some of which are covered in the case studies or exercises, but the binomial and Poisson distributions are sufficient for many applications. Their main limitation is that they cannot model dependence between trials or occurrences. The number of bacteria in jars of water taken from a source may be well modelled by a Poisson distribution, but larvae may tend to cluster (due to social instincts perhaps). If overall numbers of bacteria and larvae are similar, the counts of larvae in jars will be more spread out than counts of bacteria. This is more than a nice distinction: assessments of environmental hazards may be very sensitive to what is or is not assumed about clustering.

Table 4.1 The number of α-particle emissions from a polonium source in 7.5 s intervals

Number of emissions (x)	Number of intervals in which x particles seen	Poisson distribution probability	Poisson predicted frequency (2608 × probability)
0	57	0.0209	54.40
1	203	0.0807	210.52
2	383	0.1562	407.36
3	525	0.2015	525.50
4	532	0.1949	508.42
5	408	0.1509	393.52
6	273	0.0973	253.82
7	139	0.0538	140.32
8	45	0.0260	67.88
9	27	0.0112	29.19
10	10 ⎫		
11	4 ⎬	0.0066	17.08
12	2 ⎭		

4.1.3 Expected value

An expected value is a population average. It is not necessarily a 'most likely value'.

Example 4.6

Rutherford and his colleagues thought that the emission of α-particles from a radioactive source might be well modelled by a Poisson process. In one experiment they observed the number of emissions from a polonium source in 2608 periods of 7.5 seconds. The data (Rutherford and Geiger, 1910) are given in Table 4.1. If their theory leads to useful predictions, the relative frequencies in Table 4.1 should be consistent with Poisson probabilities. We need to choose the parameter λt of the Poisson distribution before we can calculate the probabilities and make the comparisons. If the Poisson distribution represents the imaginary population of all 7.5 s intervals that might have been observed, relative frequencies will tend towards Poisson probabilities and we can define a population mean in a similar way to the sample mean. This population mean will depend on the parameter of the distribution which can be estimated by equating population and sample means.

Let x represent the number of α-particles emitted in an interval. The mean number of α-particles emitted in the sample of 2608 intervals is

$$\bar{x} = \frac{0 \times 57 + 1 \times 203 + 2 \times 383 + \cdots + 10 \times 10 + 11 \times 4 + 12 \times 2}{2608}$$

$$= 0 \times \frac{57}{2608} + \cdots + 12 \times \frac{2}{2608}$$

$$= \sum_0^\infty x \times (\text{relative frequency of } x \text{ particles}) = 3.87$$

The population mean is

$$\mu = \sum_0^\infty xP(x)$$

If we model the population by a Poisson distribution,

$$\mu = \sum_{x=0}^\infty x \frac{e^{-\lambda t}(\lambda t)^x}{x!}$$

and as the first term is zero so we can equally well start with x at 1 and write

$$\mu - = \sum_{x=1}^\infty x \frac{e^{-\lambda t}(\lambda t)^x}{x!}$$

The next move is to factor out λt and write

$$\mu = \lambda t \sum_{x=1}^\infty \frac{e^{-\lambda t}(\lambda t)^{x-1}}{(x-1)!}$$

Now change the variable from x to y, where $y = x - 1$.

$$\mu = \lambda t \sum_{y=0}^\infty \frac{e^{-\lambda t}(\lambda t)^y}{y!}$$

The summation is the sum of Poisson probabilities from 0 to ∞ and therefore equals 1, so

$$\mu = \lambda t$$

This is consistent with the definition of λ as the average rate per unit time. The Poisson probabilities in the third column of Table 4.1 are calculated by taking λt equal to 3.87 particles per period.

The average value of the variable X, formally called the **expected value** of X or mean value of X is denoted by $E[X]$ or μ, and defined by

$$E[X] = \sum_{\substack{\text{all possible} \\ \text{values of } x}} xP(x)$$

The average value of the squared deviation from the mean, formally called the expected value of $(X - \mu)^2$ or **variance** of X is denoted by $E[(X - \mu)^2]$ or

σ^2, and defined by

$$E[(X - \mu)^2] = \sum_{\substack{\text{all possible} \\ \text{values } x}} (x - \mu)^2 P(x).$$

The average value of an arbitrary function ϕ of X, $\phi(X)$, is

$$E[\phi(X)] = \sum_{\substack{\text{all possible} \\ \text{values of } x}} \phi(x)P(x)$$

Similar arguments to those used in Example 4.6 lead to the following results:

	Mean	Variance	Std. dev.
Binomial	np	$np(1-p)$	$\sqrt{np(1-p)}$
Poisson	λt	λt	$\sqrt{\lambda t}$

The last result can be put to immediate use. If the Poisson distribution is a good model for the α-particle emissions the sample mean and variance should be similar. The variance is 3.68, which is close to the mean of 3.87.

4.2 Continuous probability distributions

Flood defence structures for residential areas are often designed to protect against annual floods with a 100-year return period. The average time between floods as large as, or larger than, the 100-year flood is 100 years, so the probability of such a flood in any one year in 0.01. The first step in the design process is to estimate this flood. If we had records for thousands of years, and assumed that nothing relevant had changed, we could estimate the 100-year flood from a histogram or, more conveniently, from the cumulative frequency polygon. In practice, records exceeding 50 years are relatively uncommon, and even if they are available there may have been some climatic or geomorphic changes. One solution to the problem is to assume some general smooth shape towards which histograms of annual floods would tend if we had enough data. The choice of general shape can be made by looking at histograms from many sites, especially those with long records, or by theoretical arguments which rely on simplifying assumptions. The details of the shape are determined by parameters which can be related to location, spread, skewness and kurtosis. The location, spread and, when necessary, skewness and kurtosis can be estimated for specific sites from relatively short records. This approach is subject to many inaccuracies, but is more reliable than using the largest flood in the past 100 years (try Exercise 4.21), if such a record were available, and feasible with fewer data!

A similar problem arises with the pavers for the container park. The port authority really wishes to know the proportion of pavers which have a cement content out of specification, rather than the population mean and standard deviation. If a bell-shaped distribution is assumed for the population, this proportion is determined by the mean and standard deviation which can be estimated from the sample of 24 pavers. The estimate of the proportion will be

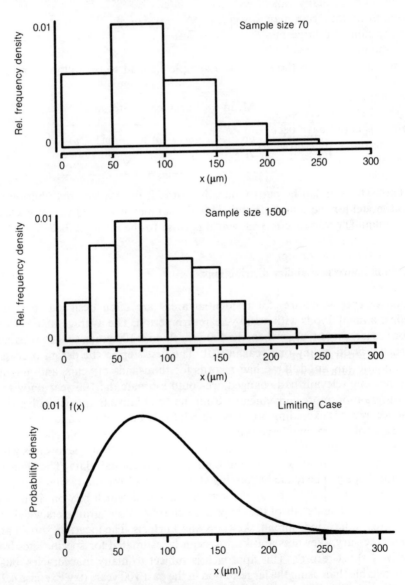

Fig. 4.6 Histograms of trough-to-peak heights (hull roughness data) tending to a smooth curve as the number of transects increases.

inaccurate because it is sensitive to the values of the population mean and standard deviation, which are themselves estimated from a fairly small sample. It is nevertheless the best that can be made without testing more pavers, and the manufacturer of the pavers would no doubt prefer to negotiate a discount on the basis of this estimate than have 100 000 pavers rejected. The bell-shaped distribution, usually called the **normal distribution**, has been found to give a good fit to data from many industrial processes. It can be derived theoretically as the sum of a large number of errors, none of which predominates. This is the renowned **central limit theorem**, and a proof is given in Appendix A2.

It is helpful to establish a few general points about histograms tending to smooth curves before looking at specific distributions. Refer back to the data on ship hull roughness. There were 550 readings and a histogram was drawn with 13 class intervals. If many more readings had been available more class intervals could have been used. It is reasonable to suppose that as the sample size becomes very large the histogram will look more like a smooth curve (Fig. 4.6). The histogram is a plot of 'relative frequency density' against the variable (x), in this case maximum trough-to-peak height over a 50 mm transect length. The smooth curve it is imagined to tend towards is known as a **probability density function** (pdf). The curve is thought of as a model for the histogram of the imaginary infinite population of all such heights, and can be expressed by a formula $f(x)$. The area under the histogram between any two numbers a and b gives the proportion of sample heights in this range. Similarly, the area under the pdf between a and b represents the probability that a randomly selected height (X) will be between them. Formally,

$$\Pr(a < X < b) = \int_a^b f(x)\,dx$$

This is illustrated in Fig. 4.7. The cumulative frequency polygon for the roughness data showed the proportion less than any particular value x. As the

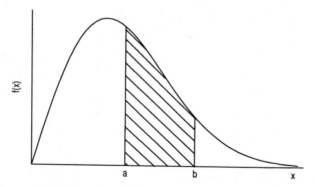

Fig. 4.7 If the variable X has the pdf $f(x)$, the area under the curve between a and b represents $\Pr(a < X < b)$.

histogram tends towards the pdf, the cumulative frequency polygon tends towards the **cumulative distribution function** (cdf), which gives the proportion of the population less than x. The cdf is denoted by $F(x)$, where

$$F(x) = \Pr(X < x) = \int_{-\infty}^{x} f(\theta)\,\mathrm{d}\theta$$

The cdf is the integral of the pdf, and, conversely, the derivative of the cdf is the pdf:

$$\frac{\mathrm{d}F(x)}{\mathrm{d}x} = f(x)$$

The cdf, its relationship with the pdf, and the corresponding relationship between the cumulative frequency polygon and histogram are shown in Fig. 4.8. Any function which is never negative and has an area underneath it of 1 is a possible pdf, but there are a few commonly used distributions which allow for a very wide range of shapes.

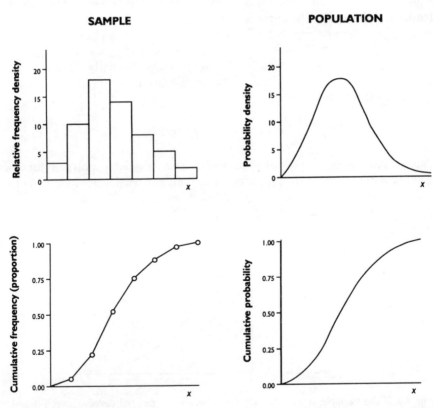

Fig. 4.8 Relationships between pdf, histogram, cdf and cumulative frequency polygons.

4.2.1 The normal distribution

The normal distribution is usually attributed to Abraham de Moivre (1667–1754), who essentially derived the pdf as an approximation to the sum of terms from a binomial distribution with a large number of trials, but Hilary Seal suggests, in a letter to the editors of the *Journal of the Institute of Actuaries Students' Society*, reproduced in Kendall and Plackett (1977), that Thomas Simpson was the first to draw attention to the physical significance of the result in 1757. Kendall and Plackett (1977) also reproduce a bibliographic note (Daw and Pearson 1972) on de Moivre's derivation. The normal distribution is also commonly known as the **Gaussian distribution** after Carl Friedrich Gauss (1777–1855). Gauss's first description of the distribution was in an aside in his 1809 text *Theoria Motus Corporum Celestium*. However, as was common at the time, he gave no reference, so it is hard to establish whether he derived it independently or was quoting what was generally known. The pdf is

$$f(x) = \frac{1}{\sigma\sqrt{2\pi}} e^{-((x-\mu)/\sigma)^2/2} \qquad \text{for } -\infty < x < \infty$$

It has a bell shape which is centred on μ, and is nearly all contained within 3σ of μ. The parameters μ and σ are the mean and standard deviation of the distribution, and are the values the sample mean and standard deviation tend towards as the sample size increases (demonstrated in section 4.2.4). The pdf has a point of inflexion when x is a distance σ from μ, (Fig. 4.9). While $f(x)$ is formally defined for any value of x, it is negligible if x is more than 4σ from μ, and the distribution is a good approximation for many variables, such as weights or volumes dispensed during filling processes, dimensions of machined items and so on. A normal distribution, with the same mean and variance, is shown superimposed on the histogram of the compressive strengths of concrete blocks in Fig. 3.11.

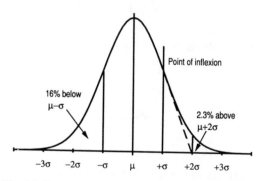

Fig. 4.9 Some properties of the normal distribution.

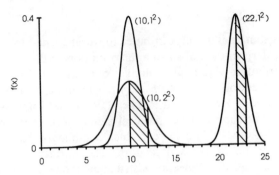

Fig. 4.10 Normal distributions with means 10, 22 and standard deviations of 1 and 2. Both shaded areas are 0.34.

The cdf

$$F(x) = \int_{-\infty}^{x} \frac{1}{\sigma\sqrt{2\pi}} e^{-((\theta - \mu)/\sigma)^2/2} \, d\theta$$

cannot be expressed exactly as a simple function of x and has to be evaluated numerically. However, μ and σ simply determine the location and scale (Fig. 4.10), so it is only necessary to produce tables for one choice of these parameters, $\mu = 0$ and $\sigma = 1$ being the most convenient. This leads to the definition of the **standard normal distribution** as a normal distribution with a mean of 0 and a standard deviation of 1. It is conventional to use z for a standard normal variable and the Greek letter ϕ for its distribution, so its pdf is written

$$\phi(z) = \frac{1}{\sqrt{2\pi}} e^{-z^2/2}$$

and its cdf (Fig. 4.11), Table E.2 in Appendix E, is

$$\Phi(z) = \frac{1}{\sqrt{2\pi}} \int_{-\infty}^{z} e^{-\theta^2/2} \, d\theta$$

The **inverse** cdf(Φ^{-1}) is defined by:

$$z = \Phi^{-1}(p)$$

if and only if

$$\Phi(z) = p$$

Example 4.7

From Table E.2, $\Phi(1)$ equals 0.8413. The distribution is symmetric about 0 so the area between -1 and 1 is 0.6826 (Fig. 4.12). Since σ is a scale factor (proved

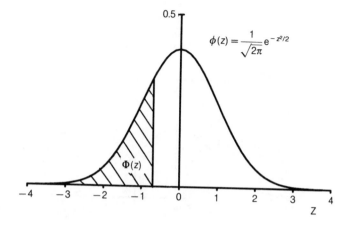

$$\phi(z) = \frac{1}{\sqrt{2\pi}}e^{-z^2/2}$$

$\Phi(z)$

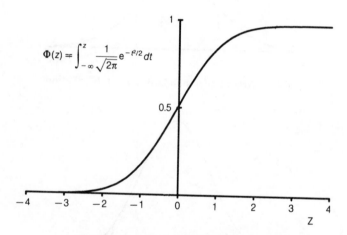

$$\Phi(z) = \int_{-\infty}^{z} \frac{1}{\sqrt{2\pi}}e^{-t^2/2}\,dt$$

Fig. 4.11 Standard normal pdf and cdf.

formally below) we can deduce that 68%, approximately two-thirds, of the distribution lies within one standard deviation of the mean. The value of $\Phi(2)$ is 0.9772, and it follows that 0.9544, approximately 95%, of the distribution lies within two standard deviations of the mean. The value of $\Phi(4)$ is outside the range of Table E.2, but it exceeds 0.99997. This is why the area beyond four standard deviations from the mean is nearly always thought of as negligible.

Percentage points of the standard normal distribution are often needed and the standard notation for the upper $\alpha \times 100\%$ point (the value such that a proportion α lies above it) is z_α (Fig. 4.13). By definition,

$$\Phi(z_\alpha) = 1 - \alpha$$

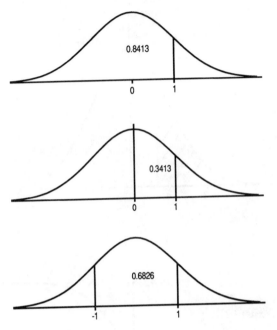

Fig. 4.12 Areas under standard normal distribution.

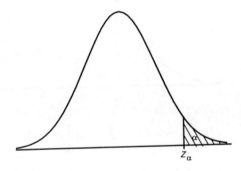

Fig. 4.13 $\alpha \times 100\%$ percentage point of standard normal distribution.

They can be obtained by interpolating within Table E.2, but it is convenient to have a separate table of common values (Table E.3). Both areas and percentage points can be obtained from MINITAB:

Calc ▶ Probability Distributions ▶ Normal...

Example 4.8

From Table E.3, $z_{0.05}$ equals 1.645, so 90% of the distribution lies between -1.645 and 1.645. As an approximation to this, the number 1.7 is easy to remember. Using the notation for the inverse cdf,

$$1.645 = \Phi^{-1}(0.95)$$

Scaling normal distributions

If the variable X has a normal distribution with mean μ and variance σ^2 we write

$$X \sim N(\mu, \sigma^2)$$

The variable Z has a standard normal distribution, so

$$Z \sim N(0, 1).$$

The scaling property can be expressed as,

$$\Pr(X < x) = \Pr\left(Z < \frac{x - \mu}{\sigma}\right)$$

The proof follows from substituting

$$z = \frac{\theta - \mu}{\sigma}$$

in the integral for $\Phi(x)$. That is

$$\Pr(X < x) = \int_{-\infty}^{x} \frac{1}{\sigma\sqrt{2\pi}} e^{-((\theta - \mu)/\sigma)^2/2} \, d\theta$$

$$= \int_{-\infty}^{(x - \mu)/\sigma} \frac{1}{\sqrt{2\pi}} e^{-z^2/2} \, dz$$

$$= \Pr\left(Z < \frac{x - \mu}{\sigma}\right)$$

Another way of expressing this result follows from noting that

$$\Pr(X < x) = \Pr\left(\frac{X - \mu}{\sigma} < \frac{x - \mu}{\sigma}\right)$$

since the inequalities are equivalent, and observing that $(X - \mu)/\sigma$ therefore has the same distribution as Z. In symbols

$$\frac{X - \mu}{\sigma} \sim N(0, 1)$$

Example 4.9

The specification, in BS6717, for the cement content of pavers is $380 \, \text{kg m}^{-3}$, equivalent to 16.9% cement if dry concrete density is taken as $2250 \, \text{kg m}^{-3}$. The mean and standard deviation of the cement contents of the 24 blocks were 17.19% and 1.79%, respectively. Let X represent the percentage cement content

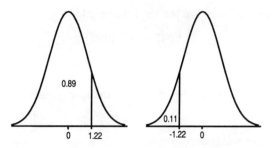

Fig. 4.14 Using the symmetry of the normal distribution to calculate areas.

of a paver. No limits around 16.9% are given in the standard, but an engineer representing the port authority would be concerned if many pavers had cement contents below 15%. If we assume that cement contents have a normal distribution with a mean of 17.19 and a standard deviation of 1.79, that is

$$X \sim N(17.19, 1.79^2)$$

then

$$\Pr(X < 15) = \Pr\left(Z < \frac{15 - 17.19}{1.79}\right)$$

$$= \Pr(Z < -1.22)$$

This is not tabled directly but can easily be found from $\Phi(1.22)$ by using the fact that the normal distribution is symmetric and has a total area of 1 (Fig. 4.14). The estimated proportion is 0.11, but there are several sources of error. Cement content analyses are not exact measurements of the actual cement contents of the selected pavers. Even if they were, the sample mean and standard deviation will not exactly equal the population values and the population will not be exactly normal. Assessment of the accuracy of such estimates will be covered in later chapters.

Example 4.10

A distillery sells bottles of whisky with declared contents of 750 ml. The 1979 Weights and Measures Act stipulates that:

(i) the average contents must exceed 750 ml;
(ii) no more than $2\frac{1}{2}\%$ of output must contain less than 735 ml;
(iii) no bottles must contain less than 720 ml.

Suppose that the volumes of whisky are normally distributed with a mean of 770 ml and a standard deviation of 18 ml. Will the distillery comply with the Act?
    We first check condition (ii):

$$\Pr(X < 735) = \Pr\left(Z < \frac{735 - 770}{18} \right)$$

$$= \Pr(Z < -1.944) = 0.026$$

and condition (iii):

$$\Pr(X < 720) = \Pr(Z < -2.78) = 0.003$$

Condition (i) is obviously satisfied, but we have an estimated 2.6% of bottles containing less than 735 ml and an estimated 0.3% containing less than 720 ml, thus infringing conditions (ii) and (iii) and representing a slight risk of prosecution.

Now suppose the distillery buys new bottling machinery with a standard deviation of 10 ml. The mean volume dispensed can be adjusted. What volume would you recommend setting for the mean?

Condition (i) requires:

$$750 < \mu$$

Condition (ii) requires:

$$735 + 1.96\sigma < \mu \Rightarrow \mu > 754.6$$

Condition (iii) is less specifically defined in the context of a normal distribution. To be on the safe side

$$720 + 4\sigma < \mu \Rightarrow \mu > 760$$

I would suggest setting the mean at a target value of 760 ml.

Process capability

Many industrial variables have distributions which are approximately normal. The fact that nearly all the normal distribution lies within three standard deviations of the mean leads to a definition of the **process capability index** as

$$C_p = \frac{\text{Upper Specification Limit} - \text{Lower Specification Limit}}{6\sigma}$$

It is calculated by replacing σ with its estimate, and values less than 1 (low capability) are usually considered unsatisfactory. If a process has low capability it is well worthwhile checking that the specification limits really do need to be that close before attempting to reduce variability! The process capability index is appropriate provided the process mean is at the mid-range. However it may be at some other value, in which case the **process performance index**, defined as

$$C_{pk} = \text{minimum of} \begin{cases} \dfrac{\text{Upper Specification Limit} - \text{Mean}}{3\sigma} \\[2ex] \dfrac{\text{Mean} - \text{Lower Specification Limit}}{3\sigma} \end{cases}$$

is more relevant. This can also be used with one-sided specification limits which are set for impurities.

The interpretation of C_p is sensitive to the assumption of normality (Exercise 4.16), but markedly non-normal distributions might themselves indicate problems. If out-of-specification items are rejected or reworked, and records are only kept of measurements made on items that are within the specification, we might expect rather flat empirical distributions with sharp cut-off points!

Example 4.11

A small engineering company makes clutch assemblies for a car manufacturer. One of the components is a steel rod. The lengths of these rods can be approximated by a normal distribution with mean 100.4 mm and standard deviation 0.8 mm. Each rod costs 12p to make and is immediately usable if its length is between 99 mm and 101 mm. Rods shorter than 99 mm cannot be used but have a scrap value of 2p. Rods longer than 101 mm can be shortened to an acceptable length at an extra cost of 4p. Find the average cost of a usable rod. It has been suggested that all rods which are not immediately usable should be scrapped. If the mean of the process can be altered without affecting the standard deviation,

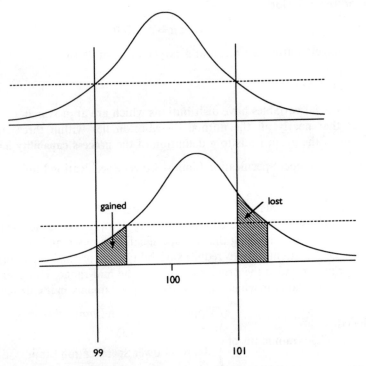

Fig. 4.15 Number of rods outside specification minimized by setting mean at mid-point.

what value would minimize the scrap? What would the average cost of a rod be now? What value of the mean would minimize the cost of a rod? What are the process capability and performance indices?

Let X be the length of rod. Then

$$\Pr(X > 101) = \Pr(Z > 0.75) = 0.2266$$

and

$$\Pr(X < 99) = \Pr(Z < -1.75) = 0.0401$$

Hence the average cost of a usable rod is given by

$$\frac{12 + (0.2266)4 - (0.0401)2}{1 - 0.0401} = 13.36p$$

We now investigate the proposal that all rods which are outside the specified range be scrapped. The scrap will be minimized if the mean is set at 100 mm (Fig. 4.15), but the average cost of a usable rod will increase to 14.68p.

We find the minimum cost by trying a few other values for the mean. The costs per rod with the mean at 100.3 mm and 100.2 mm are 13.35p and 13.40p, respectively. The minimum cost per rod will be obtained if the mean is between 100.3 mm and 100.4 mm.

The lower and upper specification limits are 99 and 101. Substitution into the definitions gives

$$C_p = \frac{2}{6 \times 0.8} = 0.42$$

$$C_{pk} = \frac{0.6}{3 \times 0.8} = 0.25$$

The process is not running satisfactorily and the cost of reducing the standard deviation might be offset by direct savings of up to 1.36p per rod and the 'hidden' costs associated with the return or rejection of rods when the clutches are assembled.

4.2.2 The uniform distribution

Suppose tube trains are separated by 10-minute intervals. If passengers arrive at the platform with no knowledge of the timetable, their waiting times, X (minutes), will be **uniformly distributed** between 0 and 10. The pdf is

$$f(x) = 0.1 \qquad \text{for } 0 \leqslant x \leqslant 10$$

and the cdf is

$$F(x) = 0.1x \qquad \text{for } 0 \leqslant x \leqslant 10$$

These are shown in Fig. 4.16. The distribution generalizes to any interval $[a, b]$:

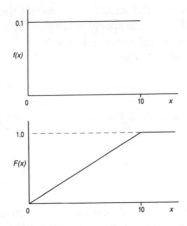

Fig. 4.16 Uniform distribution pdf and cdf.

the pdf becomes

$$f(x) = \frac{1}{b-a} \qquad \text{for } a \leqslant x \leqslant b$$

If X has a uniform distribution on $[a, b]$, we write

$$X \sim U[a, b]$$

The sequence of decimals, $\{r_j\}$, between 0 and 1 produced by the algorithm of Exercise 2.16 should have a uniform distribution. Uniform random numbers from $U[0, 1]$ can be transformed to give a random sample from any other distribution (Example 4.13).

4.2.3 The exponential distribution

The Poisson distribution is the distribution of the number of events over a length of time t in a Poisson process with an average rate of λ per unit time. The number of events is a discrete random variable associated with the process. A continuous variable associated with the same process is the time between events.

Let T represent the time until the first event occurs in a Poisson process with a rate λ. We now concentrate on a fixed length of time t, and write down an expression for the occurrence of no event by time t.

$$\Pr(T > t) = \Pr(\text{no event in time } t) = e^{-\lambda t}$$

Notice that this probability does not depend on when we start timing because occurrences in a Poisson process are assumed to be independent. The cdf is by definition

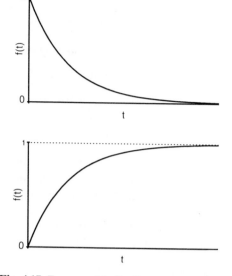

Fig. 4.17 Exponential distribution pdf and cdf.

$$F(t) = \Pr(T < t) = 1 - e^{-\lambda t} \qquad \text{for } 0 < t$$

and the pdf follows by differentiation

$$f(t) = \lambda e^{-\lambda t} \qquad \text{for } 0 < t$$

These are shown in Fig. 4.17. Since occurrences in a Poisson process are independent, this exponential distribution models the distribution of time from any 'starting time' until the next event, and in particular the times between events. This 'lack of memory' property of the exponential distribution is an essential feature of Markov processes, named after Andrei Andreevich Markov (1856–1922), which are used, for example, to model queues. As λ is the average number of events per unit time, the average time between events must be $1/\lambda$. If T has an exponential distribution with parameter λ, we write, in tribute to Markov,

$$T \sim M(\lambda)$$

Example 4.12

A computer manager knows that jobs are submitted to a mainframe machine at an average rate of 3 per minute. Compare the mean time between jobs with the median time. What is the probability of no new job being submitted in a minute?

The mean is $\frac{1}{3}$ minute (20 s). To find the median, denoted v, note that

$$\Pr(T < v) = 1 - e^{-\lambda v} = 0.5$$

Hence

$$v = -\ln(0.5)/\lambda = 0.23 \text{ minutes (14 s)}$$

The probability of no new job being submitted in a minute is

$$\Pr(T > 1) = e^{-3 \times 1} = 0.050$$

Example 4.13

I have written a computer routine that generates uniform (pseudo)random numbers on $[0, 1]$. I want to transform these to be a random sample from an exponential distribution with parameter λ, as part of a simulation for a complex queueing process.

Let r be a typical uniform random number. The required transformation is to make t the subject of the formula,

$$F(t) = r$$

For the exponential distribution, the equation

$$1 - e^{-\lambda t} = r$$

can be rearranged to give

$$t = (-\ln(1 - r))/\lambda$$

The values of t corresponding to the values of r given by the routine will be a random sample from the exponential distribution.

A similar construction can be used to obtain a random sample from any distribution. The reason why it works is that, for any random variable X with cdf $F(x)$,

$$\Pr(a < X < b) = F(b) - F(a)$$

If $R \sim U[0, 1]$

$$\Pr(F(a) < R < F(b)) = F(b) - F(a)$$

But the inequality on the left-hand side can be rearranged as

$$\Pr(F^{-1}(F(a)) < F^{-1}(R) < F^{-1}(F(b)))$$

which equals

$$\Pr(a < F^{-1}(R) < b)$$

Therefore $F^{-1}(R)$ has the same distribution as X. It is immediately useful if F^{-1} can be expressed as a formula. The graphical construction is shown in Fig. 4.18.

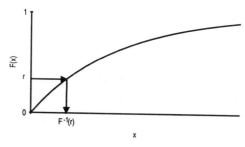

Fig. 4.18 Generating exponential random numbers from uniform random numbers.

4.2.4 Expected value

The general argument is the same as for discrete distributions, with summation replaced by integration. The starting point is the expressions for $\bar{x}$ and s calculated from grouped data. As the number of class intervals tends to infinity, the approximation involved become negligible and the sum becomes an integral (Fig. 4.19).

Example 4.14

The mean of the exponential distribution is given by

$$E[X] = \int_0^\infty t\lambda e^{-\lambda t}\,dt$$

Integration by parts gives $E[X] = 1/\lambda$.

The expected value of X (its mean) for any continuous distribution, denoted $E[X]$ or μ as for the discrete case, is given by

$$\mu = \int_{-\infty}^\infty xf(x)dx$$

while the variance, which we can again write as $E[(X - \mu)^2]$ or σ^2, is given by

$$\sigma^2 = \int_{-\infty}^\infty (x - \mu)^2 f(x)dx$$

The average value of an arbitrary function ϕ of X, $\phi(X)$, is

$$E[\phi(X)] = \int_{-\infty}^\infty \phi(x)f(x)dx$$

For example, skewness is

$$\gamma = \int_{-\infty}^\infty (x - \mu)^3 f(x)dx/\sigma^3$$

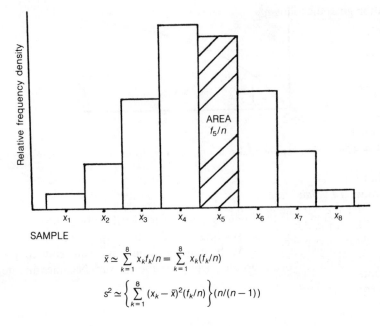

$$\bar{x} \simeq \sum_{k=1}^{8} x_k f_k / n = \sum_{k=1}^{8} x_k (f_k / n)$$

$$s^2 \simeq \left\{ \sum_{k=1}^{8} (x_k - \bar{x})^2 (f_k / n) \right\} (n/(n-1))$$

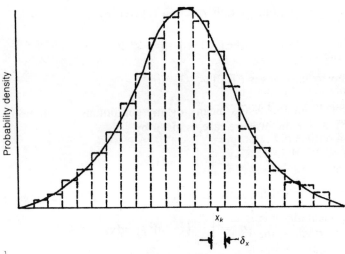

POPULATION modelled by *pdf* $f(x)$. Approximate by histogram with a large number of class intervals c and large frequencies f_k

$$\sum_{k=1}^{c} x_k (f_k / n) \simeq \sum_{k=1}^{c} x_k f(x_k) \delta x \simeq \int_{-\infty}^{\infty} x f(x) \, dx = \mu$$

$$\sum_{k=1}^{c} (x_k - \bar{x})^2 (f_k / n) \simeq \sum_{k=1}^{c} (x_k - \bar{x})^2 f(x_k) \delta x \simeq \int_{-\infty}^{\infty} (x - \mu)^2 f(x) \, dx = \sigma^2$$

Fig. 4.19 Mean and variance of a continuous distribution.

Integration gives the following results for the continuous distributions considered so far:

	Mean	Variance	Std. Dev.	Skewness	Kurtosis
Normal	μ	σ^2	σ	0	3
Uniform	$(a+b)/2$	$(b-a)^2/12$	$(b-a)/\sqrt{12}$	0	2
Exponential	$1/\lambda$	$1/\lambda^2$	$1/\lambda$	2	9

Note that the relationship between the mean and variance of the exponential distribution is not the same as for the Poisson distribution.

In practice the parameters are usually considered unknown, and have to be estimated from the data. For the normal distribution use $\bar{x}$ and s in place of μ and σ, and for the exponential distribution $1/\bar{x}$ can be substituted for λ. However, the uniform distribution is often used in contexts where the values of the parameters can be assumed rather than estimated. If they do have to be estimated, equating sample and population means and variances is not very satisfactory (Exercise 4.18).

4.2.5 Predicting the 100-year flood (extreme value type I distribution)

None of the distributions we have looked at so far would be plausible for annual floods on the Thames. Dodd (1923) and Fisher and Tippett (1928) proved that the distribution of the maximum in random samples of size m from a distribution with an unbounded upper tail, decaying at least as fast as an exponential distribution, tends to a type I extreme value distribution of greatest values (EVGI) – also known as the **Gumbel** or **double exponential distribution** – as m tends to infinity. An outline of the proof is given in Appendix A3. Annual floods are the maximum of 365 daily discharges, but the daily discharges are certainly not independent and this has the effect of reducing m from 365 (Fisher and Tippett's result can be generalized to non-independent sequences). The assumption that daily discharges come from a distribution with an unbounded upper tail which decays exponentially seems plausible enough, but m is a long way short of infinity! Nevertheless, the EVGI distribution often provides a good empirical fit to yearly maxima and is widely used in the UK. It also has an advantage of being relatively simple. The EVGI has cdf

$$F(x) = e^{-e^{-(x-\xi)/\theta}} \qquad \text{for } -\infty < x < \infty$$

and a pdf which can be obtained by straightforward, if slightly awkward, differentiation

$$f(x) = \frac{1}{\theta} e^{-(x-\xi)/\theta} e^{-e^{-(x-\xi)/\theta}} \qquad \text{for } -\infty < x < \infty$$

The parameter ξ is the mode of the distribution and θ, which must be positive, is a scale factor proportional to the standard deviation. The features of the distribution in terms of these parameters are

median	$\xi + 0.366\,51\theta$
mean (μ)	$\xi + 0.577\,22\theta$
standard deviation (σ)	$1.282\,55\theta$
skewness	1.14
kurtosis	5.4

The skewness of the Thames' annual maxima is 1.08, and as this is so close to the theoretical value for the EVGI we might expect a good fit. The easiest way to estimate the parameters is to equate the sample mean and standard deviation with the population values. The mean and standard deviation of the Thames' maxima are $324.2\,\mathrm{m}^3\,\mathrm{s}^{-1}$ and $119.5\,\mathrm{m}^3\,\mathrm{s}^{-1}$, respectively. The parameter estimates are the solutions of the equations,

$$324.2 = \hat{\xi} + 0.577\,22\hat{\theta}$$
$$119.5 = 1.282\,55\hat{\theta}$$

and are 270.4 for $\hat{\xi}$ and 93.2 for $\hat{\theta}$. The EVGI pdf with these parameter values is superimposed on the histogram in Fig. 4.20, and there is a remarkably good correspondence.

Suppose our objective is to estimate the upper 1% point ($x_{0.01}$) of the distribution. By definition

$$F(x_{0.01}) = \exp(-\exp(-(x_{0.01} - \xi)/\theta)) = 0.99$$

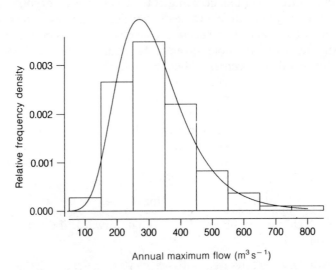

Fig. 4.20 EVGI distribution fitted to Thames' annual maxima.

If $x_{0.01}$ is made the subject of the formula

$$x_{0.01} = \xi + (-\ln(-\ln(0.99)))\theta$$

The final step is to substitute in our estimates to obtain

$$\hat{x}_{0.01} = 270.4 + 4.600 \times 93.2 = 699$$

One alternative method for estimating the parameters is from a probability plot. This has the advantage of providing an indication of how good the fit is, and is particularly useful for smaller data sets. The plot is well worth doing for this reason alone. It is convenient to introduce a common notation for an ordered sample. A sample of size n is written as

$$x_1, \ldots, x_n$$

The subscripts often represent the chronological order, and if they are a random sample this is irrelevant from a statistical point of view. When they are assumed to be a random sample, as are the Thames data, you should check that there are no significant patterns over time. The first step in a probability plot is to place the data into ascending order. We write

$$x_{(1)} < x_{(2)} < \cdots < x_{(n)}$$

and the brackets round the subscript indicate that they have been ordered. Notice that in general $x_{(i)}$ will not usually be the same as x_i, $x_{(i)}$ is known as the *i*th order statistic, which is slightly shorter than the '*i*th largest in the sample'. Now let F be the cdf of any distribution. The expected value of the *i*th order statistics, $E[X_{(i)}]$, is the average value of the *i*th largest value in a very large number of samples, each of size n. These expected values can be found for different distributions in terms of i, n and the distribution parameters. However, the following approximation is usually adequate:

$$F(E[X_{(i)}]) = \frac{i - \varepsilon}{n + 1 - 2\varepsilon}$$

for some ε between 0 and 0.5. It is intuitively sensible, as it states that the proportion below the average value of the *i*th largest among n is about $i/(n+1)$. The value of ε which gives the best approximation for the EVGI distribution is 0.4.

So,

$$\exp(-\exp(-(E[X_{(i)}] - \xi)/\theta)) = (i - 0.4)/(n + 0.2)$$

and taking natural logs twice, followed by slight rearrangement, gives

$$E[X_{(i)}] = \xi + (-\ln(-\ln((i - 0.4)/(n + 0.2))))\theta$$

The probability plot is a plot of $x_{(i)}$ against $-\ln(-\ln((i - 0.4)/(n + 0.2)))$. The slope and intercept of a line drawn through the points are estimates of θ and

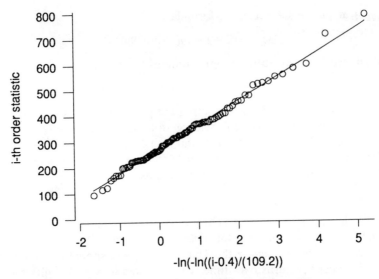

Fig. 4.21 EVGI plot of Thames annual maxima at Kingston.

ξ, respectively. Any obvious curvature suggests that the EVGI distribution is not really appropriate. The line can be drawn by eye, but try not to pay too much attention to the end points because the variances of order statistics increase dramatically as i approaches 1 or n. The plot and the line I drew are shown in Fig. 4.21. My estimates of ξ and θ from the line are 270 and 95. There is no evidence of any curvature in the plot, which again confirms that the EVGI is a good empirical fit to these data. The MINITAB commands for the probability plot, assuming the Thames's data are in C1, are

```
RANK   C1   C2
LET   C3 = − LOGE( − LOGE((C2 − .4)/(N(C1) + .2)))
PLOT   C1   C3
```

The command RANK puts the number 1 next to the smallest value, 2 next to the second smallest and so on. Use

```
PRINT   C1   C2
```

to see the effect. Ties are assigned the average rank. If you use the mouse:

Calc ▶ Mathematical Expressions
   Variable: C2
   Expression: Rank (C1)
Calc ▶ Mathematical Expressions
   Variable: C3
   Expression: − Loge(− Loge((C2 − .4)/(N(C1) + .2)))

Graph ▶ Plot
   Graph Variables: Y   X
                   C1   C3
File ▶ Display Data
   Columns, constants and matrices to display: C1   C2

Estimates of skewness from small data sets are unreliable so the fixed skewness of the EVGI, which has been found to give a good fit to many long records of annual maxima and also has some theoretical basis, is often an advantage when estimating floods with high return periods from short records in the UK. However, these estimates can be very sensitive to the form of distribution assumed and the method of estimating its parameters (see Exercise 4.21). Some attempts have been made to 'regionalize' skewness and kurtosis values, especially in the USA with its varied climate, and to use distributions with more parameters, such as the Wakeby distribution. More recently, variations based on L-moments have been an active research topic (Vogel *et al.*, 1993). Another important application of extreme value theory is in the design of sea defences (for an example, see Tawn, 1988).

4.2.6 Describing uncertainty in engineering schemes (lognormal distribution)

Shortly before the water authorities were required to produce their first asset management plans, the Water Research Centre had published its *Sewerage Rehabilitation Manual* (1986). Part of this manual covered unit costs of engineering work and a typical formula, for the unit cost (X, in pounds per metre) of an *in-situ* relining technique for a sewer of diameter D and length L, was

$$X = 18.03 L^{-0.232} D^{0.619}$$

where the negative fractional exponent of L represents the fact that larger schemes generally have a lower cost per metre. This formula was based on data supplied by all the water authorities in the UK and was supplemented by a statement that 80% of unit costs were expected to be within factors of 0.6 and 1.5 of the values given by the formula. These limits assume that X has a lognormal distribution.

A non-negative random variable X has a **lognormal distribution** if

$$\ln X \sim N(a, b^2)$$

The mean and variance of X are

$$\mu = e^{a + b^2/2}$$
$$\sigma^2 = \mu^2 (e^{b^2} - 1)$$

and the pdf is

$$f(x) = \frac{1}{xb\sqrt{2\pi}} e^{-(1/2b^2)[\ln(x) - a]^2} \qquad \text{for } 0 \leqslant x$$

Notice that the mean of X is not exponential of the mean of $\ln X$ and, conversely, the mean of $\ln X$ is not $\ln \mu$. This is a particular case of the general result that $E[\phi(X)]$ is not the same as $\phi(E[X])$, unless ϕ is a linear function. The lognormal distribution is a consequence of the central limit theorem if a large number of errors, none of which predominates, are multiplied together. It can also be justified, as can any other distribution, by comparison with histograms or, in the absence of data, as a plausible shape! The simplest way to fit the distribution to a set of data is to take natural logarithms of the data and set a and b equal to the mean and standard deviation of these transformed data.

The formula for unit cost was derived by fitting the model

$$\ln X = \beta_0 + \beta_1 \ln L + \beta_2 \ln D$$

to the data supplied by the water authorities using the techniques described in Chapter 9. The coefficients β_1 and β_2 were estimated as -0.232 and 0.619, respectively. The coefficient β_0 was estimated as 2.824 and the standard deviation of the differences between the natural logarithms of costs and their fitted values was 0.37. These differences, and hence $\ln X$ for given L and D, appeared to be approximately normally distributed. So the WRc statistician calculated the factor of 18.03 in the formula for X from the equation for μ in the last paragraph, with a and b equal to 2.824 and 0.37 respectively. The 80% limits follow from 80% limits for $\ln X$, which is assumed to have a normal distribution, by taking the exponential. Since 80% of a normal distribution is within 1.282 standard deviations of the mean (Table E.3),

$$\Pr(a - 1.28b < \ln X < a + 1.28b) = 0.80$$

It follows that

$$\Pr(e^{a-1.28b} < X < e^{a+1.28b}) = 0.80$$
$$\Pr(e^{a+b^2/2}\, e^{-1.28b - b^2/2} < X < e^{a+b^2/2}\, e^{1.28b - b^2/2}) = 0.80$$
$$\Pr(\mu e^{-1.28b - b^2/2} < X < \mu e^{1.28b - b^2/2}) = 0.80$$

Substitution of 0.37 for b gives the quoted limits of

$$\Pr(0.6\mu < X < 1.5\mu) = 0.80$$

Example 4.15

Parts of the water company's asset management plan (AMP) are made up from a large number of relatively small schemes. Each scheme involves a quantitative assessment of the work to be done and an indication of the uncertainty about this. For example, the length of a sewer which needs replacement might be estimated from a closed-circuit TV survey as 100 m. The estimate of uncertainty is rather more subjective and engineers were asked to choose from a set of symmetric and asymmetric 90% prediction intervals based on normal and lognormal distributions, respectively. In the AMP a large number of schemes

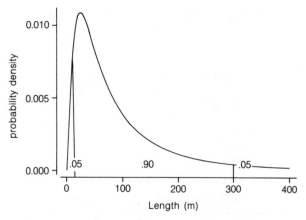

Fig. 4.22 A lognormal distribution (with mean 100 and standard deviation of 118); areas under the curve correspond to probabilities.

are added together and the uncertainty in the total is mainly due to other sources, particularly inaccuracies in unit costs and any tendencies consistently to over- or underestimate the size of schemes. This is the subject of Chapter 7.

Figure 4.22 shows a lognormal distribution for a length X, with a mean of 100 and a standard deviation of 118. The corresponding values of a and b, for the distribution of ln X, are the solutions of

$$b = \sqrt{(\ln(\sigma^2/\mu^2 + 1))}$$
$$a = \ln \mu - b^2/2$$

and equal 4.169 and 0.934, respectively. There is a 90% probability that ln X is within

$$4.169 \pm 1.645 \times 0.934$$

which equals

$$[2.633, 5.705]$$

This corresponds to a 90% probability that X will be within 14 and 300 (with equal areas of 0.05 in both tails). The median of the distribution is $e^a = 65$. The mode of the distribution is the value of x for which $f(x)$ has a maximum. Straightforward differentiation and substitution into the general result gives a value of 27. Similar calculations lead to the prediction intervals shown in Table 4.2.

4.2.7 Predicting fatigue damage (the Weibull distribution)

Our version of Miner's rule is a rather naive model for the lifetime of joints, and modern computer software can be used for much more realistic simulations.

Table 4.2 Asymmetric 90% prediction intervals for lognormal distribution

Mean	90% Confidence band	Standard deviation	Mode
μ	$[-10\%, \ +10\%]$ or $[0.90\mu, 1.10\mu]$	0.06μ	(0.99μ)
μ	$[-21\%, \ +25\%]$ or $[0.79\mu, 1.25\mu]$	0.14μ	(0.97μ)
μ	$[-38\%, \ +50\%]$ or $[0.62\mu, 1.50\mu]$	0.27μ	(0.90μ)
μ	$[-50\%, \ +75\%]$ or $[0.50\mu, 1.75\mu]$	0.42μ	(0.79μ)
μ	$[-60\%, +100\%]$ or $[0.40\mu, 2.00\mu]$	0.52μ	(0.70μ)
μ	$[-76\%, +150\%]$ or $[0.24\mu, 2.50\mu]$	0.81μ	(0.47μ)
μ	$[-86\%, +200\%]$ or $[0.14\mu, 3.00\mu]$	1.18μ	(0.27μ)

For comparison, if a normal distribution is assumed:

Mean	90% Confidence band	Standard deviation	Mode
μ	$[-100\%, +100\%]$ or $[0, 2\mu]$	0.61μ	(μ)

An M.Sc. student, Edward Sarfo-Karikari, experimented with a routine which required the applied stress distributions to be described in terms of the Weibull distribution.

This distribution is named after the Swedish physicist, Waloddi Weibull, who popularized it in various papers (from 1939 onwards). It can be given some theoretical justification, using extreme-value arguments similar to those for the EVGI, for certain applications, but it can also be used (simply) because it provides a good fit to observed data. The cdf is

$$F(x) = 1 - e^{-(x/\beta)^{\alpha}} \qquad \text{for } x > 0$$

and the pdf follows from differentiation:

$$f(x) = \frac{\alpha}{\beta^{\alpha}} x^{\alpha - 1} e^{-(x/\beta)^{\alpha}} \qquad \text{for } x > 0$$

The mean, variance and skewness are rather complicated expressions which involve the gamma function:

$$\mu = \beta \Gamma(1 + 1/\alpha)$$

$$\sigma^2 = \beta^2 [\Gamma(1 + 2/\alpha) - \Gamma^2(1 + 1/\alpha)]$$

$$\gamma = \frac{\Gamma(1 + 3/\alpha) - 3\Gamma(1 + 2/\alpha)\Gamma(1 + 1/\alpha) + 2\Gamma^3(1 + 1/\alpha)}{[\Gamma(1 + 2/\alpha) - \Gamma^2(1 + 1/\alpha)]^{3/2}}$$

It can exhibit a wide range of shapes (Fig. 4.23), but with only two parameters it is not possible to choose μ, σ and γ independently. A three-parameter Weibull distribution can be constructed by introducing a lower bound l, and treating $X - l$ as a two-parameter Weibull variable. The same thing can also be done for the lognormal distribution, and any other distribution with a fixed lower limit. The snag is that reliable estimation of l can be a problem and its value may have important consequences (Exercise 4.21). Stresses, caused by waves,

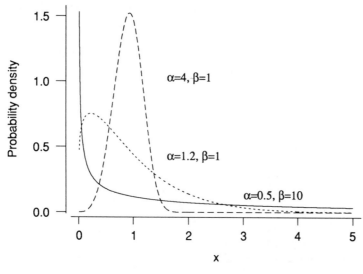

Fig. 4.23 Weibull pdf.

on joints of an offshore structure can reasonably be assumed to have a lower bound of 0.

A simple way of fitting the distribution, which also gives an indication of its applicability, is by a probability plot. A value of ε equal to 0.4 in the approximation for $F(E[X_{(i)}])$ is reasonable. You should check that a line through a plot of $\ln(x_{(i)})$ against $\ln(-\ln(1-(i-0.4)/(n+0.2)))$ will have a gradient of $1/\alpha$ and an intercept of $\ln \beta$. For grouped data there will be points corresponding only to ends of intervals.

Example 4.16

We will fit a Weibull distribution to the stresses caused by waves in the N–S direction on the same joint of the offshore drilling platform that we considered in section 3.3. The frequencies are found by adding the first and fifth columns in Table 3.3. The stresses corresponding to the right-hand end points of the wave height grouping intervals were calculated by interpolating the second column in Table 3.10. For example, a wave height of 0.95 corresponds to a stress of approximately

$$26.3 + \frac{0.95 - 0.475}{1.45 - 0.475} \times (78.8 - 26.3) = 52$$

The next eleven values are calculated in a similar way, but as 1.95 is mid-way between 1.45 and 2.45, and so on, the arithmetic is easier. The final stress, which corresponds to a wave height of 12.95 m, is found by extrapolation. So we have:

Cumulative number of waves (i) causing stress less than $x_{(i)}$	Stress (Pa) $x_{(i)}$
1 400 330	52
1 753 410	100
1 821 512	144
1 837 255	190
1 841 610	234
1 824 992	281
1 843 477	334
1 843 654	393
1 843 719	457
1 843 743	522
1 843 752	589
1 843 755	662
1 843 756	737

A plot of $\ln(x_i)$ against $\ln(-\ln(i-(i-0.4))/1\,843\,756.2))$ is shown in Fig. 4.24. I found it convenient to fit a least-squares line (see Chapter 8 on regression) through the points, but a line fitted by eye would do as well. The gradient and intercept were 1.157 and 3.356, respectively. The estimates of the parameters

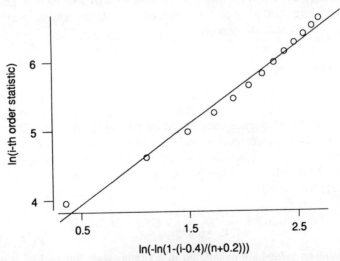

Fig. 4.24 Plot to estimate parameters of Weibull distribution for stresses.

are the solutions of the equations,

$$1.157 = 1/\hat{\alpha}$$
$$3.356 = \ln \hat{\beta}$$

and it follows that

$$\hat{\alpha} = 0.86 \quad \hat{\beta} = 28.67$$

A histogram of the stresses and the fitted Weibull pdf are shown in Fig. 4.25. You may have noticed that there was some evidence of curvature in the probability plot. A more complex distribution could give a slightly better fit but a Weibull seems adequate given the other approximations and simplifications involved in the modelling of the crack propagation.

The computer package allows the parameters of the Weibull distributions for stresses at a joint to vary during a simulation. This facility can be used to take some account of different directions and differences between years, as well as sampling and modelling errors for the single year's record. The model for the damage in shallow water was not affected by the time order of the stresses, and the simulation used simple random samples from the Weibull distribution. In deep water the frequency composition of wave forces becomes important, and the time order of the applied stresses is an essential part of the modelling (an overview is given in section 11.3). Even in shallow water, there is some evidence that high stresses early in a structure's life are more detrimental than if they occur later. An explanation is that they open up cracks that would

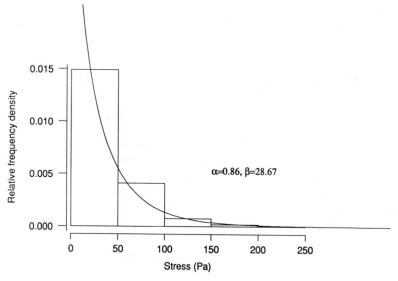

Fig. 4.25 Weibull distribution fitted to stresses at a particular joint due to waves from South East.

otherwise be unaffected by small stresses. Time-series models for stresses can be used to give more accurate descriptions of the effects of the sea environment in which the platforms are expected to operate.

The modelling of crack growth in random environments is a challenging and lively area of research. It is of great practical value and brings people from various disciplines together. An important aspect of the work is to compare theoretical predictions with data from test structures and operational platforms.

4.2.8 Normal score plot

We can approximate the probability that a variable X is less than the average value of the ith largest datum in a random sample of n by

$$F(E[X_{(i)}]) = \frac{i - 0.375}{n + 0.25}$$

This is a special case of the general approximation suggested in section 4.2.5 with ε set equal to 0.375 which gives a slightly better approximation for the normal distribution. A practical difference is that F is no longer available as a formula and we have to use tables. Now

$$Z = (X - \mu)/\sigma$$

is the standard normal variable and

$$E[Z_{(i)}] = E\left[\frac{X_{(i)} - \mu}{\sigma}\right] = \frac{E[X_{(i)}] - \mu}{\sigma}$$

Rearrangement gives

$$E[X_{(i)}] = \mu + \sigma E[Z_{(i)}]$$

We can either calculate approximations to the $E[Z_{(i)}]$ using the inverse cdf of the standard normal distribution,

$$E[Z_{(i)}] = \Phi^{-1}((i - 0.375)/(n + 0.25))$$

or look up the accurate $E[Z_{(i)}]$ which are called **normal scores** directly (see, for example, Lindley and Scott, 1984). Approximate normal scores are also available as a MINITAB command (NSCO). A **normal score plot** is a plot of the ordered sample $x_{(i)}$ against $E[Z_{(i)}]$, and any systematic curvature is evidence of discrepancies from the normal distribution. It is shown for the cubes in Fig. 4.26(a), and complements the superposition of the pdf on the histogram. We again see discrepancies in the tails. Probability plots for small data sets are inevitably less informative, but it is still sometimes possible to distinguish obvious curvature in plots from samples as small as 15, despite the scatter about any notational line. You can investigate this for yourself by generating random samples from various distributions using MINITAB. Normal score plots for

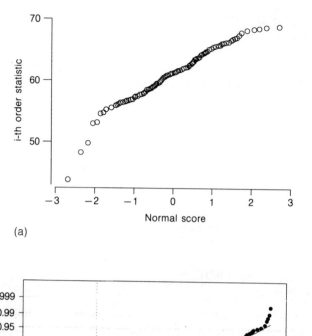

(a)

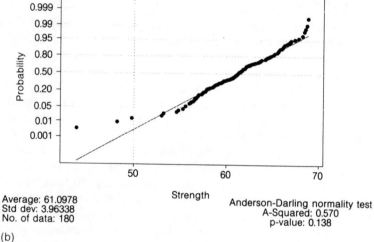

Average: 61.0978
Std dev: 3.96338
No. of data: 180

Strength

Anderson-Darling normality test
A-Squared: 0.570
p-value: 0.138

(b)

Fig. 4.26 (a) Normal score plot for compressive strengths of 180 concrete cubes. (b) Normal probability plot.

very large random samples from different distributions show the characteristic shapes you might look for. The following sequence of commands generates a random sample of 1000 from a uniform distribution on $[0, 1]$ and also converts this to corresponding random samples from normal ($N(0, 1)$) and exponential $M(1)$ distributions. The normal score plots should be an S-shape, straight line, and J-shape, respectively.

```
RANDOM   1000   C1;
  UNIFORM   0   1..
INVCDF   C1   C2;
  NORMAL   0   1.
LET   C3 = − LOGE(1 − C1)
NSCO   C1   C9
PLOT   C1   C9
PLOT   C2   C9
PLOT   C3   C9
```

Or, using the mouse:

Calc ▶ Random Data ▶ Uniform...
Calc ▶ Probability Distributions ▶ Normal...
Calc ▶ Mathematical Expressions...
Calc ▶ Mathematical Expressions...
  Variable: C9
  Expression: Nscores (C1)
Graph ▶ Plot...

If you reduce the sample size to 15 and repeat the exercise several times (STORE the commands into a macro file) you will get an impression of the discriminating potential of normal score plots with small samples.

An equivalent procedure to a normal score plot is to plot the ordered data $x_{(i)}$ against $(i − 0.375)/(n + 0.25)$ on **normal probability paper**. This special paper has a probability scale (it is usually $1000 \times$ probability) such that the cdf of the normal distribution becomes a straight line. The special probability scale could be constructed by putting $1000 \times (i − 0.375)/(n + 0.25)$ alongside the $E[Z_{(i)}]$. The only difference in the plots is that the probability axis is usually vertical on the special paper. It is very convenient for calculating the mean and standard deviation, because the mean corresponds to a probability of 0.5 and two standard deviations either side correspond to probabilities of 0.025 and 0.975 respectively. A normal probability plot for the cubes, using the grouped data in Table 3.7 rather than all the points, is shown in Fig. 4.26(b). Estimates of percentage points can be read off the plot directly. Probability papers for other distributions used to be sold, but they are becoming harder to find because most analyses are now computer-based. MINITAB Release 9 includes normal and Weibull plotting routines:

Graph ▶ Normal Plot...

4.3 Modelling rainfall

The purpose of this section is to show how a combination of Poisson processes can be used to model a complex phenomenon such as rainfall. Rainfall distribution is essential information for designers of water resource projects ranging from flood protection to irrigation schemes. Ideally, and provided there were no long-term climate changes, statistics could be calculated from long records over an extensive network of rain gauges. In practice, rain gauge networks are often sparse or non-existent and, even in countries with good coverage, records for periods exceeding 50 years are relatively uncommon. Furthermore, records usually consist of daily rainfall totals and for some purposes, such as assessment of the hydraulic performance and pollution impact of sewers, finer resolution, down to five-minute rainfall totals, is needed. Even if suitable data are available for a project they will not necessarily be freely available. The Meteorological Office in the UK usually charges for data. It follows that a reliable method for generating long realistic rainfall time series would be valuable.

Deterministic rainfall models are not particularly appropriate because probability distributions of their parameter values would still be needed. In any case the physics of atmospheric processes is not completely understood and deterministic rainfall models tend to be of little practical value in engineering design problems (Cho and Chan, 1987). Much work on the theoretical development of **stochastic** (random) rainfall models has been completed, and the following is a relatively simple approach, the **Neyman–Scott rectangular pulses** (NSRP) rainfall model, for rainfall at a single site. It is less restrictive than it might at first appear because many hydraulic flow models require rainfall inputs in this form, and it has recently been developed and tested for this application (Cowpertwait *et al.*, 1991). It could also be used as a first step, before using some method of spatial disaggregation, when rainfall distribution patterns over an area are required. The NSRP model is built up from five Poisson processes.

Storm origins, which can correspond to the leading edges of cold fronts, occur in the NSRP rainfall model as a Poisson process with a rate λ per hour. A typical realization is shown on the first line of Fig. 4.27. Each storm origin has a random number (C) of rain cells associated with it. The number of rain cells is generated via a Poisson distribution, specifically ($C - 1$) is a Poisson random variable with mean ($v - 1$). The reason for modelling ($C - 1$), rather than C, as a Poisson random variable is to ensure that all storms have rain. The parameter v is the average number of rain cells per storm. The next stage is to position the (rain) cell origins. The waiting times from the storm origin to the cell origins are independent exponential random variables with a mean of $1/\beta$ hours. The second line of Fig. 4.27 shows the cell origins, together with the original storm origins. The cells are assumed rectangular and their size depends on duration and intensity. The cell durations are modelled as independent exponential random variables with a mean of $1/\eta$ hours, and the intensities as independent exponential random variables with a mean of $1/\xi$ mm per hour. The cells themselves are included in the third line of Fig. 4.27 and summed to give the rainfall intensity at any time, on the bottom line. It follows that the model is built up from five independent Poisson processes and is defined by five parameters λ, β, v, η, and ξ. It is an example of a **clustered point process model**.

Storm origins arrive according to a Poisson process

Each origin generates a random number of rain cells with cell origins at ✳

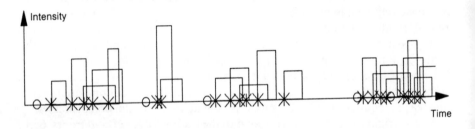

The intensity and duration of each rain cell follow exponential distributions – the intensity is constant throughout the duration

The total intensity at any point in time is the sum of the intensities of all active rain cells at that point

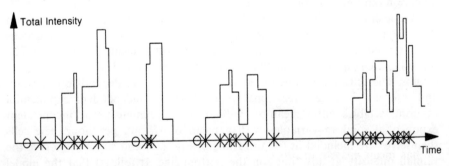

Fig. 4.27 A schematic representation of the Neyman–Scott Rectangular Pulses Rainfall Model.

Having defined the model, the next step is to estimate the parameters. Rainfall has to be accumulated in some way if it is to be recorded, and rainfall time series of daily totals are available for many sites. A relatively straightforward method of estimating the five parameters of the model is to compare statistics calculated from these time series with their long-term average values (expected values), provided expressions for the latter in terms of the model parameters are known (details are given in Rodriguez-Iturbe *et al.*, 1987); Cowpertwait 1991). One of the objectives of our project was to develop a procedure for simulating rainfall at ungauged sites. This was done by relating parameter estimates to geographical variables such as distance from coast, latitude, height, east or west coast, and time of year. Regression techniques, which are described in a later chapter, were used.

4.4 Summary

Population and sample

Sample	Population
Discrete variables	
line chart	probability function $P(x)$
Continuous variables	
histogram	pdf $f(x)$
cumulative frequency polygon	cdf $F(x)$
Discrete and continuous	
average	expected value
$\bar{x}, s^2, s$	μ, σ^2, σ
$\hat{\gamma}, \hat{\kappa}$	γ, κ

Binomial distribution

X is the number of successes in n trials with constant probability of success p. $\Pr(X = x)$ is denoted $P(x)$ and given by

$$P(x) = {}_nC_x p^x (1 - p)^{n-x} \qquad \text{for } x = 0, 1, \ldots, n$$

X has mean np, variance $np(1 - p)$ and standard deviation $\sqrt{(np(1 - p))}$. We write:

$$X \sim \text{Bin}(n, p)$$

Poisson process

Occurrences are random and independent with underlying rate λ per unit of continuum. The discrete variable X is the number of occurrences in length t of continuum, and has the Poisson distribution.

The time between occurrences T is a continuous variable with the exponential distribution.

Poisson distribution

$$P(x) = e^{-\lambda t}(\lambda t)^x/x! \qquad \text{for } x = 0, 1, \ldots$$

X has mean λt, variance λt and standard deviation $\sqrt{(\lambda t)}$. We write:

$X \sim \text{Poisson } (\lambda t)$

Exponential distribution

$$F(t) = 1 - e^{-\lambda t} \qquad \text{for } 0 < t$$
$$f(t) = \lambda e^{-\lambda t} \qquad \text{for } 0 < t$$

T has mean $1/\lambda$, variance $1/\lambda^2$ and standard deviation $1/\lambda$. We write

$T \sim M(\lambda)$

Normal distribution

$$f(x) = \frac{1}{\sigma\sqrt{2\pi}} e^{-((x-\mu)/\sigma)^2/2} \qquad \text{for } -\infty < x < \infty$$

X has mean μ, variance σ^2 and standard deviation σ. We write

$X \sim N(\mu, \sigma^2)$

The standard normal distribution has the pdf

$$\phi(z) = \frac{1}{\sqrt{2\pi}} e^{-z^2/2} \qquad \text{for } -\infty < z < \infty$$

and has mean 0, variance 1 and standard deviation 1.

Approximately two-thirds (68.3%) of a normal distribution is within one standard deviation from the mean, 95% within two standard deviations, and nearly all (99.7%) within three standard deviations. The cdf of the standard normal distribution is

$$\Pr(Z < z) = \Phi(z)$$

The upper $\alpha \times 100\%$ point z_α is such that an area α lies to the right and $(1-\alpha)$

to the left,

$$\Phi(z_\alpha) = 1 - \alpha$$

Uniform distribution

$$F(x) = \frac{x - a}{b - a} \qquad \text{for } a < x < b$$

$$f(x) = \frac{1}{b - a} \qquad \text{for } a < x < b$$

X has mean $(a + b)/2$, variance $(b - a)^2/12$ and standard deviation $(b - a)/\sqrt{12}$. We write

$$X \sim U[a, b]$$

EVGI distribution

$$F(x) = e^{-e^{-(x - \xi)/\theta}} \qquad \text{for } -\infty < x < \infty$$

$$f(x) = \frac{1}{\theta} e^{-(x - \xi)/\theta} e^{-e^{-(x - \xi)/\theta}} \qquad \text{for } -\infty < x < \infty$$

X has mode ξ, median $\xi + 0.366\,51\theta$, mean $\xi + 0.577\,22\theta$, variance $1.645\theta^2$, standard deviation $1.282\,55\theta$, skewness 1.14, and kurtosis 5.4.

Lognormal distribution

A non-negative variable X has a lognormal distribution if

$$\ln X \sim N(a, b^2)$$

The expected value of X is $e^{a + b^2/2}$. If we define $c = (e^{b^2} - 1)$, then X has variance $\mu^2 c$, standard deviation $\mu\sqrt{c}$ and skewness $c^3 + 3c$.

Weibull distribution

$$F(x) = 1 - e^{-(x/\beta)^\alpha} \qquad \text{for } 0 < x$$
$$f(x) = \alpha\beta^{-\alpha}x^{\alpha - 1}e^{-(x/\beta)^\alpha} \qquad \text{for } 0 < x$$

X has mean $\beta\Gamma(1 + 1/\alpha)$ and variance $\beta^2[\Gamma(1 + 2/\alpha) - \Gamma^2(1 + 1/\alpha)]$.

Random numbers from distributions

(i) Where the cdf F is a known formula:

- Generate a random number (r) from $U[0, 1]$ (Exercise 2.16).

- If the required distribution has a cdf $F(x)$,

$$x = F^{-1}(r)$$

is a random number from it.

(ii) Normal distributions:

- Generate m (which should be at least 10) random numbers from $U[0, 1]$.
- The average of these m numbers is, approximately, a random variate from $N(0.5, 1/(12m))$ and can be scaled to any other normal distribution. The result relies on the central limit theorem (Chapter 5), and becomes more accurate as m increases.

(iii) Discrete distributions:

The obvious analogue to the procedure for discrete distributions. Define a cumulative probability function $C(x)$ by

$$C(x) = \sum_{k \leq x} P(k)$$

Then for a given r the random number x is such that

$$C(x - 1) < r < C(x).$$

(iv) Using MINITAB:

The following command and subcommand would put 1000 normal random numbers with mean 500 and standard deviation 10 into column 1.

```
RANDOM   1000  C1;
   NORMAL   500   10.
```

Using the mouse:

```
Calc ▶ Random Data ▶ Normal
   Generate: 1000
   Store: C1
   Mean: 500
   Standard deviation: 10
```

Alternative subcommands cover a wide range of discrete and continuous distributions, but not the EVGI. However, the uniform distribution on $[0, 1]$ can be used if F^{-1} is available as a formula. So 1000 random variates from an EVGI, with ξ and θ in K1 and K2, can be obtained from:

```
RANDOM   1000  C1;
   UNIFORM   0   1.
LET   C1 = - LOGE( - LOGE(C1))
LET   C1 = K2*C1 + K1
```

Using the mouse:

```
Calc ▶ Random Data ▶ Uniform...
```

Calc ▶ Mathematical Expressions...
Calc ▶ Mathematical Expressions...

Exercises

4.1 Fifteen per cent of 'Cheap 'n' Cheerful' Roman candles (a fictitious brand of firework) are single-coloured rather than multicoloured. Find the probability of more than 3 single-coloured Roman candles in a box of 12.

4.2 The probability that a fluorescent light has a life of over 500 hours is 0.9. Find the probabilities that among 11 such lights:

(i) exactly 8 last for more than 500 hours,
(ii) at least 8 last for more than 500 hours,
(iii) at least 2 do not last for more than 500 hours.

4.3 An inspector wants to check that at least 95% of water samples from the public supply contain less than the maximum specified level for nitrates. He takes a random sample of 15 jars from the water supply each week. If the nitrate content of any jar exceeds the specified level he files a complaint. The legislation permits a maximum of 5% of jars above the specified level. Find the probability that he files a complaint if:

(i) only 2% of the jars that could conceptually be filled from the supply exceed the level;
(ii) 5% of all such jars exceed the level;
(iii) 10% of all such jars exceed the level.

Repeat the exercise if he takes random samples of 30 and files a complaint if more than 2 jars exceed the limit.

4.4 Hypergeometric distribution

A population of N electric motors contains B that do not satisfy the specification ('defectives'). A random sample of n motors is taken without replacement. Let the variable X represent the number of defectives in the sample.

(i) The probability that the first motor selected is defective is B/N. Explain why X does not have an exact binomial distribution with p equal to B/N.

(ii) Explain why X has the following distribution,

$$P(x) = \frac{{}_BC_x \, {}_{N-B}C_{n-x}}{{}_NC_n} \quad \text{for } x = 0, \ldots, n$$

which is known as the hypergeometric distribution.

(iii) Compare the binomial approximation to the hypergeometric distribution for N, B and n equal to 20, 6, and 3, respectively.

4.5 A batch of 120 tape recorders contains five that are faulty. If three are randomly selected and sent to a retailer, find the probability that he gets one faulty tape recorder. Approximate this probability with the binomial distribution and find the percentage error involved in this approximation. Find the probability that he gets more than one faulty tape recorder.

4.6 Sheets of metal have plating faults which occur randomly and independently at an average rate of 1 per square metre. What is the probability that a sheet 1.5 m by 2 m will have at most one fault?

4.7 Z is a standard normal random variable. Find:

  (i) $Pr(Z > 2.3)$,
  (ii) $Pr(Z < 1.75)$,
  (iii) $Pr(|Z| < 2.0)$,
  (iv) $Pr(|Z| > 3.0)$,
  (v) b such that $Pr(Z < b) = 0.90$,
  (vi) c such that $Pr(-c < Z < c) = 0.98$,
  (vii) k such that $Pr(|Z| > k) = 0.01$.

The mean and standard deviation of the normal variable X is given in the following. Find:

  (viii) $Pr(X < 5.8 | X \sim N(4, (0.8)^2))$,
  (ix) $Pr(X > 7.5$ or $X < 3.9 | X \sim N(5.7, (1.1)^2))$,
  (x) $Pr(82 < X < 158 | X \sim N(120, 380))$.

4.8 Resistors have a nominal rating of $100\,\Omega$, but those from supplier A have resistances which are approximately normally distributed with mean $100.6\,\Omega$ and standard deviation $3\,\Omega$. Find the percentage of resistors that will have resistances:

  (i) higher than the nominal rating;
  (ii) within $3\,\Omega$ of the nominal rating.

Another supplier, B, produces resistors whose resistances are approximately normally distributed with mean $100\,\Omega$ and standard deviation $4\,\Omega$. What percentage of these resistors will be within $100 \pm 3\,\Omega$?

4.9 Failure stresses of many standard test pieces of Norwegian pine have been found to be approximately normally distributed with a mean of $25.44\,\mathrm{N\,mm^{-2}}$ and a standard deviation of $4.65\,\mathrm{N\,mm^{-2}}$. The 'statistical minimum failing stress' is defined as the value such that 99% of test results may be expected to exceed it. What is the value in this case?

4.10 A machine stamps out can tops whose diameters are normally distributed with standard deviation 0.8 mm. At what mean diameter should the machine be set so that 5% of the tops produced have diameters exceeding 74 mm?

4.11 The British Standard BS6717 describes a sampling procedure for selecting pavers for compressive strength tests. Split each consignment, of up to 5000 pavers, into eight groups. Randomly select two pavers from each group. The standard states that the pavers are acceptable if both:

- average compressive strength exceeds 49 N mm^{-2}; and
- no pavers have compressive strengths less than 40 N mm^{-2}.

(The standard does not describe an acceptance procedure for cement contents. Analyses of cement contents are far more expensive than crushing tests.)

Suppose a manufacturer produces pavers with a mean strength of 49 N mm^{-2} and a standard deviation of 3 N mm^{-2}. What is the probability that a sample of 16 satisfies the second criterion? Give an approximate probability that a consignment would be acceptable. Another manufacturer produces pavers with a mean strength of 55 N mm^{-2} and a standard deviation of 6 N mm^{-2}. What is the probability that a sample of 16 satisfies the second criterion (it is almost certain to pass the first)?

4.12 Suppose that annual maximum floods (X) have a distribution with cdf F. The upper 1% point of this distribution is $x_{0.01}$ such that

$$F(x_{0.01}) = 0.99$$

The probability $x_{0.01}$ is exceeded in any one year is 0.01 and the return period is the reciprocal of this probability, that is 100 years.

(a) Find the probabilities that the '100 year flood' ($x_{0.01}$) is exceeded at least once in:

   (i) 10 years;
   (ii) 50 years;
   (iii) 100 years.

(b) Find the number of years for which the probability of at least one exceedance of the '100 year flood' is 0.5.

(c) Write down the probability of at least one exceedance in n years of the flood with return period of T years, in terms of n and T.

4.13 Show that

$$E[(X - a)^2]$$

is a minimum if a equals μ.

4.14 On average three lorries arrive each hour, of the working day, to be unloaded at a warehouse. Assume that they arrive at random and independently. Find the probability that the time between arrivals of successive lorries is less than 5 minutes.

4.15 Prove the results for the means and variances of the uniform and exponential distributions.

4.16 A computer-controlled machine cuts printed circuit boards to a wide variety of shapes. The sequence of shapes is variable and programmed at the start of each shift. The errors in length about the nominal dimension are approximately uniformly distributed between $-10\,\mu m$ and $+10\,\mu m$ because of a rounding operation in the programme. Assume the specification for length is within $10\,\mu m$ of nominal and that errors are uniformly distributed over this range. Calculate the process capability index C_p and comment.

4.17 A power station in a tropical country can provide between 0 and A megawatts of power.

(a) From past records the demand X on a day can be approximated by the pdf

$$f(x) = \alpha x(A - x) \qquad \text{for } 0 \leqslant x \leqslant A$$

(i) What must α equal, in terms of A, if $f(x)$ is a pdf?
(ii) Sketch $f(x)$.
(iii) Now suppose the profit on x units is $2x^2$ per day in local currency. Find the expected annual profit in terms of A.
(iv) What would your answer be if you assume the profit on x units is kx per day, where k is chosen so that the linear and quadratic profit formulae coincide when x equals $A/2$?

(b) The plant manager claims that the demand pdf is modelled just as well by an equilateral triangle. Repeat (a) with this different pdf.

4.18 Suppose we wish to fit a $U[a, b]$ distribution to a set of data. Explain why setting

$$\bar{x} = (\hat{a} + \hat{b})/2$$
$$s^2 = (\hat{b} - \hat{a})^2/12$$

and solving for $\hat{a}$ and $\hat{b}$ is not a very satisfactory procedure for estimating a and b. Preferable estimates for a and b are given by

$$(x_{(1)} + x_{(n)})/2 \pm [(n + 1)/(n - 1)][(x_{(n)} - x_{(1)})/2].$$

4.19 A safety device on a chemical reactor can be reset, after sounding a warning alarm, up to k times. It is then replaced. If alarm situations occur randomly and independently according to a Poisson process with mean λ per day, find the pdf for the useful lifetime of such a device. What is this distribution if $k = 1$?

4.20 Relationship between EVGI and Weibull distributions

The variable W has an EVGI distribution with cdf

$$F(w) = e^{-e^{-(w-\xi)/\theta}}$$

Define Y and $-W$ and find the cdf of Y by completing the following argument.

$$\Pr(W < w) = e^{-e^{-(w-\xi)/\theta}}$$

and therefore

$$\Pr(-W > -w) = e^{-e^{-(w-\xi)/\theta}}$$

But $-W$ is Y from the definition, and setting

$$-w = y$$

we can write

$$\Pr(Y > y) = \cdots$$

The variable Y has an EVLI distribution, which models the distribution of the least values in large samples from distributions whose lower tail decays at least as fast as a negative exponential function. Now define

$$X = e^Y$$

and show that X has a Weibull distribution. It follows that $-\ln X$ has an EVGI distribution, and this can be used as an alternative method for fitting the Weibull distribution.

4.21 (a) In section 4.2.5 we estimated the flood with a 100-year return period at Teddington as $699 \text{ m}^3 \text{ s}^{-1}$. This estimate was based on an assumption that annual floods have an EVGI distribution. Compare it with estimates based on:

(i) a normal distribution;
(ii) a lognormal distribution with a lower bound of 0;
(iii) a lognormal distribution with a lower bound of one less than the sample minimum: $x_{(1)} - 1$;
(iv) a Weibull distribution with a lower bound of 0,
(v) a Weibull distribution with a lower bound of $x_{(1)} - 1$.
(vi) Repeat (iv) by assuming $-\ln X$ has an EVGI distribution.
(vii) Compare skewness of the fitted distributions (you could generate a large random sample from a Weibull distribution with the fitted parameter values to obtain a good estimate of the skewness) and probability plots. Which distribution do you think is the most realistic? Why do you think this procedure is likely to give a more precise estimate than the cumulative frequency polygon?

(b) The annual maxima sea levels at Lowestoft, UK (cm above sea level) for 1953–83 are:

$$
\begin{array}{ccccccccccc}
332 & 249 & 193 & 181 & 183 & 190 & 188 & 189 & 229 & 227 & 161 \\
187 & 212 & 188 & 200 & 189 & 253 & 195 & 224 & 182 & 243 & 178 \\
180 & 266 & 190 & 228 & 168 & 188 & 195 & 190 & 278 & &
\end{array}
$$

(i) Plot the data against time.

(ii) Estimate the high tide with a 1000 year return period.

4.22 Geometric distribution and negative binomial distribution

A recent modification to the NSRP rainfall model has been to use the geometric distribution for the number of cells per storm. The probability function is

$$P(x) = p(1-p)^{x-1} \qquad \text{for } x = 1, 2, 3, 4, \ldots$$

(a) (i) Show that $P(x)$ is a probability function, that is

$$\sum P(x) = 1$$

(ii) Differentiate both sides of the above equation with respect to p and deduce that the mean value of X is $1/p$ (the variance is $(1-p)/p^2$).

(b) A sequence of Bernoulli trials is a sequence of trials, each of which has two possible outcomes (success and failure) and a constant probability of success (p). If there are exactly n such trials we have a binomial distribution, but in the following there is no fixed end to the sequence. Explain why the geometric distribution gives the distribution of the number of trials up to and including the first success.

Now let Y be the number of failures before the mth success occurs. Explain why Y has the negative binomial distribution:

$$P_Y(y) = {}_{m+y-1}C_{m-1}\, p^m (1-p)^y \qquad \text{for } y = 0, 1, 2, \ldots$$

The mean and variance of Y are $m(1-p)/p$ and $m(1-p)/p^2$. Check that these results for $m = 1$ are consistent with those for the geometric distribution. (Note the difference in definition between X and Y, even for $m = 1$.)

The negative binomial distribution is often used to model a discrete random variable with a variance greater than the mean. For example, it allows for some bunching in traffic flow simulations.

4.23 Lognormal distribution

(i) Let $Y = \ln X$. Suppose

$$Y \sim N(a, b^2).$$

Then

$$\Pr(Y < y) = \int_{-\infty}^{y} \frac{1}{b\sqrt{2\pi}} e^{-((\theta - a)/b)^2/2}\, d\theta$$

But this equals

$$F(x) = \Pr(X < x)$$

where $X = e^Y$, $x = e^y$ and $y = \ln x$. Hence derive the pdf of the lognormal distribution.

(ii) The mean of the lognormal distribution is given by

$$\mu = \frac{1}{b\sqrt{2\pi}} \int_0^\infty x \frac{1}{x} e^{-(\ln x - a)^2/(2b^2)} \, dx$$

Make the substitution $y = \ln x$ to obtain

$$\frac{1}{b\sqrt{2\pi}} \int_{-\infty}^\infty e^y e^{-(y-a)^2/(2b^2)} \, dy$$

Now note that the exponent

$$y - (y - a)^2/2b^2 = [-(y - (a + b^2))^2 + 2ab^2 + b^4]/2b^2$$

by the device of completing the square. Explain why the result

$$\mu = e^{a + b^2/2}$$

follows.

4.24 The Rayleigh distribution

This distribution is named after Lord Rayleigh (J.W. Strutt) who derived it in 1919 in the context of a problem in acoustics. It is often used to model the distribution of wave heights. The cdf is

$$F(x) = 1 - e^{-x^2/(2\theta^2)} \qquad \text{for } 0 \leqslant x$$

The mean and variance are $\sqrt{\pi/2}$ and $\frac{1}{2}(4 - \pi)\theta^2$, respectively, the skewness is 0.6311 and the kurtosis 3.2451. The significant wave height (H_s) is defined as four times the root mean square height (the square root of the mean of the squared wave heights). Show that if waves do have an approximate Rayleigh distribution about one-third exceed H_s. Start by deducing the result that for any variable X,

$$E[X^2] = \sigma_X^2 + \mu_X^2$$

This follows directly from the definitions of σ_X^2 and μ_X.

4.25 In Fig. 4.28 the radioactive source emits α-particles at angles θ which are uniformly distributed over $[-\pi/2, \pi/2]$. A screen is set up at a distance a from the source.

(i) Show that the intercepts on the screen X have a **Cauchy distribution** which has pdf

$$f(x) = \frac{a}{\pi(a^2 + x^2)} \qquad \text{for } -\infty < x < \infty$$

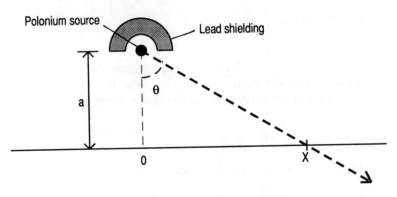

Fig. 4.28 A variable with a Cauchy distribution.

(ii) Show that Cauchy distribution has an infinite variance if the mean (which is not properly defined) is taken as 0. The central limit theorem does not hold for the Cauchy distribution. The ratio of two standard normal random variables has a Cauchy distribution, and ratios for which the denominator can be very close to zero will behave similarly.

4.26 **Censored data**
Ten pieces of electric cable were subjected to an accelerated life test for 100 days. Two lasted 100 days, but the rest failed after

$$30 \quad 39 \quad 46 \quad 51 \quad 58 \quad 66 \quad 80 \quad 97$$

days, respectively. Use graphical methods to fit a Weibull distribution.

4.27 Think of a pdf as a lamina of constant density. Show, by taking moments about the origin, that its mean corresponds to the x-coordinate of its centre of gravity. Hence deduce that the mean of a positively skewed distribution is greater than its median.

5

Combining variables

5.1 Introduction

Part of the headset bearing assembly on a bicycle is a 'crown race' which is pressed on to a 'seat' on the front fork. The interference recommended in *Sutherland's Handbook for Bicycle Mechanics*, the industry standard, is 0.1 mm (Fig. 5.1). A manufacturer of bicycles reported that crown races made by company A were frequently fracturing when fitted to front forks for mountain bikes. The following description is loosely based on an investigation by Jan Urbanowicz, who was an M.Sc. student on an industrial placement with the company at the time. He found that crown races made by company B also fractured in significant numbers during the production process whereas those manufactured by company C appeared satisfactory. The cost of a fracture is considerably more than that of the crown race itself. There is the labour cost of prising the fractured race off the seat and repeating the pressing operation with another race. The production line may be disrupted and then costs escalate. A crown race which is slightly too large is almost as inconvenient. If failures occur after bicycles leave the factory the costs are much greater. A highly stressed crown race could fracture in use, conceivably causing an accident, and a loose fit will soon result in noticeable play at the handlebars. The direct costs of customers' complaints, and repairs through dealers if faults occur during the guarantee period, are increased by the 'hidden' costs associated with losing a good reputation. The manufacturer was anxious for the problem to be identified and rectified quickly.

The starting point for such an industrial investigation is to collect relevant data. Urbanowicz took random samples of 35 forks and 35 of each of the three types of crown race from production. The means and standard deviation are given in Table 5.1. We need to know how to combine this information to find the means and standard deviations of interferences. It would not be very convenient to have to pair forks and crown races randomly, and then measure the interferences directly. The relevant formulae are special cases of simple results that have wide-ranging applications.

Now let $\{x_i\}$ be the diameters of a random sample of n seats and $\{y_i\}$ be the diameters of a random sample of n crown races. The interferences are

$$d_i = x_i - y_i$$

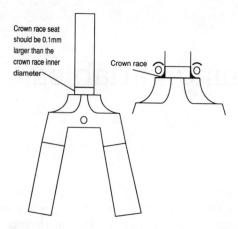

Fig. 5.1 Bicycle Fork (*Sutherland's Handbook*).

Table 5.1 Means and standard deviations of seats (on front forks) and crown races

	Sample size	Mean (mm)	Standard deviation (mm)
Front fork seat	35	27.060	0.034
Crown race A	35	26.923	0.017
Crown race B	35	26.990	0.027
Crown race C	35	26.950	0.019
Front fork seat	20	27.037	0.022

The means are simply related by

$$\bar{d} = \bar{x} - \bar{y}$$

All you need do to prove this result is sum both sides over i from 1 to n:

$$\sum d_i = \sum (x_i - y_i) = \sum x_i - \sum y_i$$

and divide by n.

The relationship between the variances is only slightly more complicated. From the definition of the variance, we have

$$s_d^2 = \sum (d_i - \bar{d})^2/(n-1)$$

The next step is to express d as the difference between x and y:

$$= \sum ((x_i - y_i) - (\bar{x} - \bar{y}))^2/(n-1)$$

$$= \sum ((x_i - \bar{x}) - (y_i - \bar{y}))^2/(n-1)$$
$$= \sum (x_i - \bar{x})^2/(n-1) + \sum (y_i - \bar{y})^2/(n-1) - 2\sum (x_i - \bar{x})(y_i - \bar{y})/(n-1)$$
$$= s_x^2 + s_y^2 - 2\sum (x_i - \bar{x})(y_i - \bar{y})/(n-1)$$

If the seats and crown races are paired at random we can expect the last term to be close to zero, for any reasonable sample size (justified in the next section), and then

$$s_d^2 = s_x^2 + s_y^2$$

Using this result, the mean interference with crown race A is estimated by

$$27.060 - 26.923 = 0.137 \, \text{mm}$$

The estimated standard deviation of these interferences is

$$\sqrt{0.034^2 + 0.017^2} = 0.038 \, \text{mm}$$

Normal score plots for all the samples were compatible with an assumption that the corresponding populations are approximately normal. It follows (Appendix A2, Theorem 3) that the interferences are also near-normally distributed. Two of the plots are shown in Fig. 5.2. The distribution of seat diameters is slightly odd, possibly a mixture of two normal distributions from different machines, but this is unlikely to have a significant effect on our assumption that about 95% of the distribution is within two standard deviations of the mean and nearly all within three standard deviations. *Sutherland's Handbook* gives no indication of tolerance limits for the 0.1 mm interference. It is high compared with British Standard recommendations for interference fits (BS4500 Part 1, 1969), but mass-produced bicycles are not precision engineered. The claimed tolerance for seats on the forks is ± 0.05 mm, and the estimated standard deviation of 0.034 mm implies poor process capability (C_p equals 0.49).

A second random sample of 20 forks was taken, and the mean and standard deviation were found to be 27.037 and 0.022. The combined estimates of mean and standard deviation are:

$$\text{mean} = (35 \times 27.060 + 20 \times 27.037)/55 = 27.052$$
$$\text{standard deviation} = (35 \times 0.034 + 20 \times 0.022)/55 = 0.030.$$

It would be preferable to recalculate the standard deviation directly from the 55 data but the above quick calculation is adequate for our purposes. This more precise estimate of standard deviation still implies poor process capability. The estimated mean interferences, with their corresponding standard deviations bracketed, using the three types of crown race are now

crown race A   0.129   (0.034)
crown race B   0.062   (0.040)
crown race C   0.102   (0.036)

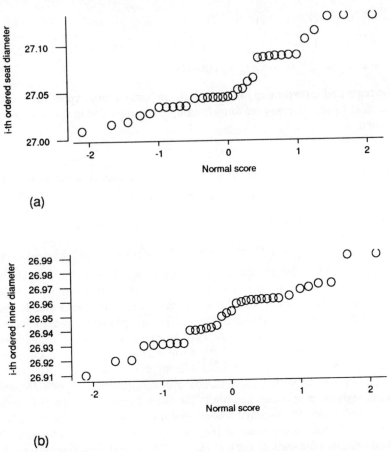

(a)

(b)

Fig. 5.2 Normal score plots for (a) seat diameters; (b) crown race C diameters.

At this stage Urbanowicz carried out a metallurgical investigation of a few crown races. Type C appeared to be less brittle, and this would account for the fact that none had been observed to fracture. Approximate calculations indicated that even their fracture strain is likely to be exceeded if the interference is more than 0.20 mm. The only effective way to reduce the standard deviation of the interferences would be to reduce the standard deviation of the seats, and 0.020 mm would be more in line with the fork manufacturer's claims. Standard deviations as low as this have been achieved on other types of fork by the same manufacturer. It would also be best to use Type C crown races, although they are more expensive, and the standard deviation of interferences, if the seat standard deviation could be reduced to 0.020 mm, would be 0.028 mm. If a normal distribution of interferences is assumed, almost all would be between 0.00 mm and 0.20 mm.

If we assume the estimates of means and standard deviations are reasonably precise (Chapter 6) we could advise manufacturer A to increase the diameter of crown races by 0.03 mm. This should result in some decrease in fractures, and any decrease in the standard deviation of seats would also help. Any further increase in the diameter of crown races would be likely to result in a small proportion of clearance fits. This might be acceptable if they could be detected before the pressing operation and another crown race substituted. We should also advise manufacturer B to decrease the diameter by about 0.04 mm and to reduce the variability of diameters.

The precision of our estimates could be improved by taking more random samples from stock. We should certainly monitor the effects of our requests to manufacturers by taking random samples from incoming batches. This would not only enable us to improve estimates of process parameters, but also give us information about the variability within and between batches.

You should note that the variance of the differences between randomly selected components is the sum of the variances of the components. This is sensible: we cannot expect deviations to cancel if components are selected at random! We could reduce the variance of the interferences by carefully matching crown races with forks, but this is not practical in mass production.

5.2 Sample covariance and correlation

We might expect some relationship between the compressive strengths and cement contents of pavers, and Fig. 2.1 allows us to make a visual assessment. Over this range of cement content there appears to be some tendency for the compressive strength to increase as the cement content increases, but it is not very marked. **Correlation** is a measure of the strength of linear association, and it lies on a scale from -1 to 1.

A first step in obtaining the correlation is to define the covariance. The sample **covariance** is

$$\widehat{cov} = \sum (x_i - \bar{x})(y_i - \bar{y})/(n - 1)$$

To understand exactly what this is, we divide the swarm of points in Fig. 2.1 into four quadrants by drawing the lines $x = \bar{x}$ and $y = \bar{y}$. Now look at a typical point (x_i, y_i), which I happen to have taken in the upper left quadrant, shown in Fig. 5.3. Since

$$x_i - \bar{x} < 0 \quad \text{and} \quad y_i - \bar{y} > 0$$

the product

$$(x_i - \bar{x})(y_i - \bar{y}) < 0$$

Similarly for all other points in the upper left quadrant and the lower right quadrant. Points in the other two quadrants will make positive contributions

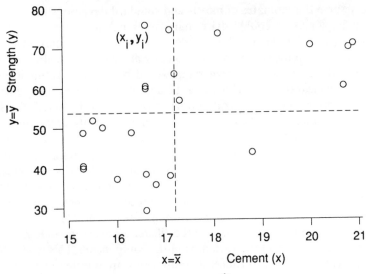

Fig. 5.3 Calculating $\widehat{cov}(x, y)$.

to the sum in the numerator of $\widehat{cov}$. If there is no association between the two variables the positive and negative contributions will tend to cancel out and, after division by $(n-1)$, $\widehat{cov}$ will be negligible. In contrast to this, if there is a tendency for y to increase as x increases most of the contributions will be positive, and $\widehat{cov}$ will be 'large' and positive. Similarly, if there is a tendency for y to decrease as x increases $\widehat{cov}$ will be 'large' and negative. The unit of $\widehat{cov}$ is the product of the unit of x with the unit of y, so the interpretation of 'large' depends on the choice of units. The covariance of the pavers is 13.4 (N mm^{-2} $\times$ % cement) and could be increased by a factor of a million by changing measurements from millimetres to metres.

The sample correlation (r) is a dimensionless quantity obtained from the covariance by:

$$r = \frac{\widehat{cov}}{s_x s_y}$$

It is shown in Example 5.5 that

$$-1 \leqslant r \leqslant 1$$

The correlation will take its extreme value of -1 if the points lie on a straight line with a negative slope, and 1 if they lie on a line with a positive slope. If points are scattered equally over all four quadrants, so that there is no linear relationship, the correlation will be close to 0. The correlation between the compressive strengths and cement contents is 0.53, and it should be noted that despite the scatter the correlation is more than half way along the scale between

Table 5.2 Amount of additive
and drying time for 11 dishes of
varnish

Additive $(mg\,g^{-1})$	Drying time (hours)
0	12.0
1	10.5
2	10.0
3	8.0
4	7.0
5	8.0
6	7.5
7	8.5
8	8.0
9	9.5
10	11.0

0 and 1. However, correlation is only a measure of linear association and points displaying an obvious 'quadratic' association can have correlations close to zero. The data in Table 5.2 are the drying time in hours for 11 dishes of varnish which had a different amount of additive mixed in. A quadratic association is evident from the scatter plot (Fig. 5.4), and the correlation is -0.25. The quadratic association will only be obvious if you plot the data, as you always

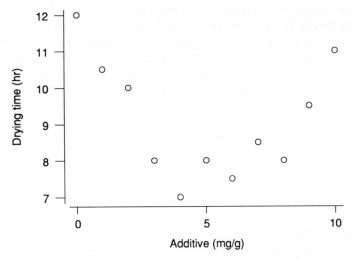

Fig. 5.4 Drying time against amount of additive for 11 dishes of varnish.

should! If you are using MINITAB and your data pairs are in C1 and C2 type

> PLOT C2 C1
> CORRELATION C1 C2

or use the mouse:

> Graph ▶ Plot...
> Stat ▶ Basic Statistics ▶ Descriptive Statistics ▶ Correlation
>   Variables: C1 C2

You should also note that an association between two variables does not necessarily mean that one causes the other. During a recession a chemical company has steadily improved its range of timber preserving products by reducing drying times. Throughout this period sales (seasonally adjusted) have been falling and the correlation between drying times and sales is 0.94, but reverting to the older products will not lead to increased sales! The company obtains more reliable information about the benefits of its research efforts by monitoring market share, but even this needs to be tempered by allowing for other changes in the market such as major competitors going out of business or making drastic price reductions.

5.3 Joint probability distributions

We often need to consider more than one variable for each member of a population, and any relationships between the variables may be of particular importance. Records of sea states are valuable for shipping engineers, shipping companies and the offshore industry. Data are often obtained from wave buoys which are fitted with instruments and transmit to data loggers. Data from a site in the North Sea are summarized in Table 5.3. They have been extracted from 38 850 fifteen-minute wave records. For each record the significant wave height (in metres), the height exceeded by about one-third of waves (Exercise 4.24), and average zero crossing period (in seconds) have been noted. If we sum the columns we obtain the distribution of crossing period (known as a **marginal distribution** if it happens to have been obtained by summing over other variables). We can find the distribution of heights by summing along rows. However, these two distributions would not tell us anything about the relationship between amplitude and frequency. Both variables affect the response of structures in deep water and any relationship between them is of great practical importance. The structure is relatively sensitive to specific frequencies, and if these tended to coincide with higher-amplitude waves then the safe operating life would be reduced. The approximate correlation can be calculated from grouped data, by taking mid-points of grouping intervals and multiplying by the frequencies (section 5.4). Bivariate data can be represented by a 3-D histogram (Fig. 5.5), the heights of the blocks being relative frequency divided by the area

Table 5.3 Significant wave height against zero crossing period for 15-minute records from a location in the North Sea*

Significant wave height (m)	0+	22	194	360	261	119	31	10	0+	0+	0+	1000
8.0	0	0	0	0	0	0	0	0	0	0+	0	0+
7.5	0	0	0	0	0	0	0+	0+	0+	0	0	0+
7.0	0	0	0	0	0	0	0+	0+	0	0	0	0+
6.5	0	0	0	0	0	0+	0+	0+	0+	0	0	0+
6.0	0	0	0	0	0	0+	0+	0+	0+	0+	0	0+
5.5	0	0	0	0	0	0+	1	1	0+	0+	0	2
5.0	0	0	0	0	0+	1	1	1	0+	0	0	3
4.5	0	0	0	0	0+	3	2	1	0+	0+	0	6
4.0	0	0	0	0	1	6	3	1	0+	0	0	11
3.5	0	0	0	0+	6	11	3	1	0+	0	0	21
3.0	0	0	0	2	21	15	3	1	0+	0	0	42
2.5	0	0	0+	18	42	19	4	1	0+	0	0	84
2.0	0	0	4	69	65	22	5	1	0+	0	0	166
1.5	0	0+	39	125	66	23	4	1	0+	0	0	258
1.0	0	9	110	113	49	14	4	1	0+	0+	0	300
0.5	0+	13	41	33	11	5	1	0+	0+	0+	0+	104
0.0												

Zero crossing period (s): 1 2 3 4 5 6 7 8 9 10 11 12

*The proportion of records in each cell, expressed per thousand (0+ entries correspond to less than 0.5 parts per thousand and are taken as 0.3 in calculations). The marginal distributions of zero crossing period and height are given along the top of the table and down the right-hand column, respectively.

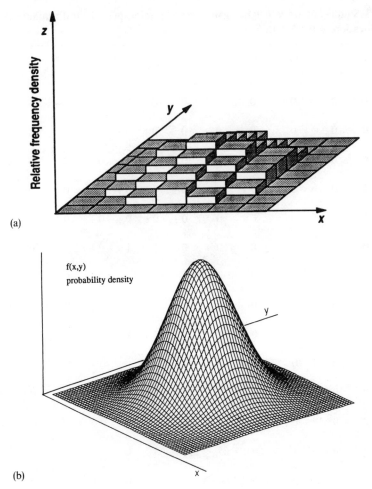

Fig. 5.5 (a) 3-D histogram. (b) Joint (bivariate) pdf.

of the class cell. As the sample size increases we imagine this tending towards a smooth surface given by a formula

$$z = f(x, y) \qquad \text{for } x, y > 0$$

such that the volume underneath it equals 1. That is,

$$\iint f(x, y)\, \mathrm{d}x\, \mathrm{d}y = 1$$

This is known as a **joint probability density function**, and the idea generalizes to more than two variables, although the geometric interpretation is lost. The **marginal probability density function** of X, which is just the pdf of X when information about y is ignored, is obtained by summing over the y-values and is given by the formula

$$f_X(x) = \int f(x, y)\mathrm{d}y$$

The marginal distribution of Y is defined in an analogous way. The marginal distributions $f_X(x)$ and $f_Y(y)$ will, in general, be different functions and, although I have emphasized this here by writing X and Y as subscripts, I shall sometimes rely on the arguments and context to make the distinction. A **joint cumulative distribution function**, $F(x, y)$, is defined by

$$F(x, y) = \Pr(X < x \text{ and } Y < y)$$

$$= \int_{-\infty}^{x} \int_{-\infty}^{y} f(\zeta, \eta)\mathrm{d}\zeta\,\mathrm{d}\eta$$

It follows that

$$f(x, y) = \frac{\partial^2 F}{\partial x \partial y}$$

Definitions for discrete variables are analogous.

5.3.1 Independence

The concept of independent events, in the probability context, carries over to the definition for variables to be independent. The variables X and Y are independent if and only if

$$f(x, y) = f_X(x)f_Y(y)$$

To see the justification for this definition, start with the probability of being close to any point (x_p, y_p) which is shown in Fig. 5.6:

$\Pr(X$ and Y lie within a rectangular area $\delta x\,\delta y$ centred on $(x_p, y_p))$
$= \Pr(X$ lies within an interval δx centred on x_p and Y lies within an interval δy centred on $y_p)$
$\simeq f(x_p, y_p)\delta x\,\delta y \qquad$ if δx and δy are small

Now if X and Y are independent the above probability is equal to

$\Pr(X$ lies within an interval δx centred on $x_p)$
$\times \Pr(Y$ lies within an interval δy centred on $y_p)$
$\simeq f_X(x_p)\delta x f_Y(y_p)\delta y$

The definition follows by comparing the two expressions and noting that the product $\delta x\,\delta y$ cancels. Since δx and δy are arbitrary, and the approximations

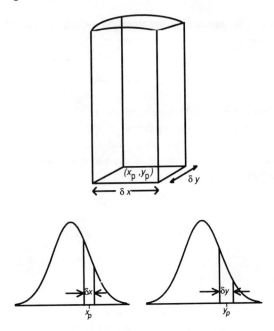

Fig. 5.6 Probabilities as volumes and areas.

are exact as δx and δy tend to 0, the definition does not depend on any approximations.

The fictitious data in Table 5.4 are fluoride contents and nitrate level, both measured in parts per million (ppm), for 764 jars of water. Fluoride is now added to the water supply in some areas of the UK as a public health measure to reduce dental decay. The addition of fluoride is contentious, although higher levels occur naturally in some waters, and the target value is 1.0 ppm. There is no reason to suspect an association between the variables, and none is apparent from the distributions.

Example 5.1

Apprentices take two practical and two theory exams. Let X, Y be the number of practical and theory exams passed, respectively. The probability distribution $P(x, y)$ based on extensive records is shown in the following table:

	x 0	1	2
y			
0			0.1
1		0.1	0.1
2	0.1	0.2	0.4

Table 5.4 Fluoride content and nitrate level for 764 jars of water

Fluoride content (ppm)	3	22	126	270	192	80	50	16	1	3	0	1	764
1.3				2									2
1.2			2	8									10
1.1		1	2	4	6	1							14
1.0		1	10	44	28	17	9	1					110
0.9	1	6	26	80	81	29	13	6	1	1			244
0.8	2	10	54	90	60	24	18	7		2		1	268
0.7		3	28	39	15	8	6	1					100
0.6		1	4	3	2	1	4	1					16
	0	1	2	3	4	5	6	7	8	9	10	11	

Nitrate level (ppm)

For example,

$$\text{Pr(an apprentice passes 3 or 4 exams)}$$
$$= P(1, 2) + P(2, 1) + P(2, 2) = 0.7$$

The marginal distributions are

$$P_X(x) = \begin{cases} 0.1 & x = 0 \\ 0.3 & x = 1 \\ 0.6 & x = 2 \end{cases}$$

$$P_Y(y) = \begin{cases} 0.1 & y = 0 \\ 0.2 & y = 1 \\ 0.7 & y = 2 \end{cases}$$

X and Y are not independent, for example

$$P_{XY}(0, 0) = 0 \quad \text{yet} \quad P_X(0)\, P_Y(0) = 0.01$$

Example 5.2

X and Y are the amounts of cement and plaster sold in a week by a small builders' merchant (units are tens of tonnes). The merchant consults his past three years' records and finds that sales of cement and plaster, between March and October, can be reasonably modelled by the pdfs

$$f_X(x) = \tfrac{1}{2}x \qquad \text{for } 0 \leqslant x \leqslant 2$$
$$f_Y(y) = \tfrac{1}{2}(1 + 3y^2) \qquad \text{for } 0 \leqslant y \leqslant 1$$

respectively. He calculates the sample correlation between weekly sales of cement and plaster and obtains a value of -0.07. He decides to assume the variables are independent because such a small correlation is of little practical significance and the estimate is not precise enough to contradict an assumption of independence. It is reasonable because bad weather is likely to affect cement sales but not plaster sales.

If the two distributions are independent the joint probability density function is their product,

$$f(x, y) = x(1 + 3y^2)/4 \qquad \text{for } 0 \leqslant x \leqslant 2, 0 \leqslant y < 1$$

and is shown in Fig. 5.7(a).

(a) Confirm that $f(x, y)$ is a joint pdf.

Slice the volume under the surface parallel to the x-axis and let $A(y)$ be a typical face area (Fig. 5.7(b)). Then

$$A(y) = \int_0^2 x(1 + 3y^2)/4 \, dx$$

$$= \frac{1 + 3y^2}{4} \int_0^2 x \, dx = \frac{1 + 3y^2}{4} [\tfrac{1}{2} x^2]_0^2$$

$$= (1 + 3y^2)/2$$

The volume V is the sum of elements of volume $A(y) \, \delta y$. That is,

$$V = \int_0^1 A(y) dy = \int_0^1 \frac{1 + 3y^2}{2} dy = [\tfrac{1}{2}y + \tfrac{1}{2}y^3]_0^1 = 1$$

So the total probability is 1 and the function is a pdf.

(b) Demonstrate that the marginal probability density functions of X and Y are

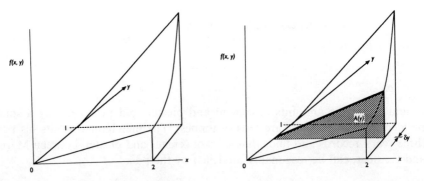

Fig. 5.7 Bivariate distribution of cement and plaster sales: (a) pdf; (b) a typical element of volume.

just the pdf he started with.

$$f_X(x) = \int_0^1 \tfrac{1}{4}x(1 + 3y^2)dy = \tfrac{1}{2}x \qquad\qquad \text{for } 0 \leqslant x \leqslant 2$$

$$f_Y(y) = \int_0^2 \tfrac{1}{4}x(1 + 3y^2)dx = \tfrac{1}{2}(1 + 3y^2) \qquad \text{for } 0 \leqslant y \leqslant 1$$

(c) In the past he has always started the week with 20 tonnes of cement and 10 tonnes of plaster, but he wishes to know the probability that stocks of 10 tonnes of cement and 5 tonnes of plaster would suffice for some week in the future. He requires

$$\Pr(X < 1 \text{ and } Y < 0.5) = \int_0^1 \int_0^{0.5} f(x, y)\, dx\, dy$$

$$= 0.25 \int_0^1 x\, dx \int_0^{0.5} (1 + 3y^2) dy = 0.078$$

Now find the probability that he would run out of both cement and plaster.

$$\Pr(1 < X \text{ and } 0.5 < Y) = \int_1^2 \int_{0.5}^1 f(x, y)\, dx\, dy$$

This equals 0.516, and you should convince yourself why it is not 1 minus the last probability!

We shall look at joint distributions, especially the bivariate normal, in more detail in Chapter 8. For the moment our main interest is in means, variances and covariances.

5.4 Population covariance and correlation

We can calculate an approximate mean for the wave heights in Table 5.3 from

$$\bar{y} = \sum_{i=1}^{11} \sum_{j=1}^{16} y_{ij} f_{ij}/n$$

where y_{ij} is the y-coordinate of the centre, and f_{ij} is the relative frequency, of the (i, j)th cell. The calculation leads to a $\bar{y}$ of 1.37 m, and similar calculation gives $\bar{x}$ equal to 4.94 s. If n becomes large the 3-D histogram tends towards the joint pdf $f(x, y)$ and relative frequencies tend towards probabilities $f(x, y)\, dx\, dy$. So

$$E[Y] = \int \int y f(x, y)\, dx\, dy$$

$$= \int y f_Y(y)\, dy = \mu_Y$$

The mean of Y in the joint distribution is the mean of Y in the marginal distribution, as it should be. Similarly for μ_X, σ_X^2, and σ_Y^2. An approximate value for the covariance between wave height and period is

$$\widehat{cov} = \sum_{i=1}^{11} \sum_{j=1}^{16} (x_{ij} - \bar{x})(y_{ij} - \bar{y}) f_{ij}/n$$

The values of s_x, s_y and cov calculated for the data in Table 5.3 using the above formulae are 1.22 s, 1.81 m and 1.28 m s, respectively. Hence the sample correlation (r) is approximately 0.58. The definition of population covariance, $cov(X, Y)$, is defined analogously to $\widehat{cov}$, as

$$cov(X, Y) = E[(X - \mu_X)(Y - \mu_Y)] = \int \int (x - \mu_X)(y - \mu_Y) f(x, y) \, dx \, dy$$

The population correlation, ρ, is defined by

$$\rho = \frac{cov(X, Y)}{\sigma_X \sigma_Y}$$

If X and Y are independent the covariance is zero because the defining integral factorizes to

$$\int (x - \mu_X) f(x) dx \int (y - \mu_Y) f(y) dy$$

both terms of which are themselves zero. The converse is not necessarily true because X and Y may have a nonlinear association, or the range of Y may depend on the value of X (see Exercise 5.16).

We are now in a position to state and prove a most useful result.

5.5 Linear combinations of random variables

Let X and Y be variables with means μ_X, μ_Y and variances σ_X^2, σ_Y^2, respectively. Now construct a variable which is a linear combination of X and Y:

$$W = aX + bY$$

where a and b are constants. Then

$$\mu_W = a\mu_X + b\mu_Y$$

and

$$\sigma_W^2 = a^2\sigma_X^2 + b^2\sigma_Y^2 + 2ab \, cov(X, Y)$$

The proof involves squaring a bracket, and remembering that the integral of a sum is the sum of the integrals. The result for the mean is immediate.

$$\mu_W = E[W] = E[aX + bY]$$
$$= aE[X] + bE[Y] = a\mu_X + b\mu_Y$$

The result for the variance is

$$\sigma_W^2 = E[(W - \mu_W)^2] = E[((aX + bY) - (a\mu_X + b\mu_Y))^2]$$
$$= E[(a(X - \mu_X) + b(Y - \mu_Y))^2]$$
$$= E[a^2(X - \mu_X)^2 + b^2(Y - \mu_Y)^2 + 2ab(X - \mu_X)(Y - \mu_Y)]$$
$$= a^2\sigma_X^2 + b^2\sigma_Y^2 + 2ab\,\text{cov}(X, Y)$$

Notice that the final term disappears if X and Y are independent. The result extends to any number of variables.

Example 5.3

An oil company has a survey vessel which measures the depth of the sea bed with two instruments which give readings X and Y. Both instruments have been carefully calibrated and give unbiased estimates of depth. That is, if the depth is θ,

$$E[X] = E[Y] = \theta$$

However, the second instrument is more precise and

$$\sigma_X^2 = 2\sigma_Y^2$$

A surveyor intends averaging the two results, but she thinks some weighted average will be better than $(X + Y)/2$.

First, assume the errors are independent and find the mean and variance (in terms of σ_Y^2) of

$$0.5X + 0.5Y$$

and

$$0.2X + 0.8Y$$

Both have a mean of θ. The variance of the first linear combination is

$$0.25\sigma_X^2 + 0.25\sigma_Y^2 = 0.75\sigma_Y^2$$

The variance of the second combination is $0.72\sigma_Y^2$.

These results suggest that there might be a best (unbiased with smallest variance) combination somewhere between taking 0.5 of X and taking 0.2 of X. Define

$$W = aX + (1 - a)Y$$

The fact that the weights, a and $(1 - a)$, sum to 1 is a requirement for W to be unbiased. We express σ_W^2 in terms of a and σ_Y^2, and then find the value of a that minimizes this variance.

$$\sigma_W^2 = 2a^2\sigma_Y^2 + (1 - a)^2\sigma_Y^2 + (1 - 2a + 3a^2)\sigma_Y^2$$

Setting the derivative with respect to a equal to zero gives a minimum of $\frac{2}{3}\sigma_Y^2$ when a equals $\frac{1}{3}$.

Now suppose we know from experiments on a test rig in Loch Ness that the errors have a correlation of 0.4. What are the mean and variance of $\frac{1}{3}X + \frac{2}{3}Y$?

The mean is unaffected by the correlation and still equals θ. The variance is

$$(\tfrac{1}{3})^2 \times 2\sigma_Y^2 + (\tfrac{2}{3})^2\sigma_Y^2 + 2(\tfrac{1}{3})(\tfrac{2}{3})0.4(\sqrt{2}\sigma_Y)\sigma_Y$$

which equals $0.92\sigma_Y^2$. You are left to find the best linear combination under these circumstances.

Example 5.4

A company manufactures walls for timber-frame houses. A particular design is shown in Fig. 5.8. The distributions of heights of components are known. The company needs to know the likely differences in heights of finished walls, if components are selected at random, before houses are built! We need the result (Appendix A2.2) that a linear combination of normally distributed random variables has a normal distribution. If X_1, X_2 are distributed $N(3.0, 0.20^2)$ and Y_1, Y_2 are distributed $N(2.5, 0.16^2)$, find the probability that the difference in heights, D, exceeds 0.5 m when components are selected at random.

$$D = (X_1 + Y_1) - (X_2 + Y_2) = X_1 + Y_1 - X_2 - Y_2$$
$$E[D] = E[X_1] + E[Y_1] - E[X_2] - E[Y_2]$$
$$= 3.0 + 2.5 - 3.0 - 2.5 = 0$$

It is convenient to use var(D) for σ_D^2 and so on, because it avoids two subscript levels.

$$\text{var}(D) = \text{var}(X_1) + \text{var}(Y_1) + \text{var}(X_2) + \text{var}(Y_2)$$
$$= 0.20^2 + 0.16^2 + 0.20^2 + 0.16^2$$
$$= 0.1312$$

Taking the square root gives a standard deviation of 0.362. (The third decimal place is not reliable but we will round only at the end of the problem.)

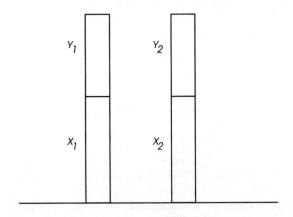

Fig. 5.8 Front and back walls of timber frame houses.

We have now established that

$$D \sim N(0, 0.362^2)$$

We require

$$\Pr(D < -0.5 \quad \text{or} \quad 0.5 < D)$$
$$= 1 - \Pr(-0.5 < D < 0.5)$$
$$= 1 - \Pr(-1.38 < Z < 1.38)$$

where, in standard notation, Z is $N(0, 1)$. Reference to tables of the standard normal distribution gives the corresponding probability as 0.17.

Example 5.5 Correlation lies between -1 and 1

Prove that $-1 \leqslant \rho \leqslant 1$. We have seen that

$$\text{var}(aX + bY) = a^2 \, \text{var}(X) + b^2 \, \text{var}(Y) + 2ab \, \text{cov}(X, Y)$$

Then set a equal to σ_Y and b equal to σ_X, and use the fact that $\text{var}(aX + bY)$ must be positive to obtain ρ greater that -1. Repeat with b equal to $-\sigma_X$ to obtain ρ less than 1. An analogous argument with sample quantities gives

$$-1 \leqslant r \leqslant 1$$

Example 5.6

A helicopter operating in the North Sea can carry ten passengers. Passengers' weights are normally distributed with a mean of 80 kg and a standard deviation of 14 kg. Let X represent passenger weight. Then

$$X \sim N(80, 14^2).$$

Find the following probabilities.

(i) Pr(a randomly selected passenger exceeds 90 kg)

$$\Pr(X > 90) = \Pr(Z > (90 - 80)/14)$$
$$= \Pr(Z > 0.71) = 0.24$$

(ii) Pr(total weight of 10 passengers selected at random exceeds 900 kg). Define

$$T = X_1 + \cdots + X_{10}$$
$$\mu_T = 80 + \cdots + 80 = 800$$
$$\sigma_T^2 = 14^2 + \cdots + 14^2 = 1960$$
$$\sigma_T = 44.27$$
$$T \sim N(800, (44.27)^2)$$
$$\Pr(T > 900) = \Pr(Z > (900 - 800)/44.27)$$
$$= \Pr(Z > 2.26) = 0.012$$

(iii) Pr(mean weight of 10 passengers selected at random exceeds 90 kg)

$$\bar{X} = T/10$$
$$\mu_{\bar{x}} = 800/10 = 80$$
$$\sigma_{\bar{x}} = 44.27/10 = 4.427$$

As the event is the same as in (ii), the probability is the same.

Notice that the standard deviation of the mean of ten passengers' weights is $\sigma/\sqrt{10}$. We generalize this important result in section 5.6.

Example 5.7

A large civil engineering company specializes in contract work in Africa and operates a helicopter which can carry ten passengers. The weights of male employees are distributed with a mean of 80 kg and a standard deviation of 13 kg. The weights of female employees are distributed with a mean of 60 kg and a standard deviation of 10 kg. We have been asked to find the probability that the total weight of ten passengers exceeds 800 kg, if they are selected at random from all employees of the company and are equally likely to be male or female. This problem is different from the last because of the mixture of two distributions. We will tackle it in two parts. We first let M, W and Y represent the weights of men, women and a randomly selected passenger respectively. With these definitions,

$$\mu_Y = E[Y] = \tfrac{1}{2}E[M] + \tfrac{1}{2}E[W]$$
$$= 70 \text{ kg}$$

The variance of Y can be found from the following identity,

$$\sigma_Y^2 = E[(Y - \mu_Y)^2] = E[(Y^2 - 2\mu_Y Y + \mu_Y^2)]$$
$$= E[Y^2] - 2\mu_Y E[Y] + E[\mu_Y^2]$$
$$= E[Y^2] - 2\mu_Y^2 + \mu_Y^2$$
$$= E[Y^2] - \mu_Y^2$$

This is a useful general result, and is often rearranged as

$$E[Y^2] = \sigma_Y^2 + \mu_Y^2$$

Half the time Y will equal M and the other half it will equal W. Therefore the average value of Y^2 is given by:

$$E[Y^2] = \tfrac{1}{2}E[M^2] + \tfrac{1}{2}E[W^2]$$

Now

$$E[M^2] = \sigma_M^2 + \mu_M^2 = 6569$$

and similarly,

$$E[W^2] = 3700$$

so $\sigma_Y^2 = 234.5$.

Note that we cannot say

$$Y = \tfrac{1}{2}M + \tfrac{1}{2}W$$

even if the helicopter always carries equal numbers of males and females!

To continue with the problem, let T be the total weight, that is

$$T = \sum_{i=1}^{10} Y_i$$

Because simple random sampling is used it is reasonable to assume the Y_i are independent and σ_T^2 is 10 times σ_Y^2, which is 2345. The expected value of T is 700, provided men and women are equally likely to be selected.

If we now use the central limit theorem (section 5.6)

$$\Pr(800 < T) = \Pr\left(\frac{800 - 700}{\sqrt{2345}} < Z \right)$$

and the required probability is

$$\Pr(2.065 < Z) = 0.02$$

5.6 Distribution of the sample mean

Let $\{X_i\}$ be a random sample of size n from any distribution with a mean μ and finite variance σ^2. Define the sample total, T, by

$$T = X_1 + \cdots + X_n$$

Using the results for a linear combination of variables and the fact that randomization makes an assumption of independence valid,

$$\mu_T = \mu + \cdots + \mu = n\mu$$

and

$$\sigma_T^2 = \sigma^2 + \cdots + \sigma^2 = n\sigma^2$$

Hence

$$\sigma_T = \sigma\sqrt{n}$$

Now the sample mean is by definition

$$\bar{X} = T/n$$

a simple scaling by the reciprocal of n. In general, if W is a scaling of X

$$W = aX$$

and the general result of section 5.5 with $b = 0$ gives

$$\sigma_W^2 = a^2 \sigma_X^2 \quad \text{and} \quad \sigma_W = a \sigma_X$$

as it must from a dimensional argument. Making use of this gives

$$\mu_{\bar{X}} = \mu_T/n = n\mu/n = \mu$$
$$\sigma_{\bar{X}}^2 = \sigma_T^2/n^2 = n\sigma^2/n^2 = \sigma^2/n$$

and finally

$$\sigma_{\bar{X}} = \sigma/\sqrt{n}$$

In Example 5.6 passengers weights were distributed with a mean of 80 kg and a standard deviation of 14 kg. We can now write down the mean and standard deviation of the average weight of ten randomly selected passengers, in one step, as 80 kg and $14/\sqrt{10} = 4.427$ kg. The above results give us the mean and standard deviation of the distribution of $\bar{X}$. The following theorem tells us the form of that distribution.

Central limit theorem

Let $\{X_i\}$ be a random sample of size n from a distribution with mean μ and finite variance σ^2. Then

$$\frac{\bar{X} - \mu}{\sigma/\sqrt{n}} \text{ is approximately distributed as } N(0, 1)$$

The approximation inproves as n increases, and is usually good for $n > 30$. If the population itself is nearly normal the approximation is excellent for any value of n. The distribution of $\bar{X}$ is an example of a **sampling distribution**.

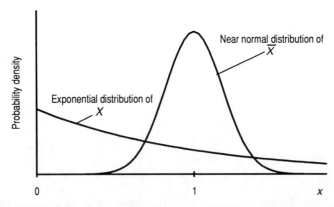

Fig. 5.9 Distribution of sample mean when taking random samples of 30 from an exponential distribution.

Imagine taking millions of samples of size n from our population, replacing after each selection unless you think of an infinite population, and recording the means. We could then draw a histogram of all these means, and work out their mean and standard deviation. The histogram would look like a normal distribution, the mean of the means would be μ and the standard deviation of the means would be $\sigma/\sqrt{n}$. The term **standard error** is often used for standard deviation of the mean. Figure 5.9 shows the distribution of X when taking random samples of 30 from an exponential distribution with a mean of 1. The central limit theorem is proved in Appendix A2.

Example 5.8

This example is part of a case study of the application of statistical techniques in shipyards by Kattan (1993). One of his aims was to reduce rework by simplifying designs. The panel shown in Fig. 5.10 is made up of five plates (6 m × 1.5 m), eight stiffeners and three webs. The tolerance for the overall width is -7 mm to $+3$ mm. Kattan asked the welders to record process data at their workstations over two months. They provided the following information for the axis OX.

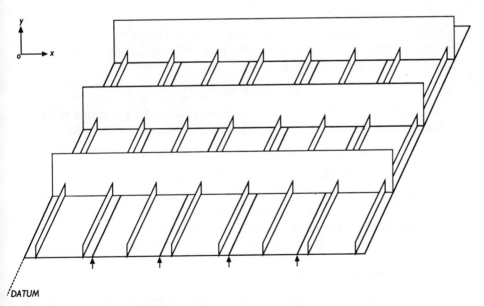

Fig. 5.10 Built-up stiffened plate panel.

Operation	Average discrepancy (mm)	Standard deviation of discrepancies (mm)
Plate cutting	+0.5	0.9
Plate alignment	+1	0.8
Butt welding shrinkage	−4	1.1
Shrinkage per stiffener	−1	1.0
Shrinkage per web	−1.5	1.2

A welder makes four butt welds, and welds eight stiffeners and three webs on each panel. If we also allow for the four alignments and five plate cuttings we have a mean error in overall width of

$$(5 \times 0.5) + (4 \times 1) + (4 \times -4) + (8 \times (-1)) + (3 \times (-1.5))$$
$$= -22 \, \text{mm}$$

This result suggests that the designer should make an additional allowance of 20 mm to bring the average panel width to the middle of the tolerance range (−2 mm). The variance of the error, if we assume the individual discrepancies are independent, is

$$5 \times 0.9^2 + 4 \times 0.8^2 + 4 \times 1.1^2 + 8 \times 1.0^2 + 3 \times 1.2^2$$
$$= 23.8 \, \text{mm}^2$$

and the standard deviation is 4.9 mm. If we assume the average width has been brought to the middle of the range, the probability of having to rework the panel because it is out of tolerance is

$$2\Pr(Z > 5/4.9) = 0.31$$

Now suppose the designer uses three plates (6 m × 2.5 m), reducing the butt welds and alignments to two on each panel. The variance will be reduced to

$$3 \times (0.9)^2 + 2 \times (0.8)^2 + 2 \times (1.1)^2 + 8 \times (1.0)^2 + 3 \times (1.2)^2$$
$$= 18.45 \, \text{mm}^2$$

The standard deviation is now 4.3 mm, and the probability of rework is reduced to 0.25. This is a worthwhile reduction, but we could do better still if the number of stiffeners could be safely reduced. Kattan reports that the number was reduced to six, with an increase in cross-sectional area to compensate.

The finite population correction

So far in this chapter we have assumed random sampling from an imaginary infinite population. This is adequate unless the sample is a substantial proportion of the population, in which case the variance of $\bar{X}$ will be appreciably reduced. Suppose $X_1, \ldots, X_n$ is a random sample of size n from a finite population of size N with mean μ and variance σ^2. The population mean μ is defined by

$$\mu = \sum_{i=1}^{N} x_i/N$$

and the population variance is defined as

$$\sigma^2 = \sum_{i=1}^{N} (x_i - \mu)^2/N$$

Let T be the sample total and $\bar{X}$ the sample mean. Then

$$E[T] = n\mu \qquad \text{var}(T) = n\sigma^2(1 - (n/N))(N/(N-1))$$
$$E[\bar{X}] = \mu \qquad \text{var}(\bar{X}) = (\sigma^2/n)(1 - (n/N))(N/(N-1))$$

The factor of $N/(N-1)$ is usually negligible. The more interesting factor of $1 - (n/N)$, which reduces the variance from the infinite population case, is called the **finite population correction** (FPC) and is negligible if N is much larger than n. The distribution of T, and hence X, is well approximated by a normal distribution if n is reasonably large or if the population is nearly normal. I have not included a proof (for a straightforward account, see Barnett, 1974), but the variance formula does reduce to the result for an infinite population as N tends to infinity and gives 0, as it should, if the sample is the entire population (n equals N).

Example 5.9

The most unusual application of the finite population I have come across is to do with the price of rocks from a quarry. Brian Taylor (Ward Robinson, Newcastle upon Tyne) told me about this when it arose during computer software development for the quarrying company. The management wished to know the distribution of errors if it adopted an approximate charging policy for a small proportion of sales.

Rocks are transported from the quarry to a wharf. Here they are stacked in distinct piles within bunkers designated for specific mass bands (Fig. 5.11). The rocks are weighed by a load cell during the lifting operation, so the company knows the exact masses of all the rocks.

However, the rocks are taken off by barges and the skippers often finish with only part of a pile, either because the barge is fully loaded or because they want to catch the tide. The cranes on the barges are not equipped with load cells.

A particular skipper loads a whole number of piles and then finishes with 30 from a pile of 40 rocks in a 5–7 tonne range. The mean and standard deviation of masses for the pile of 40 rocks were recorded as 6.12 and 0.56 tonnes, respectively. The normal distribution will be an excellent approximation to the distribution of T (it is reasonable for $2 \leqslant n \leqslant 38$). There is therefore a 95% chance

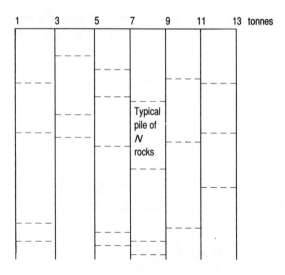

Fig. 5.11 Piles of rocks in bunkers.

that the total mass of the 30 rocks taken will be within

$$30 \times 6.12 \pm 1.96 \times \sqrt{30 \times 0.56^2 \times (1 - (30/40))}$$

which after some arithmetic works out as

[180.6, 186.6]

This assumes that rocks are loaded at random, which is reasonable as speed of loading is the skipper's main concern. The rocks are sold by mass and the error in the price for the fractional pile is unlikely to be more than 2%, and will be a negligible percentage of the total price.

5.7 Statistical process control charts

A company makes a range of burners for domestic gas appliances. The process starts with steel plates, perforates them, rolls them into tubes, and flanges the ends. The last two operations are carried out by a single remarkable and rather old machine which needs careful setting at the start of each run of burners of a particular type. The distance from the base of the finished burner to the first gas outlet hole is critical for burner performance. Before making a large batch, the manufacturer needs to check that the machine has been set correctly so that, provided everything stays the same, nearly all the burners will lie within specification. Furthermore, occasional checks should also be made to ensure that machine settings have not drifted or suddenly changed.

The specification length, from base to first gas outlet hole, for a particular type of burner is that it should be between 67.50 mm and 68.50 mm. The standard deviation of this length had been estimated as 0.10 mm, by measuring a random sample from a large batch when the machine had been 'working well'. A postgraduate student, Medhi Akrami, investigated the use of some simple **statistical process control** (SPC) techniques, also known as **statistical quality control** (SQC) techniques, with this process, and checking the estimate of standard deviation was one objective. An experienced operator had set the machine up and was preparing for a run of several thousand. Akrami decided to take five samples of 5 off the production line, the samples separated only by the time taken to measure five burners, followed by samples of 5 at approximately, but not exactly, half-hour intervals. Exactly equal time intervals should be avoided in case they coincide with any periodicities of the process. It is also preferable to avoid taking 5 consecutive items off the line when sampling, because they may be more similar than 5 selected haphazardly from a minute or so of production. Before describing his work we need some definitions.

5.7.1 Mean chart

Suppose that the process produces lengths that are approximately normally distributed with a standard deviation which is known from past experience. The mean can be adjusted, and the target value is τ. Samples of n are taken every so often, and we shall assume they are random samples from a process with an underlying mean of μ. Then, given all our assumptions,

$$\bar{X} \sim N(\mu, (\sigma/\sqrt{n})^2)$$

If the mean is on target – that is, if $\mu = \tau$ – there will be a probability of 0.001 that $\bar{X}$ exceeds

$$\tau + 3.09\sigma/\sqrt{n}$$

and the same probability that $\bar{X}$ is less than

$$\tau - 3.09\sigma/\sqrt{n}$$

The procedure for the mean chart is to plot sample means against time and take action if a point falls outside action limits drawn at

$$\tau \pm 3.09\sigma/\sqrt{n}$$

The idea dates back to the work of Shewhart and others in the 1920s (Shewhart, 1931) and the mean chart is also commonly known as a **Shewhart chart**. The rationale behind it is that action will be taken unnecessarily, when μ is on target, at an average rate of 1 in 500 samples but that this is offset by the advantages of detecting changes in μ and taking suitable action. Provided the action is appropriate, and this will be taken up in the course of our example, it is undoubtedly a useful technique. Lockyer and Oakland (1981) give an example of a company in the paint industry, turning over about £35 million a year, which saved more than £250 000 by installing Shewhart mean and range charts (section 5.7.2) to monitor filling processes and ensure conformity with the 1979 Weights and Measures Act.

5.7.2 Range chart

Any changes in process variability are as important to detect as changes in the mean. Unfortunately they are likely to be more difficult to correct. Increased variability may be caused by worn machinery or inexperienced operators. A sudden decrease is likely to be due to faults in measuring equipment. Table E.5

Table 5.5 (a) Lengths from base of burner to first gas outlet hole (mm). (b) Means and ranges for a further 35 samples of size 5, in chronological order

(a)

	Sample 1	Sample 2	Sample 3	Sample 4	Sample 5
	68.20	68.14	68.16	68.01	68.19
	68.10	68.29	68.03	67.97	68.30
	68.10	67.97	68.17	68.03	68.25
	68.13	68.11	68.05	68.02	68.19
	68.01	68.11	68.10	68.05	67.95
Mean	68.108	68.124	68.102	68.016	68.176
Range	0.190	0.320	0.140	0.080	0.350
Standard deviation	0.068	0.114	0.063	0.030	0.134

(b)

68.070	68.052	68.122	68.124	68.080
0.320	0.300	0.190	0.140	0.250
67.998	68.088	67.980	68.116	68.088
0.150	0.320	0.340	0.760	0.290
68.020	68.048	68.104	68.038	68.164
0.300	0.170	0.230	0.220	0.390
67.962	68.034	68.036	67.948	68.010
0.240	0.310	0.220	0.170	0.210
68.022	68.106	68.014	67.966	67.990
0.310	0.200	0.240	0.260	0.120
67.994	68.052	67.992	68.080	68.068
0.200	0.190	0.200	0.150	0.100
68.048	68.010	68.012	68.032	67.940
0.120	0.140	0.250	0.210	0.210

gives factors of the standard deviation from which to calculate upper (1 in 1000) and lower (1 in 1000) action lines for the range of samples of size n. These are sensitive to the assumption of normality of individual lengths, unlike the mean chart which is quite robust (central limit theorem). In general, if the variable for which they are constructed has a kurtosis greater than 3 (which is the kurtosis of the normal distribution) the upper factor should be multiplied by

$$\sqrt{(1 + (\kappa - 3)(n - 1)/(2n))}$$

and the lower factor should be divided by this (Wetherill and Brown, 1991).

The results that Akrami obtained for the burners are summarized in Table 5.5. Only the first five samples, which were taken in rapid succession, are given in full. The means and ranges are plotted in Fig. 5.12. The target length was taken as the middle of the specified range, 68.00 mm, and the fifth sample falls above the upper action line. What action should be taken? All the points are above the target and there is no reason to suspect any sudden change. The mean of all 25 data is 68.1052. It seems that the operator has set the machine

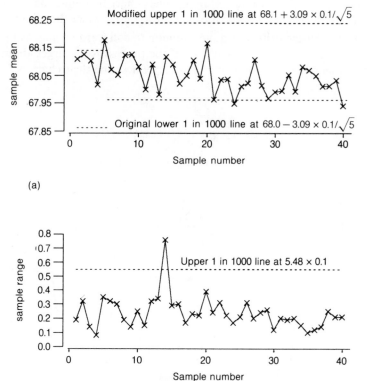

(a)

(b)

Fig. 5.12 (a) Mean chart (Shewhart chart). (b) Range chart.

up for a mean of about 68.10 mm rather than 68.00 mm. However a mean of 68.10 mm will do, provided the standard deviation is only 0.10 mm, because nearly all the burners will be within $68.10 \pm 3 \times 0.10$ mm, which is itself well within specification. The machine is difficult to set up and any attempted fine adjustment could lead to a greater discrepancy. The action taken was to recalculate action lines about the 'acceptable setting' of 68.10 mm, take a sixth sample immediately, and then continue to monitor the process at approximate half-hour intervals. The only other cause for doubt about the process was a single high range value. The only action taken was to take another sample immediately.

Even if the estimate of standard deviation was reliable, the lack of action in this example would have been justified by the facts that the process has high capability, fine adjustment is difficult, and the burners do not work noticeably better with a length of exactly 68.00 mm than anywhere else within their specification. In Lockyer and Oakland's (1981) paint example any deviation from the target has an associated cost – the company either gives away paint or runs an increased chance of prosecution. In this case action involved adjusting the process (Fig. 5.13). Before setting up their chart they must have estimated the standard deviation of volume of paint dispensed when the filling machine was working satisfactorily. It is quite common to add warning lines at $1.96\sigma/\sqrt{n}$ from the target value, and they did so. If the process is on target about 5% of points will lie outside the warning lines. They describe their chart as follows.

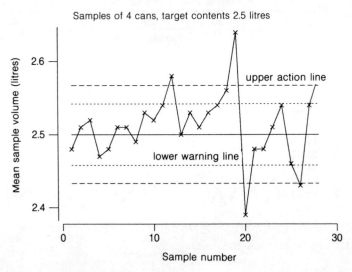

Fig. 5.13 Lockyer and Oakland's example of a mean chart in the paint industry (courtesy of *Management Today*).

The mean chart shows action and warning lines together with the average or mean volume of samples of four tins. The process is well under control (within the action lines) until sample 12, when the mean rises above the upper action line; corrective action must be taken. Its action brings the process back into control again until sample 18, where the mean is in the warning zone – another sample should be taken immediately. This sample (19) gives a mean well above the action line; corrective action forthwith. But action results in over-correction and sample mean 20 is below the lower action line. The process continues to drift upwards out of control between samples 21 to 24 and from 26 to 30. The process equipment was investigated as a result of this – a worn friction valve was slowly and continually vibrating open, letting too much paint into the tins. Probably this wouldn't have been identified so quickly in the absence of SQC. This simple example indicates the power of process control charts in quality control and in early warning of equipment trouble.

Adjustments may have to be made on a rather *ad-hoc* basis, and records should be kept so that the company can tend towards an optimum strategy. I think the value of SPC charts is that the process is being monitored. When producing the burners a small change in the process mean would be inconsequential and resetting the process is a last resort. Adherence to a '3σ rule' would not be helpful. At the other extreme I have seen a semi-automatic process for producing tyre side walls which can be adjusted by entering a new set point into a computer console. If the process had a tendency to wander slightly, an adjustment by a fraction of the difference between each sample mean and the target value, after every sample, would reduce the variability (see Exercise 5.18). It would admittedly be better still to stop the process wandering, but the cost of doing this might exceed the benefits. I do not advocate continual adjustment as a panacea, but it could be worth a thought in some circumstances! Despite progress in automatic, self-tuning, feedback controllers and robots, SPC still has an important role in industry (Efthimiadu *et al.*, 1993). Small companies can rarely afford such equipment. In highly complex manufacturing processes there will still be a need for human intervention, if only to check the automatic controllers. The Nashua Corporation states on its packs of ten computer diskettes: 'Literally hundreds of charts, like that shown here [picture on box], control the enemy of any manufacturing process – variation.'

5.7.3 Estimation of process standard deviation

One of the objectives of Akrami's investigation was to check whether the assumed standard deviation of lengths was reasonable. The obvious estimate with which to compare it is the standard deviation of all 200 data. This is 0.110, slightly higher than the assumed value of 0.10. An alternative estimate would be an average of the 40 estimates that can be made within each sample. As this is quite a fine comparison we should be slightly careful about how the averaging is done. If a divisor $(n-1)$ is used when calculating variances from the sample the expected value (long-term average value if you imagine taking very many such samples) will equal the population variance (σ^2). The proof is straightforward and you are asked to complete it in Exercise 5.13.

A direct consequence of this result is that the expected value of the sample standard deviation is slightly less than the population standard deviation, by a factor of $\sqrt{(1-1/(2n))}$ for a normal distribution. We should, therefore, average the 40 sample variances and then take square root, rather than average the 40 standard deviations. This gives an estimated process standard deviation of 0.102 mm. This is an estimate of local variability, and if the process mean had tended to undulate it would be significantly smaller than the standard deviation of the pooled data. There is little difference for the burners, and the process can be seen to be fairly stable from the mean chart. Akrami nevertheless recommended the use of the pooled value, 0.11, because any slight undulation that might exist in the process would be quite acceptable and no attempts to trace or remove it were going to be made.

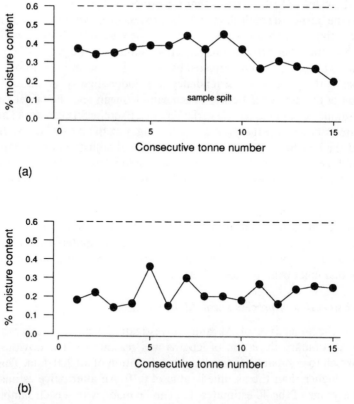

(a)

(b)

Fig. 5.14 Runs charts for (a) drier number 1 (b) drier number 3.

5.7.4 Runs chart

It is not always appropriate to take small random samples during the process. A company sells milled zirconium silicate to the ceramics industry as a refractory and glaze opacifier. The raw material is milled in a continuous process, dried and packed in one-tonne batches. The product specification is in terms of particle size and chemical composition but does not include moisture content. However, a customer had recently complained of the higher moisture content of a recent delivery (0.99% compared with previous deliveries containing 0.2% to 0.3%). A student plotted percentage moisture, against batch number for two different driers (Fig. 5.14). Monitoring of the process provides useful information, and the runs chart can be annotated if any other information is available. A formal specification limit of 0.6% maximum moisture would seem to be achievable.

5.8 Nonlinear functions of random variables

Nonlinear functions of variables are common in engineering. The flow in rivers is usually measured by a weir (Fig. 5.15). The flow (Q) is approximately related to the height (H) of the water surface above the crest, the width of the weir (L), gravity (g), and a weir discharge coefficient (K) which is introduced in an attempt to account for non-ideal flow (Dally et al., 1984), by

$$Q = \tfrac{2}{3}(2g)^{1/2}LKH^{3/2}$$

A hydrologist has estimated the standard deviation of K and H as 10% and 5%, respectively, for a particular field study, and assumes g and L are known constants. The standard deviation of K has to be assessed from expert knowledge, as she cannot investigate a random sample of K-values from this weir. The standard deviation of H may also be assessed in this way. She would like to know the standard deviation of Q, and whether it will be unbiased when K and H are.

The general theory relies on a Taylor series expansion. Let ϕ be an arbitrary

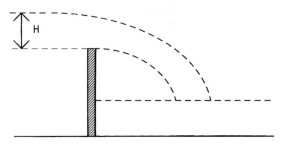

Fig. 5.15 Flow over sharp crested rectangular weir.

function of X and Y and expand ϕ about their mean values. Then

$$\phi(X, Y) = \phi(\mu_X, \mu_Y) + \frac{\partial \phi}{\partial x}(X - \mu_X) + \frac{\partial \phi}{\partial y}(Y - \mu_Y)$$

$$+ \frac{1}{2!}\left(\frac{\partial^2 \phi}{\partial x^2}(X - \mu_X)^2 + 2\frac{\partial^2 \phi}{\partial x \partial y}(X - \mu_X)(Y - \mu_Y) + \frac{\partial^2 \phi}{\partial y^2}(Y - \mu_Y)^2\right) + \cdots$$

where all the partial derivatives are evaluated at (μ_X, μ_Y). We start by considering the question of bias. Take expectation of both sides to obtain the approximate result that

$$E[\phi(X, Y)] = \phi(\mu_X, \mu_Y) + \frac{1}{2}\frac{\partial^2 \phi}{\partial x^2}\sigma_X^2 + \frac{1}{2}\frac{\partial^2 \phi}{\partial y^2}\sigma_Y^2 + \frac{\partial^2 \phi}{\partial x \partial y}\mathrm{cov}(X, Y)$$

Remember that the expected value of a constant is that constant and the expected value of a variable minus its mean, for example $(X - \mu_X)$, is 0.

The expected value of a nonlinear function of variables is not that function of the expected values of the variables – for example, the mean of $\ln X$ is not $\ln(\mu_X)$, as you may remember from the lognormal distribution.

If the variables are independent the covariance term is zero and can be dropped. An approximation to the variance of $\phi(X, Y)$ is usually based on the linear terms only. If X and Y are independent

$$\mathrm{var}(\phi(X, Y)) = \left(\frac{\partial \phi}{\partial x}\right)^2 \sigma_X^2 + \left(\frac{\partial \phi}{\partial y}\right)^2 \sigma_Y^2$$

Example 5.10

For the weir

$$Q = aKH^{3/2}$$

where a is a constant and K and H are independent variables which are unbiased estimates, that is their expected values are assumed to equal the actual quantities, with standard deviations of 10% and 5%. If we are to calculate the bias in Q and the standard deviation of Q we need:

$$\frac{\partial Q}{\partial K} = aH^{3/2}$$

$$\frac{\partial^2 Q}{\partial K^2} = 0$$

$$\frac{\partial Q}{\partial H} = \tfrac{3}{2}aKH^{1/2}$$

$$\frac{\partial^2 Q}{\partial H^2} = \tfrac{3}{4}aKH^{-1/2}$$

The bias is approximately

$$E[Q] - aE[K]E[H]^{3/2} = \tfrac{3}{8}aKH^{-1/2}\sigma_H^2 = \tfrac{3}{8}aKH^{-1/2}(0.05H)^2$$

In percentage terms,

$$100 \times \tfrac{3}{8}aKH^{-1/2}(0.05H)^2/aKH^{3/2}\% = 0.1\%$$

and this is negligible. The variance is approximately

$$(aH^{3/2})^2\sigma_K^2 + (\tfrac{3}{2}aKH^{1/2})^2\sigma_H^2$$

and after some algebra the standard deviation is found to be 12.5%.

Results based on the linear, or linear and quadratic terms, of the infinite Taylor series are approximations, but they often suffice.

5.9 Summary

Bivariate distributions

The function $f(x, y)$ is a joint pdf (or bivariate pdf in the case of two variables) if it is never negative and has a volume of 1:

$$\int\int f(x, y)\,dx\,dy = 1$$

The marginal distribution of x is the distribution of x alone. It can be obtained from the joint pdf by summing over y:

$$f_X(x) = \int f(x, y)\,dy$$

If $\phi(X, Y)$ is a function of X and Y

$$E[\phi(X, Y)] = \int\int \phi(x, y)f(x, y)\,dx\,dy$$

These definitions extend to any number of variables.

Correlation

	Sample estimate of population parameter	Population parameter
Covariance	$\widehat{\mathrm{cov}} = \sum(x - \bar{x})(y - \bar{y})/(n - 1)$	$\mathrm{cov}(X, Y) = E[(X - \mu_X)(Y - \mu_Y)]$
Correlation	$r = \widehat{\mathrm{cov}}/(s_x s_y)$	$\rho = \mathrm{cov}(X, Y)/(\sigma_X \sigma_Y)$

Linear combinations

If $W = aX + bY$, where a, b are constants, then

$$E[W] = aE[X] + bE[Y]$$
$$\text{var}(W) = a^2 \text{var}(X) + b^2 \text{var}(Y) + 2ab \text{cov}(X, Y)$$

Distribution of the sample mean

If $\{X_i\}$ is a random sample of size n from a normal distribution with mean μ and variance σ^2, then a good approximation for large n, or any n if the population is nearly normal, is

$$\bar{X} \sim N(\mu, \sigma^2/n)$$

The mean and variance are exact for any distributions used to model an infinite population, by the previous result, and the normal sampling distribution of $\bar{X}$ is a consequence of the central limit theorem. The standard deviation of $\bar{X}$, which is $\sigma/\sqrt{n}$, is commonly referred to as its standard error (SEMEAN in the MINITAB command DESCRIBE).

Finite population correction

If a sample of n is taken (without replacement) from a population of N the variance of $\bar{X}$ is reduced by a factor of $(1 - n/N)$. If sampling were to be with replacement, which is difficult to justify from a practical point of view, the population would be effectively infinite and the reduction in variance would not be obtained.

Nonlinear functions

Approximate results for the mean and variance of nonlinear functions can be obtained from a Taylor series expansion about the mean values of the variables.

Exercises

5.1   The following are pairs of measurements of the percentage carbon content (x) and the permeability index (y) of 10 sinter mixtures:

x	4.4	5.5	4.2	3.0	4.5	4.9	4.6	5.0	4.7	5.1
y	12	14	18	35	23	29	16	12	18	21

Plot the data and calculate the correlation coefficient r.

5.2   The lifetime of a spacecraft component is normally distributed with a mean of 3000 hours and a standard deviation of 800 hours. A failed component can be replaced, immediately, by a new one during the mission. What is the probability that one spare will suffice for a mission of 3000 hours if component lifetimes are assumed independent?

5.3 A chemical test necessitates measuring two independent quantities, A and B, and then expressing the result of the test as C, where

$$C = A - B/4$$

Assume that A is normally distributed with mean 10 and standard deviation 1, and that B is normally distributed with mean 34 and standard deviation 2. Calculate the proportion of C values you would expect to be negative.

5.4 A lift is rated to carry ten persons of 800 kg. If the people using the building have masses which are normally distributed with mean 75 kg and standard deviation 10 kg, what is the probability that the total mass of 10 persons selected at random will exceed 800 kg?

5.5 The diameters of rivets follow a normal distribution with mean 2.3 mm and standard deviation 0.05 mm. An independent process produces steel plates with holes whose diameters follow a normal distribution with mean 2.35 mm and standard deviation 0.1 mm. What is the probability that a randomly selected rivet will fit a hole selected at random?

5.6 A surveyor measures an angle with two instruments, A and B. Both instruments are unbiased, but the standard deviation of measurements made with instrument A is twice that of B. Due to wind effects, simultaneous measurements made with the two instruments have a correlation of 0.25. Let X and Y be simultaneous measurements of a bearing made with instrument A and instrument B, respectively. What value of α will minimize the variance of

$$\alpha X + (1 - \alpha) Y?$$

5.7 Let X represent driver reaction times ('thinking times') before applying brakes in an emergency. Assume X is distributed with mean 0.67 s and standard deviation 0.35 s. Let W represent the 'thinking distance' for cars braking from a speed of 50 km h^{-1}. Remember that distance is the product of speed and time, and hence write down the mean and standard deviation of W. The 'braking distance' for cars braking from 50 km h^{-1}, Y, is distributed with mean 14 m standard deviation 8 m. Write down the mean and standard deviation of 'stopping distance', $W + Y$, for cars braking from 50 km h^{-1} if W and Y are assumed independent. Assume stopping distance is normally distributed and find the distance that will be exceeded by 1% of such cars. Repeat the calculation if stopping distance is log-normally distributed.

5.8 Jars of water are taken from the public supply in an old quarter of a city and analysed for lead content. The variance of these measurements (M) is σ_M^2. The detection of minute amounts of lead is inevitably subject to errors (E) and the variance of measurements made on known standard solutions is σ_E^2. Express the variance σ_L^2 of actual (as opposed to measured) lead contents (L) in terms of σ_M^2 and σ_E^2.

A large sample of jars was taken and the mean and standard deviation of the measurements of lead contents were $46\,\mu g\,l^{-1}$ and $18\,\mu g\,l^{-1}$. The standard deviation of many measurements of lead content of known $40\,\mu g\,l^{-1}$ solutions is $6\,\mu g\,l^{-1}$. Find the standard deviation of actual lead contents and the correlation between errors and measurements.

5.9   A vault of toxic chemical is known to be buried somewhere within a circular area of radius 1 km. The probability of finding it by drilling a shaft of area A is $A\,f(x, y)$, where $f(x, y)$ is a bivariate uniform distribution:

$$f(x, y) = \frac{1}{\pi} \qquad \text{for } 0 \leqslant x,\, y \leqslant 1$$

Show that X and Y are uncorrelated but dependent by considering

$$\Pr(1/\sqrt{2} < X \text{ and } 1/\sqrt{2} < Y)$$

5.10 For the following joint probability function:

$$f(x, y) = x + y \qquad \text{for } 0 < x,\, y < 1$$

(i) determine $\Pr\{X < 0.5 \text{ and } Y \leqslant 0.5\}$;
(ii) find the marginal probability density functions for X and Y
(iii) determine $\Pr\{0.5 \leqslant X \text{ and } 0.5 \leqslant Y\}$.

5.11 Suppose a variable X with mean μ, and variance σ^2 is to be used as an estimator of θ. The bias, denoted B, is

$$\mu - \theta$$

and the mean squared error (MSE) is

$$E[(X - \theta)^2]$$

Show that:

(i) $MSE = \sigma^2 + B^2$
(ii) $E[X^2] = \sigma^2 + \mu^2$

5.12 Prove that if X and Y are independent variables, then

$$\text{var}(XY) = \text{var}(X)E[Y]^2 + \text{var}(Y)E[X]^2 + \text{var}(X)\text{var}(Y).$$

You should start

$$\text{var}(XY) = E[(XY - E(XY))^2]\cdots$$

and you should find part (ii) of Exercise 5.11 useful. Compare the exact result with the approximation for $\text{var}(XY)$ obtained from the linear terms of the Taylor expansion.

5.13 Start from the identity

$$\sum (X - \bar{X})^2 = \sum ((X - \mu) - (\bar{X} - \mu))^2$$

and show that

$$\sum (X - \bar{X})^2 = \sum (X - \mu)^2 - n(\bar{X} - \mu)^2$$

Take expectation of both sides and obtain

$$E\left[\sum (X - \bar{X})^2 \right] = (n - 1)\sigma^2$$

Hence deduce that $E[S^2]$ equals σ^2. Use part (ii) of Exercise 5.11 to demonstrate that S is not an unbiased estimator of σ.

5.14 Resistors are usually manufactured to a set of preferred values. Different resistance values can be obtained by combining resistors in series or in parallel. Suppose two resistors R_1 and R_2 are from populations with means of $1\,\text{k}\Omega$ and $2\,\text{k}\Omega$, respectively, and standard deviations of 1%. Find the mean and standard deviation (in percentage terms) for:

(i) $R_S = R_1 + R_2$
(ii) $R_P = (R_1^{-1} + R_2^{-1})^{-1}$

5.15 A helical gear wheel is held in a jig and measurements of distances from an origin to points on the profile are made in a rectangular Cartesian coordinate system. The standard deviation of measurements along any axis is $8\,\mu\text{m}$. The distance between two points on the gear, with coordinates (x_1, y_1, z_1) and (x_2, y_2, z_2), is of the form

$$[(x_1 - x_2)^2 + (y_1 - y_2)^2 + (z_1 - z_2)^2]^{1/2}$$

What is the standard deviation of such distances?

5.16 A bivariate pdf is defined by

$$f(x, y) = 8xy \qquad \text{for } 0 \leqslant x \leqslant 1,\, 0 \leqslant y \leqslant x$$

Find the marginal distributions of X and Y and hence show that they are not independent.

5.17 An electronic system has two different components in joint operation. Let X and Y denote the lifetimes of components of the first and second types, respectively. The joint pdf is given by

$$f(x, y) = \tfrac{1}{8} x \exp[-(x + y)/2] \qquad \text{for } 0 < x,\, y$$

Find

(i) $\Pr(1 < X$ and $1 < Y)$ and

(ii) $\Pr(X + Y < t)$ for any $t \geqslant 0$.

Hence write down in pdf of the random variable $T = X + Y$.

5.18 Regular adjustment

Let m_t denote the process mean when sample t is taken and τ be the target value. The sample sizes are n and the sample mean at time t is $\bar{X}_t$. The regular adjustment is proportional to the difference between the target value and the sample mean. So

$$m_{t+1} = m_t + \theta(\tau - \bar{X}_t)$$

for some choice of θ. If the process standard deviation is σ,

$$\bar{X}_t = m_t + E_t$$

where $E_t \sim N(0, \sigma^2/n)$ and is independent of m_t.

(i) Show that

$$m_{t+1} = (1 - \theta)m_t + \theta\tau - \theta E_t$$

(ii) The $\{m_t\}$ process is stable provided $(1 - \theta)$ is between -1 and 1, and then $\mathrm{var}(m)$ is constant. Show that

$$\mathrm{var}(m) = \theta\sigma^2/((2 - \theta)n) \qquad \text{for } |1 - \theta| < 1$$

(iii) Suppose $\theta = 0.5$, $n = 5$, and the process is on target. How much would the variability of the process be increased by implementing regular adjustment? What would the other disadvantages be? Under what circumstances might regular adjustment be worthwhile? (For suggested answers, see Metcalfe, 1992.)

6

Precision of estimates

One of the advantages of using a random sampling scheme is that the precision of an estimate can be assessed from the sample itself, and we need to have some indication of this if we are to make informed decisions.

6.1 Precision of means

6.1.1 Confidence interval for population mean when the population standard deviation is assumed known

A pharmaceuticals company markets one of its hair shampoo products in bottles with declared contents of 70 ml. The average volume dispensed by the filling machine can be adjusted, but once set up the standard deviation of shampoo dispensed has been found to be about 2.6 ml. To comply with current UK legislation the volumes dispensed should have a mean of at least 71.4 ml, if it is assumed that all the product will be within four standard deviations of the mean. At the beginning of a long run, the machine was set up and 100 bottles were filled. The volumes of a random sample of 20 of these were measured, and the sample mean was 71.68 ml. The precision of this estimate is important if we are to make a sensible decision. If it is only within ± 1 ml of the corresponding population mean we would need to increase the sample size before considering any adjustments, whereas if it is within ± 0.01 ml we might wish to reduce the setting.

The general method for assessing the precision is to construct a confidence interval. Suppose that we have a random sample of n observations from a distribution with an unknown mean μ but a known standard deviation σ. From the central limit theorem the distribution of the sample mean is, at least approximately,

$$\bar{X} \sim N(\mu, \sigma^2/n)$$

There is, for example, a 95% probability that $\bar{X}$ will be within $1.96\sigma/\sqrt{n}$ of μ. However, such a probability statement is not immediately helpful because we have already calculated a specific value, $\bar{x}$, of $\bar{X}$ and wish to say something about μ. We therefore rearrange it and say we are 95% confident μ is within

$1.96\sigma/\sqrt{n}$ of $\bar{x}$. This is formally described as a 95% confidence interval for μ, given by

$$\bar{x} \pm 1.96\sigma/\sqrt{n}$$

In applications, I tend to imagine there is a 95% chance that μ is within the interval. While 95% confidence intervals are commonly quoted, we can construct a $(1 - \alpha) \times 100\%$ confidence interval for μ as

$$\bar{x} \pm z_{\alpha/2}\sigma/\sqrt{n}$$

The more confident we insist on being, the wider the interval becomes. Intervals of 90% and 99% confidence are sometimes chosen instead of 95%.

A 95% confidence interval for the mean of the population of shampoo bottles that could be filled if the process continued on its present settings is

$$71.68 \pm 1.96 \times 2.6/\sqrt{20}$$

which is equal to

$$71.7 \pm 1.1$$

Our estimate is not very accurate and the population mean may well be significantly less than 71.4 ml. I would recommend randomly selecting a further 60 bottles, which would give a total sample size of 80 and a confidence interval of half the width, before filling many thousands of bottles. The mean and range should be monitored during the run with SPC charts.

Theoretical justification
A theoretical derivation of a 95% confidence interval for μ when σ is known demonstrates the basic argument, which can be used to justify all the confidence interval constructions given in this chapter. It follows from

$$\bar{X} \sim N(\mu, \sigma^2/n)$$

that

$$\Pr(\mu - 1.96\sigma/\sqrt{n} < \bar{X} < \mu + 1.96\sigma/\sqrt{n}) = 0.95$$

Algebraic manipulations of the inequalities will not affect the probability. Start by subtracting $(\bar{X} + \mu)$ throughout:

$$\Pr(-\bar{X} - 1.96\sigma/\sqrt{n} < -\mu < -\bar{X} + 1.96\sigma/\sqrt{n}) = 0.95$$

Now multiply by -1, remembering to change the signs of the inequalities, and finally put into ascending order to obtain

$$\Pr(\bar{X} - 1.96\sigma/\sqrt{n} < \mu < \bar{X} + 1.96\sigma/\sqrt{n}) = 0.95$$

There is therefore a 95% probability that the random interval

$$[\bar{X} - 1.96\sigma/\sqrt{n}, \; \bar{X} + 1.96\sigma/\sqrt{n}]$$

includes μ. On average, 95% of the intervals constructed in this way would include μ. In practice we will construct one such interval,

$$[\bar{x} - 1.96\sigma/\sqrt{n}, \; \bar{x} + 1.96\sigma/\sqrt{n}],$$

and will be 95% confident that it includes μ.

There is an interesting alternative approach to constructing confidence intervals, based on Bayes' theorem, for which we need the concept of conditional distributions. An example is given in Exercise 8.6, and a Bayesian analysis is used to evaluate the effects of mini-roundabouts in Chapter 11.

6.1.2 Confidence interval for population mean when the population standard deviation is unknown

It is more usual to have to estimate the population standard deviation from the sample, as well as the mean. The sampling distribution of $\bar{X}$, used above, can be written as

$$\frac{\bar{X} - \mu}{\sigma/\sqrt{n}} \sim N(0, 1)$$

We now need to modify this by replacing the unknown constant σ by the variable S. A statistician working for the Guinness Brewery in Dublin, William Sealy Gosset, investigated this and published his results (for example in *Biometrika* in 1907) under the pen name 'Student'. The relevant sampling distribution depends on the **degrees of freedom** available when calculating S, which are the sample size less 1 in this application, and is called a **t-distribution** with $n - 1$ degrees of freedom. That is,

$$\frac{\bar{X} - \mu}{S/\sqrt{n}} \sim t_{n-1}$$

The t-distributions, with three or more degrees of freedom, look similar to the standard normal distribution but are slightly more spread out (Fig. 6.1). A table of percentage points is given in Table E.4. One explanation for the use of term 'degrees of freedom' is the following. There are n observations, and if μ were known there would be n independent deviations, $(X_i - \mu)$, from which to calculate S^2. However, μ is not known and is replaced by $\bar{X}$. This imposes a constraint on the new deviations, $(X_i - \bar{X})$, which must sum to zero. Recall that a direct consequence of the definition of the mean is that

$$\sum (X_i - \bar{X}) = 0$$

The degrees of freedom are therefore reduced by 1 to $n - 1$. Equivalently, if μ were known we would have n independent bits of information to use when

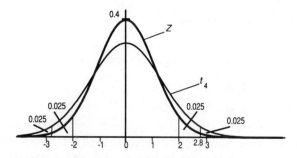

Fig. 6.1 t-distribution with four degrees of freedom compared with standard normal distribution.

calculating S but when it is replaced by $\bar{x}$ we could calculate the last deviation given the other $n-1$ deviations. It is worthwhile becoming used to this argument as it reappears in other contexts. A $(1-\alpha) \times 100\%$ confidence interval for μ is now given by

$$\bar{x} \pm t_{n-1,\alpha/2} s/\sqrt{n}$$

For large sample sizes, greater than about 30, the t-distribution is very similar to the standard normal and 1.7 and 2.0 are accurate enough for 90% and 95% confidence intervals, respectively.

Example 6.1

Fluoride is added to the water supply in the Newcastle upon Tyne area as a public health measure with the aim of reducing dental decay. There is some controversy over this policy, and it is important to keep added fluoride within naturally occurring levels. The target fluoride content is 1.0 ppm and no more than 5% of samples should contain more than 1.1 ppm. If the distribution of fluoride contents is assumed to be normal, this corresponds to a standard deviation of 0.06 ppm.

A public health inspector has just analysed 16 jars of water taken from a random sample of treatment works at random times during the past week. Her results can be summarized by

$$\bar{x} = 1.035 \text{ ppm} \quad s = 0.104 \text{ ppm}$$

She wishes to construct a 95% confidence interval for the mean fluoride content during the past week and therefore needs the upper $2\frac{1}{2}\%$ point of the t-distribution. This can be found from Table E.4 and equals 2.131. The confidence interval is given by

$$1.035 \pm 2.131 \times 0.104/\sqrt{16}$$

which becomes

$$1.035 \pm 0.055$$

She has no real evidence that the mean was too high but she is slightly concerned about the variability. A simple approximate technique for constructing a confidence interval for the population standard deviation is covered in section 6.2.

The theory behind the t-distribution assumes the sample is drawn from a normal distribution, but it is not very sensitive to this assumption, and becomes less sensitive to it as the sample size increases. Even so, in certain cases there may be reason to believe that some transformation of the variable X, such as $\sqrt{X}$ or $\ln X$, has a near-normal distribution. If the sample is small – a guideline is less than 20 – the transformed data could be analysed to produce a confidence interval for the mean of the distribution of the transformed variable. If this is done some care is needed with interpretation, because the mean of X is not equal to the inverse transform of the mean of its transform. To see why this is so use a Taylor expansion for the function of X about μ_X, and let Y denote the transformed variable (as in section 5.8).

$$Y = \phi(X) \simeq \phi(\mu_X) + \phi'(\mu_X)(X - \mu_X) + \tfrac{1}{2}\phi''(\mu_X)(X - \mu_X)^2 + \cdots$$

Taking expectation gives the following approximate result

$$E(Y) \simeq \phi(\mu_X) + \tfrac{1}{2}\phi''(\mu_X)\sigma_X^2$$

In fact, the inverse transform of the mean of Y is equal to the median of the distribution of X.

MINITAB commands
Suppose the data in C1 are a random sample from some population. If we assume a value for the population standard deviation, and put it in K2, a K1% confidence interval for the population mean is given by

    ZINTERVAL K1 K2 C1

If we use the mouse:

    Stat ▶ Basic Statistics ▶ 1-Sample Z ▶ Confidence interval

    If we estimate the population standard deviation from the sample a K1% confidence interval, based on the t-distribution, is given by
Alternatively, use the mouse.

6.2 Precision of standard deviations

If we take a random sample of n from a normal distribution, an approximate result (Exercise 6.13) is

$$S \sim N(\sigma, \sigma^2/(2n))$$

The approximation improves as n increases but the expression for the variance of S remains sensitive to the assumption that the original sample was from a normal distribution. Given this assumption, the approximation is quite reliable if n is greater than about 30 and still gives a reasonable indication with smaller sample sizes. I have not given the exact distributional result, which is just as sensitive to the assumption that the random sample was from a normal distribution, because this would mean introducing another distribution (**chi-square distribution**). A rough $(1 - \alpha) \times 100\%$ confidence interval for σ will be given by

$$s \pm z_{\alpha/2} s / \sqrt{(2n)}$$

A more formal argument which gives a slightly improved approximation is left as an exercise.

Example 6.1 (continued)

A rather approximate 90% confidence interval for σ is given by

$$0.104 \pm 1.7 \times 0.104 / \sqrt{32}$$

using 1.7 for $z_{0.025}$ because the construction does not merit any greater precision, which gives

$$0.10 \pm 0.03$$

A 95% confidence interval would be 0.10 ± 0.035. There is some evidence that the standard deviation was too high but, while the inspector would probably notify the company, it would only be of serious concern if it were a frequent occurrence. The inspector might decide to increase the sample size over the next few weeks.

6.3 Comparing standard deviations

Most manufacturers aim to reduce variability in their products. They are therefore interested in selecting suppliers whose goods are consistent. Suppose we have independent random samples of sizes n_A, n_B from normal distributions with standard deviations σ_A, σ_B, respectively. From section 6.2, we have

$$S_A \sim N(\sigma_A, \sigma_A^2/(2n_A))$$

and

$$S_B \sim N(\sigma_B, \sigma_B^2/(2n_B))$$

Since they are independent their difference

$$S_A - S_B \sim N(\sigma_A - \sigma_B, \sigma_A^2/(2n_A) + \sigma_B^2/(2n_B))$$

It follows that a rough $(1 - \alpha) \times 100\%$ confidence interval for the difference in population standard deviations is given by

$$s_A - s_B \pm z_{\alpha/2}[s_A^2/(2n_A) + s_B^2/(2n_B)]^{1/2}$$

An exact confidence interval for the ratio of the population standard deviations would involve yet another distribution (**F-distribution**).

Example 6.2

The specification for a particular tantalum capacitor in a radio is $100\,\mu F$. Two companies, A and B, produce capacitors with nominal values of $100\,\mu F$. The radio manufacturer bought samples of 30 capacitors from each company and the capacitances were recorded, with the following results:

company A $n = 30$ $\bar{x}_A = 100.3$ $s_A = 5.3$
company B $n = 30$ $\bar{x}_B = 99.6$ $s_B = 8.7$

There is little difference in price, but company B is already on the radio manufacturer's list of approved suppliers. Would you recommend buying this particular item from company A?

A usual tolerance for this type of capacitor is $\pm 20\%$ and the estimated standard deviation for B's product does seem rather high. An approximate 95% confidence interval for the difference in standard deviation is

$$(8.7 - 5.3) \pm 2.0[(8.7)^2/60 + (5.3)^2/60]^{1/2}$$

that is,

$$3.4 \pm 2.6$$

So there is some quite convincing evidence that the standard deviation of capacitances in the batches from which the samples were taken differ, with A's product being less variable. If we are prepared to assume that the sampled batches were typical we would recommend A. If we are not happy about this assumption we would need to take further samples. Buying capacitors from several different outlets would be an improvement on buying them all from one outlet and make the crucial assumption more plausible.

6.4 Comparing means

6.4.1 Independent samples

Two important measures of the performance of high-voltage circuit breakers are the 'open' and 'close' times. The type X circuit breaker has been in production for some time. A new type, Y, which is similar except for a novel switch mechanism, is at the prototype stage. The new switch mechanism is mechanically simpler, and the designers hope it will make what is already a reliable product even more robust. It may also lead to reduced open and close times but this is more a matter of speculation, and not the primary objective. The chief engineer is only prepared to market the type Y circuit breaker if she can be reasonably confident that average open and close times have not increased by more than 1% of recent type X values. There have been no changes in production methods for the type X over the past six months. Test records for the 38 produced during this period are given below.

Type X, 38 circuit breakers
average open time = 23.021 ms   standard deviation = 0.806 ms
average close time = 99.641 ms   standard deviation = 1.888 ms

Ten prototype type Y circuit breakers have been made and results of similar tests on them follow.

Type Y, 10 circuit breakers
average open time = 22.461 ms   standard deviation = 0.922 ms
average close time = 96.848 ms   standard deviation = 2.121 ms

We can provide the chief engineer with the information she requires by constructing 95% confidence intervals for the differences in mean open times, and mean close times, in the corresponding populations. These will rely on an assumption that the prototypes are a random sample of production. It will only be possible to say how realistic this is if production is started, but past experience suggests it is a reasonable working assumption.

The theory does not involve any ideas that we have not already covered, and Fig. 6.2 may help clarify the situation. Populations A and B have means μ_A and μ_B, and standard deviations σ_A and σ_B, respectively. These population means and standard deviations are unknown. Now suppose we have two independent random samples of sizes n_A and n_B from these populations. We rely on the central limit theorem, which is a good approximation if the sample is large or the population near-normal, to state that

$$\bar{X}_A \sim N(\mu_A, \sigma_A^2/n_A) \quad \text{and} \quad \bar{X}_B \sim N(\mu_B, \sigma_B^2/n_B)$$

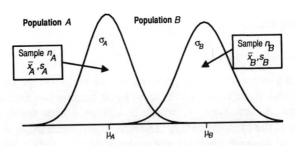

Fig. 6.2 Comparing means from independent samples.

Therefore, since the samples are independent,

$$\bar{X}_A - \bar{X}_B \sim N(\mu_A - \mu_B, \sigma_A^2/n_A + \sigma_B^2/n_B)$$

By the usual argument, a $(1 - \alpha) \times 100\%$ confidence interval for $\mu_A - \mu_B$ is given by

$$\bar{x}_A - \bar{x}_B \pm z_{\alpha/2}(\sigma_A^2/n_A + \sigma_B^2/n_B)^{1/2}$$

Usually σ_A^2 and σ_B^2 are unknown, so they are replaced by their sample estimates to obtain an approximate $(1 - \alpha) \times 100\%$ confidence interval for $\mu_A - \mu_B$ as

$$\bar{x}_A - \bar{x}_B \pm t_{v,\alpha/2}(s_A^2/n_A + s_B^2/n_B)^{1/2}$$

where

$$v = \left(\frac{s_A^2}{n_A} + \frac{s_B^2}{n_B} \right)^2 \Bigg/ \left(\frac{(s_A^2/n_A)^2}{n_A - 1} + \frac{(s_B^2/n_B)^2}{n_B - 1} \right)$$

The degrees of freedom are awkward to calculate, and the result is unlikely to be an integer, so interpolation in Table E.4 is needed. It always exceeds the smaller of the two sample sizes less 1, and if both samples are fairly large this will be close enough to the more accurate formula for the degrees of freedom.

Example 6.3

A 95% confidence interval for the difference in mean open times for the type X and type Y circuit breakers is

$$(23.021 - 22.461) \pm t_{v,0.025}(0.806^2/38 + 0.922^2/10)^{1/2}$$

where

$$v = 12.9$$

Further arithmetic gives

$$0.559 \pm 0.691$$

which is the interval

$$[-0.13, 1.25]$$

A 1% increase in open time would correspond to a difference of -0.23, so the chief engineer can be quite confident that any increase is less than 1% and have a reasonable hope of a slight decrease in open times if production begins. As an exercise you are asked to check that close times are acceptable.

A slight variant to the procedure can be used if it is reasonable to suppose the populations have the same standard deviation, and the samples do not provide evidence against this assumption. A $(1 - \alpha) \times 100\%$ confidence interval for $\mu_A - \mu_B$ is

$$\bar{x}_A - \bar{x}_B \pm t_{n_A + n_B - 2, \alpha/2} s_p[(1/n_A) + (1/n_B)]^{1/2}$$

where the pooled estimate of variance

$$s_p^2 = [(n_A - 1)s_A^2 + (n_B - 1)s_B^2]/(n_A + n_B - 2)$$

It is theoretically exact if sampling were ever to be from exact normal distributions. In fact, the approximate procedure also assumes sampling is from normal distributions but neither construction is sensitive to the assumption, and both become less so as the sample sizes increase.

Example 6.3 (continued)

If the variances are assumed the same, s_p is 0.830 and the degrees of freedom are 46. The 95% confidence interval is

0.559 ± 0.594

This is appreciably narrower, mainly because of the increased degrees of freedom, but the conclusion is unaltered.

MINITAB commands
Suppose we have independent random samples from population A and B in C1 and C2. A K1% confidence interval for the difference in the population means is given by

TWOSAMPLE K1 C1 C2

If the population variances are assumed the same a subcommand is added.

TWOSAMPLE K1 C1 C2;
    POOLED.

The equivalent procedure with the mouse is:

Stat ▶ Basic Statistics ▶ 2-Sample t ▶ Samples in different columns

The subcommand POOLED is obtained by clicking on 'Assume equal variances'.

6.4.2 Comparing means by the paired comparison procedure

Suppose we wish to check whether there is a systematic difference in masses indicated by two kitchen scales, A and B. It would not be sensible to base the comparison on the difference in average weights of a random sample of kitchen objects on scale A and an independent random sample on scale B. The variation in masses of kitchen objects would swamp any difference in the scales! The obvious way to make the comparison is to weigh the same objects on both scales and to analyse the differences. This is an example of a paired comparison experimental design. The 'pairs' are the pairs of measurements of the same object on scale A and scale B. It is usually much more efficient to plan experiments as paired comparisons if this is feasible.

If two chemical processes, A and B, are to be compared for yield it is desirable to remove, as far as is possible, variation in yield caused by factors other than the nature of the process itself. One such contributory factor might be the composition of the raw material used in the process, which could vary from batch to batch. This source of variation could be removed by using the following experimental procedure, provided the practical circumstances permit it. Take two lots from a double-size batch of raw material which has been thoroughly stirred and allocate one at random to process A, the other to process B. Measure the percentage yield from each process, x_A and x_B, respectively, and repeat for n double-size batches. Then calculate the differences in yield:

$$d_i = (x_{Ai} - x_{Bi})$$

for each of the n paired comparisons. A $(1 - \alpha) \times 100\%$ confidence interval for the difference in yield in the corresponding population is

$$\bar{d} \pm t_{n-1,\alpha/2}\, s_d/\sqrt{n}$$

The percentage yields for a medicinal product in such an experiment are given in Table 6.1. The mean and standard deviation of the six differences are

$$\bar{d} = -2.00 \quad s_d = 1.835$$

The 95% confidence interval for the mean difference in the corresponding population is

$$-2.00 \pm 2.571 \times 1.835/\sqrt{6}$$

that is,

$$-2.00 \pm 1.93$$

Despite the small sample we have some evidence that process B gives a higher yield. However, it would be prudent to follow up the experiment with some more tests. This will enable us to make a more precise estimate of the benefit, which may have to be offset against greater costs of running process B.

The paired comparisons approach to experimentation is also valuable if we want to apply our findings to a diverse population. For example, oxy-fuel gas cutting is an important process in steel fabrication. While the process is a quick, easy, and relatively inexpensive method of cutting steel, it can have undesirable

Table 6.1 Yields of a medicinal product

Batch number	1	2	3	4	5	6
x_{Ai}	60.1	57.0	57.9	58.8	60.2	58.0
x_{Bi}	63.9	60.3	57.8	61.3	59.7	61.0
d_i	−3.8	−3.3	0.1	−2.5	0.5	−3.0

effects on the cut material. For many general steel fabrication purposes, these effects are normally of little consequence but in the exacting environment of the North Sea, where safety and reliability are paramount, they are of great importance. Furthermore, the severe conditions met in the North Sea have led to the development of high-strength steels and it is these types of steel, with their enhanced chemical composition, that are more prone to the effects of the oxy-fuel gas cutting process. With this in mind Liptrot (1991) ran a series of trials to compare oxy-natural gas with the better understood oxy-propane gas cutting and to estimate the best values for other control variables in the process. It would have been possible to restrict cutting trials to test plates from the same sheet of steel. However, this would limit experimental conclusions to that type and thickness of steel. Any inferences made about other types of steel would have to rely on informed judgement, and it is always advisable to support this with empirical evidence. Liptrot decided to prepare seven pairs of test plates from high-strength steel plates of various thicknesses and grades. The problem with gas cutting is a hardening of the steel near the cut edge. After cutting each plate Liptrot ground away some of the flame-cut face, polished the exposed surface, and made 36 Vickers hardness measurements along the slope (Fig. 6.3).

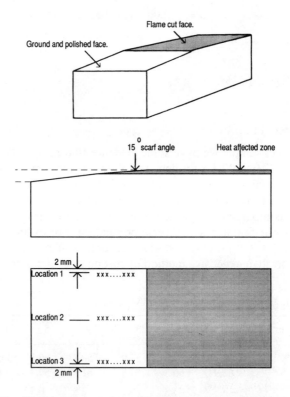

Fig. 6.3 Preparation of test plate for hardness test survey.

Table 6.2 Maximum Vickers hardness (VH10) readings on test pieces of high strength steel

Test plate	Oxy-propane	Oxy-natural gas
1	370	333
2	333	336
3	330	299
4	306	294
5	314	297
6	322	373
7	290	304

The maximum of each set of 36 hardness measurements was used in the analysis (Table 6.2). Individual differences in the maxima are unlikely to be near-normally distributed but, even with a sample as small as seven, the procedure for the confidence interval is reliable enough in this context. Calculations give

$$\bar{d} = 4.1 \quad s_d = 30.1$$

A 90% confidence interval for the difference (propane minus natural gas) in maximum hardness is

$$4 \pm 22$$

So, at this stage of the study there was no evidence of any difference in the corresponding population, but the comparison was not very precise. The results were sufficiently encouraging to continue with the investigation, as natural gas would be preferred to propane for some types of work, provided the hardening problem is not made worse. The next stage of the study was to adjust control variables, especially speed of cutting, with the aim of reducing the maximum hardness values.

6.4.3 *Preference for the paired comparison procedure*

Suppose that the company only wishes to consider oxy-natural gas cutting for a specific grade and thickness of high-strength steel. One approach to the comparison would be to prepare 14 test plates and randomly assign half to the oxy-natural gas cutter. The paired comparison approach would be to prepare pairs of test plates, each pair from the same piece of steel, and randomly assign one from each pair to the oxy-natural gas cutter. The paired comparison procedure would lead to a more accurate comparison if variations in steel extended over areas equivalent to test pieces. Even if variations were much more localized, the only disadvantage would be a loss of degrees of freedom for the t-distribution. I would choose to use the paired comparison design for this experiment.

6.5 Sample size

6.5.1 Confidence interval for a mean

Suppose we intend constructing a 95% confidence interval for a population mean from a random sample of size n. If the width of the confidence interval is to be 2δ, which corresponds to $\pm\delta$, the sample size n must satisfy the equation

$$\delta = 2\sigma/\sqrt{n}$$

which after rearrangement gives

$$n = 4\sigma^2/\delta^2$$

If the criterion is phrased in terms of a 90% confidence interval, for example, 2 would be replaced by 1.7 (as precise a value of $z_{0.05}$ as is justified, for reasons which follow). An estimate of σ is needed if this formula is to be useful. The estimate can be based on past experience, results published in the literature, or a small preliminary sample. If n is less than 30 we should, perhaps, use $t_{0.025}$ with $n-1$ degrees of freedom instead of 2 and iterate to find a final value of n. But even then, the value of n is unlikely to lead to a confidence interval with an exact width of 2δ because the sample estimate of σ will not usually coincide with the value assumed for the sample size calculation. In any case, δ is normally a somewhat arbitrary choice. I would consider including a preliminary sample with the main sample, provided it was selected at random from the same population.

Example 6.4

In section 5.1 we recommended that manufacturer A should increase the diameter of crown races by 0.03 mm if our estimate of the mean, 26.923 mm (sample size 35 and standard deviation 0.017), was reasonably precise.

A manager has decided to delay contacting the manufacturer until a random sample has been taken from a new delivery of crown races. He would like the 95% confidence interval to have a width of about ±0.005 mm. The required sample size n is given by

$$n = 4 \times 0.017^2/0.005^2$$

and equals 47 when it is rounded up. The manager will treat the new delivery as a different finite population, although all deliveries can be thought of as samples from the infinite population of all A's output. Furthermore, the variability between deliveries is of importance in itself. The sample of 47 from the new delivery will be in addition to the original sample of 35.

6.5.2 Confidence interval for a standard deviation

If the width of the 95% confidence interval is to be 2δ the sample size must satisfy the equation

$$\delta = 2\sigma/\sqrt{2n}$$

and hence

$$n = 2\sigma^2/\delta^2$$

As before, σ must be replaced by an estimate for us to decide on a sample size.

Example 6.4 (continued)

The manager also intends estimating the standard deviation of seat diameters from another delivery of forks to back up his complaint to the manufacturer. He would like the width of the 95% confidence interval to be 0.004. The sample size is given by

$$n = 2 \times 0.030^2/0.004^2$$

which rounds up to 113. He then decides he would like the same precision for the estimate of the standard deviation of seat diameters for the forks already in stock. He already has two random samples from this stock of sizes 35 and 20, respectively. I would recommend taking a third random sample of 58, to give a total of 113 forks from the stock, and calculating the standard deviation of the three samples pooled together. This is not a rigorous approach but I think it is justifiable in this situation. The general criticism is that we can continue, repeatedly constructing confidence intervals with more data, until we obtain the answer we want. Proper techniques that allow for this are described by Jennison and Turnbull (1989).

6.5.3 Confidence intervals for differences in standard deviations

To obtain maximum precision for a given sum of the two sample sizes set the ratios σ_A^2/n_A and σ_B^2/n_B equal, to ϕ say. The 95% confidence interval for the difference will have a width of 2δ if

$$\delta = 2\sqrt{[\phi/2 + \phi/2]}$$

Therefore

$$\phi = \delta^2/4$$

and the sample sizes follow by substituting estimates for σ_A^2 and σ_B^2.

6.5.4 Confidence intervals for difference in means

The paired comparison procedure is a special case of that outlined in section 6.5.1 with σ being the standard deviation of the differences for matched pairs.

The result for independent samples follows a similar argument to that for the difference in standard deviations. With the same notation

$$\delta = 2\sqrt{[\phi + \phi]}$$

and

$$\phi = \delta^2/8$$

Example 6.5

Twenty years ago the National Engineering Laboratory (NEL) in the UK proposed a 'rolling four-ball test' as a means of assessing the liability of a material to surface fatigue failure. Briefly, one ball is forced down on to a rotating cluster of three balls. Preliminary results (data from NEL, reproduced in Greenfield, 1974) for the ball lifetimes (in minutes) for samples from two types of steel are given in Table 6.3. The samples are too small to make a choice between assumptions of normal or lognormal distributions from them alone, although for the second sample, at least, the normal score plot of the logarithms appears slightly closer to being scattered about a line. Past experience suggests the lognormal distribution would be the better model. Suppose we wish the 95% confidence interval for the difference in means of the natural logarithms

Table 6.3 Lifetimes of clusters of four balls in the four-ball test (x, minutes) for samples from two types of steel: A and B; mean and standard deviation of x and $y(\ln x)$

Steel A (x_A)	Steel B (x_B)
182	272
245	171
68	375
45	170
209	205
158	348
32	202
55	363
140	151
	112
$\bar{x}_A = 126.0$	$\bar{x}_B = 236.9$
$s_{x_A} = 78.4$	$s_{x_B} = 95.8$
$\bar{y}_A = 4.617$	$\bar{y}_B = 5.392$
$s_{y_A} = 0.747$	$s_{y_B} = 0.412$

Data from National Engineering Laboratory.

of lifetimes to have a width of 0.6. Then

$$\phi = 0.3^2/8 = 0.0112$$
$$n_A = \sigma_A^2/\phi$$

and this is estimated by

$$0.747^2/0.0112 = 50$$

A similar calculation gives $n_B = 16$. These calculations are based on rather imprecise estimates of standard deviations and I would increase n_B. One of the aims of the experiment was a comparison of the lower 10% quantile of the distributions and this is continued in Exercise 6.11.

6.6 Proportions

We shall assume there is some underlying population which contains a proportion (p) with some attribute. The population could be the hypothetical output of a process, which, if present trends continue, will produce a proportion p of items outside specification. Another example is the population of all freight vehicles licensed for use on the roads, of which a proportion p are overladen. In the first example, direct measurement of p is impossible and, in the second, it is quite impractical. Yet, in both cases the attribute is undesirable and we would like to estimate the extent of the problem before deciding what action is appropriate. Assume we take a random sample of n items from the population, so that the number X with the attribute has a bionomial distribution:

$$X \sim \text{Bin}(n, p)$$

If n is large it is much more convenient to use an approximation to the discrete binomial distribution. Provided the smaller of np and $n(1 - p)$ exceeds about 5 the normal distribution with the same mean and variance,

$$X \sim N(np, np(1 - p)),$$

can be used instead. We are approximating a discrete variable by a continuous one, and in geometric terms (Fig. 6.4) this is equivalent to substituting areas under curves for heights of lines. So, the probability that $X = 3$ is close to the area under the normal curve between 2.5 and 3.5. This area has a width of 1 and an 'average height' which is roughly equal to the length of the binomial line. There would be no point in using a normal distribution for a single probability but we do gain some advantage when calculating the probability that $X \leqslant 3$.

Example 6.6

In Figure 6.4

$$X \sim \text{Bin}(13, 0.4)$$

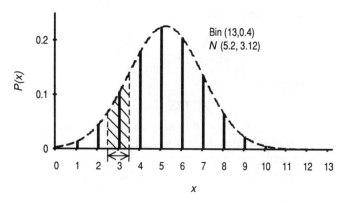

Fig. 6.4 Normal approximation to binomial distribution.

and is approximated by

$X \sim N(5.2, 3.12)$

The probability that $X \leqslant 3$

$$\Pr(X = 0, 1, 2, 3 | X \sim \text{Bin}(13, 0.4)) = 0.1686$$

is approximated by

$$\Pr(X < 3.5 | X \sim N(5.2, 3.12)) = 0.1679$$

The reason for using 3.5 rather than 3 should be apparent from the nature of the approximation. It is sometimes known as the **continuity correction**.

Example 6.7

A filter manufacturer has agreed to make a special liquid filter for a pharmaceutical company. This presents severe technical problems, but the filter manufacturer will make a profit provided at least 75% of production satisfies the final tests. In fact 117 of the first 140 proved to be satisfactory. What is the probability of such a promising start if the process is only capable of producing 75% satisfactory filters? Let X be the number of satisfactory filters in a random sample of 140.

$$\Pr(X \geqslant 117 | X \sim \text{Bin}(140, 0.75))$$

can be approximated by

$$\Pr(X > 116.5 | X \sim N(105, 26.25))$$

which equals 0.01. This assumes that the process control variables will remain the same and that defectives are produced independently rather than, for example, in cluster. In other words, we have assumed the first 140 to be a random sample from all production. This may not be realistic – it may be possible to improve the process with experience – but it provides a benchmark.

6.6.1 Confidence interval for a proportion

It is usually more appropriate to construct a confidence interval for the population proportion than to calculate probabilities as in Example 6.7. It is no longer apparent how a continuity correction should be applied, and the simplest solution is to ignore it.

Let p represent the proportion of defectives in the population. Take a random sample of size n from this population. Then the number of defectives in the sample, X, is approximately normally distributed:

$$X \sim N(np, np(1-p))$$

We can scale this to give the distribution of the sample proportion:

$$\frac{X}{n} \sim N(p, p(1-p)/n)$$

By the usual argument a $(1-\alpha) \times 100\%$ confidence interval would be given by

$$\frac{x}{n} \pm z_{\alpha/2} \sqrt{\frac{p(1-p)}{n}}$$

but this includes the unknown p in the standard deviation of X/n. This is usually estimated by

$$\sqrt{\frac{x}{n}\left(1-\frac{x}{n}\right)/n}$$

which is reasonable provided it is small compared with the lesser of x/n and $(1-x/n)$, leading to an approximate $(1-\alpha) \times 100\%$ confidence interval of

$$\frac{x}{n} \pm z_{\alpha/2} \sqrt{\frac{x}{n}\left(1-\frac{x}{n}\right)/n}$$

for large samples.

Example 6.8

A water company found that 38 out of a random sample of 200 properties had lead communication pipes to the main. A 95% confidence interval for the proportion in the corresponding population is

$$38/200 \pm 1.96 \times \sqrt{(38/200) \times (162/200)/200}$$

which gives

$$0.19 \pm 0.05$$

The company has decided that a more precise estimate is needed and wishes to halve the width of the confidence interval. This means increasing the sample size by a factor of 4, and a further 600 properties will be investigated.

6.6.2 Sample size

A company produces a successful computer software package and is considering the development of a graphics supplement. It would market this at a special 'introductory price' and cover development costs if 4% of registered users bought it. The marketing manager therefore intends sending out questionnaires to a random sample of registered users. From past experience she thinks that about one quarter of the users who express interest will buy the supplement, and the board of the company has decided to proceed with the project if at least 10% of users show enthusiasm.

Although this might not cover development costs the overall package would be improved and increased sales should justify the investment in the development. However, the board does need to be reasonably confident about the proportion of interested users, and has stipulated that the lower end of the 90% confidence interval for the proportion should exceed 0.10. A sample size is calculated on the basis that a confidence interval with a width of 0.06 could conceivably extend from 0.10 up to 0.16. Then the decision to proceed would be borderline, despite the upper end of the interval corresponding to covering costs by sales to current users alone. The sample size is found by solving the equation

$$0.03 = 1.7\sqrt{(\hat{p}(1 - \hat{p})/n)}$$

with a value of 0.13 substituted for $\hat{p}$, and turns out to be 364. She had budgeted for 500 reply-paid questionnaires and decides to increase the sample size to 500, which is about 1 in 10 registered users. She takes a random number between 1 and 10, and sends out the questionnaire to the corresponding name and every tenth thereafter in an alphabetical list. This is not a simple random sample but it is an adequate approximation to one. It is hard to believe that there is any periodicity, which corresponds to registering enthusiasm about a graphics supplement, in the list. If the list happened to be geographically ordered, and this could possibly be linked to enthusiasm about buying a graphics supplement, the list sample would actually give a desirable spread. The problem of non-response, which is a feature of surveys even if they are about more serious issues, is not a problem for the marketing manager because no response can reasonably be assumed to correspond to no interest.

6.6.3 Confidence interval for difference in proportions

Consider two large, or infinite, populations A, B and let the proportion of defectives be p_A, p_B, respectively. Draw random samples of sizes n_A, n_B from the populations and let X and Y represent the number of defectives in each sample (Fig. 6.5). Then

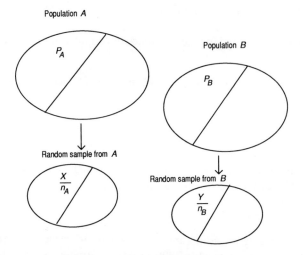

Fig. 6.5 Comparing proportions.

$$\frac{X}{n_{\mathrm{A}}} \sim N\left(p_{\mathrm{A}}, \frac{p_{\mathrm{A}}(1-p_{\mathrm{A}})}{n_{\mathrm{A}}}\right)$$

$$\frac{Y}{n_{\mathrm{B}}} \sim N\left(p_{\mathrm{B}}, \frac{p_{\mathrm{B}}(1-p_{\mathrm{B}})}{n_{\mathrm{B}}}\right)$$

and since the samples are independent

$$\frac{X}{n_{\mathrm{A}}} - \frac{Y}{n_{\mathrm{B}}} \sim N\left(p_{\mathrm{A}} - p_{\mathrm{B}}, \frac{p_{\mathrm{A}}(1-p_{\mathrm{A}})}{n_{\mathrm{A}}} + \frac{p_{\mathrm{B}}(1-p_{\mathrm{B}})}{n_{\mathrm{B}}}\right)$$

where the approximations will be good provided $n_{\mathrm{A}}p_{\mathrm{A}}$, $n_{\mathrm{A}}(1-p_{\mathrm{A}})$, $n_{\mathrm{B}}p_{\mathrm{B}}$, $n_{\mathrm{B}}(1-p_{\mathrm{B}})$ all exceed 5. Provided n_{A}, n_{B} exceed about 30 an approximate $(1-\alpha) \times 100\%$ confidence interval for $p_{\mathrm{A}} - p_{\mathrm{B}}$ is given by

$$(x/n_{\mathrm{A}} - y/n_{\mathrm{B}}) \pm z_{\alpha/2}\sqrt{(x/n_{\mathrm{A}})(1 - x/n_{\mathrm{A}})/n_{\mathrm{A}} + (y/n_{\mathrm{B}})(1 - y/n_{\mathrm{B}})/n_{\mathrm{B}}}$$

Example 6.9

A chemical company manufactures a transparent plastic in $1\,\mathrm{m} \times 2\,\mathrm{m}$ sheets. Sheets quite often contain 'flaws', which means part of the sheet has to be cut off and recycled. Changes have recently been made to the process in the hope of improving matters. Random samples, stratified over shifts, were taken during the week before and the week after the changes: 23 out of 145 sheets were flawed before the changes, while 7 out of 94 sheets were flawed after the changes. We will ignore the stratification when evaluating the change, because samples from within each stratum are rather small and there is no obvious difference between shifts.

A 95% confidence interval for the difference in proportion in the corresponding population is:

$$(23/145 - 7/94) \pm 2.0\sqrt{((23/145)(122/145)/145 + (7/94)(87/94)/94)}$$

which gives

$$0.08 \pm 0.08$$

There is some evidence that the changes have reduced the proportion of flawed sheets. In principle, allowing for the stratification in the analysis should reduce the width of the confidence interval slightly.

6.6.4 McNemar's Test

Matched pairs can again sometimes be used to increase the precision of a comparison. A random sample of 200 motorists was asked whether or not it is advisable to avoid drinking any alcohol before driving. The same motorists were followed up after an advertising campaign, designed to encourage motorists to abstain. The results are given in Table 6.4. McNemar's test focuses attention on those people who change their minds, 75 of them in our example. If the campaign had no effect, people who changed their minds would be equally likely to change from 'avoid' to 'not avoid' or vice versa. Let X be the number among the 75 who change from 'not avoid' to 'avoid'. One way of presenting the analysis is to argue that if the campaign has no effect, then $X \sim \text{Bin}(75, 0.5)$. The probability $\Pr(X \geqslant 50)$ can be approximated by

$$\Pr(Z > ((49.5 - 37.5)/\sqrt{(75 \times 0.5 \times 0.5)}) = \Pr(Z > 2.77) = 0.003$$

An alternative approach is to construct a confidence interval for the proportion (p) of motorists in the underlying population of all motorists who change their minds, who would change their minds from 'not avoid' to 'avoid'. A 95% confidence interval is 0.67 ± 0.11. Either way, we have evidence that the campaign has had some effect. However, 65 'not avoid' motorists were unaffected and it would be interesting to compare it with other campaigns.

Table 6.4 Motorists' views on abstaining if driving: before and after a publicity campaign

		After campaign	
		Avoid	Not avoid
Before campaign	Avoid	60	25
	Not avoid	50	65

Table 6.5 Drink drive figures for England and Wales during the period from 19 December to 1 January

	Breath tests administered	Positive breath tests	Percentage positive	Injury accidents
1989	68 519	5910	8.63	5715
1990	78 783	5294	6.72	6157
1991	64 118	4921	7.67	4581
1992	59 069	4248	7.19	4958

Source: The Guardian, 3 January 1991; 4 January 1993.

6.7 Random breath tests?

The statistical methods we have covered assume random samples. Whenever possible we should draw random samples, but we are often presented with data that have not been obtained in this way. The results of breath tests shown in Table 6.5 were taken from *The Guardian* of 3 January 1991 and 4 January 1993. The following is an article by Lawrence Donegan (1993) which appeared in the latter.

Senior police officers said yesterday they would continue to press for random breath tests after it was revealed that the Government's anti-drink driving campaign had little impact on the number of motorists arrested over the Christmas period.

Figures released by the Association of Chief Police Officers (ACPO) over the weekend showed that 300 motorists a day gave positive breath tests between December 19 and January 1. The number of fatal road accidents rose by almost 6 per cent, despite what police described as a 'shocking' campaign showing a dying accident victim.

Walter Girven, chairman of the ACPO traffic committee, said he was sorry the Government's publicity drive had not been more effective.

'I thought the video produced by the Department of Transport was extremely hard-hitting. It is difficult to visualize how much further one could go in terms of publicity.' The ACPO believed police officers should have the right to stop any driver at any time, he said.

The number of positive breath tests was down from 4,921 last year to 4,248, but this was matched by a fall in the number of tests carried out – 59,069, compared with 64,118 last year.

Mr Girven, who is also chief constable of Wiltshire, said police were reaching a hard core of drivers who were prepared to continue taking risks.

'Nationally over the 14-day period, more than 300 motorists a day supplied a positive breath test. This is a slight reduction on last year but not enough to indicate room for optimism. The police will continue their efforts to target the anti-social activities of a selfish minority of motorists during 1993.'

Road minister Kenneth Carlisle welcomed the fall in positive tests, which he claimed showed the Government's campaign against drink driving was working. But the figures prompted calls from a number of MPs for tougher action.

Andrew MacKinlay, a Labour member of the transport select committee, said he would be tabling Commons questions to the Transport Secretary calling for higher penalties. 'I am bitterly disappointed that despite the horrific pre-Christmas anti-drink-drive publicity the figures are still what they are. I will not be satisfied until this callous disregard for other people's lives is eliminated,' he added.

The AA said: 'These are obviously encouraging figures, but there are still people out there who insist on having one for the road. Hopefully the hard core will continue to decrease.'

A spokesman for the Campaign Against Drink-Driving said the figures showed there was still a 'reckless' hard core of motorists, 'and we can't seem to get to these people. We're still talking about 700 people a year killed in drink-related crashes. If we can't get below this figure then the Government is going to have to give the police extra powers.'

My thoughts include the following.

- Injury accidents will probably depend on the weather, although it is possible that there are several conflicting mechanisms. For example, bad weather makes driving more hazardous but may also reduce the number of people on the roads, and so on.
- A generous interpretation of the effect of the 1992 campaign is as follows. Police will tend to follow drivers who they suspect of being 'over the limit' and stop them for any infringement. They may also breathalyse any driver committing an offence, but these people, who may be less likely to give a positive test, account for the difference in the number of tests administered. The reduction in positive tests from 4921 to 4248 would then suggest that the situation has improved, slightly, from last year. A confidence interval based on an assumption of Poisson occurrences could be constructed by the method described in Exercise 6.10. But assuming positive tests occur with a Poisson distribution which is unaffected by the reduced number of tests is very optimistic, and represents a 'best scenario'. A formal confidence interval constructed in this way would imply a precision which is unjustified.
- A less optimistic, and probably more realistic, approach is to compare the proportions, pointing out that if anything the reduced number of tests administered in 1992 will tend to increase the proportion of positive tests. If this is done the 95% confidence interval for the reduction in percentage is

0.48 $\pm$ 0.30. There is some evidence of a small improvement, but the size of the reduction is disappointing.

- Any reduction would not necessarily be due only to the campaign. Publicity throughout the year and changes in public attitudes may also play a part. A statistical evaluation of the campaign itself would involve broadcasting it in some regions only, and comparing results of random breath tests with those from other regions. Testing would need to be done at similar times of day and in similar locations, between 10 p.m. and midnight near bars for example, and would require a very careful design. Despite the issue of random tests, it is hard to justify witholding the campaign from some regions when it is expected to help reduce accidents.

6.8 Summary

Precision

A precise estimate is not necessarily reliable. Replicate measurements of the speed of light by the same physicist using the same apparatus may be consistent but biased, because of calibration errors in equipment or systematic errors introduced by the method. A confidence interval for the mean of the 'corresponding population' will be a confidence interval for the mean of the imaginary population of all measurements this physicist might have made with this apparatus. The current accepted speed of light, $299\,792\,458\,\mathrm{m\,s^{-1}}$, is the result of a consensus between many physicists and the possibility of bias is much less.

In engineering applications it is often impractical for different people to make measurements with different equipment. So it is particularly important that measuring equipment is properly calibrated at regular intervals, and that staff are trained to use it correctly. If, despite these precautions, bias remains a potential source of inaccuracy, attempts will have to be made to reduce it to insignificant levels. Sensible judgements have to be made. I think it would be reasonable to have the same person measure all the volumes of shampoo in the sampled bottles, but if I were sending pavers for chemical analysis I would rather use two or more commercial laboratories than send them all to one. If I were comparing the cement content of pavers from two manufacturers I would ensure both sets of pavers were treated in the same way. I would still prefer to use two laboratories, but I would send half the number of pavers from each manufacturer to one laboratory, and the remaining pavers to another laboratory. Inter-laboratory trials are covered in detail in British Standard BS5497.

It is also important to distinguish between repeated measurements, at steady state, of one run of a process and measurements made on a different run under the same nominal operating conditions. You are unlikely to be able to replicate conditions exactly, and there may be unidentified impurities in different batches of raw material. A very precise measurement of water purity from a single run of an effluent plant may not be a reliable estimate of the mean of all such runs.

Random sample from a single population

A $(1 - \alpha) \times 100\%$ confidence interval for μ when σ is assumed known is

$$\bar{x} \pm z_{\alpha/2}\sigma/\sqrt{n}$$

and when σ is estimated by s this becomes

$$\bar{x} \pm t_{n-1, \alpha/2}s/\sqrt{n}$$

A rather approximate $(1 - \alpha) \times 100\%$ confidence interval for σ is given by

$$s \pm z_{\alpha/2}\sqrt{(s^2/(2n))}$$

This is sensitive to the assumption that the population is approximately normal.
A $(1 - \alpha) \times 100\%$ confidence interval for p is

$$x/n \pm z_{\alpha/2}\sqrt{[(x/n)(1 - x/n)/n]}$$

Percentage points

The upper $(\alpha/2) \times 100\%$ points for the standard normal distribution can be found in Table E.3 or the last row of Table E.4. The upper $(\alpha/2) \times 100\%$ points for t are given in Table E.4. However, the values for 95% confidence and 90% confidence are so commonly used that we give them here. For 90% confidence use $z_{0.05}$ which equals 1.645 (for which 1.7 is an adequate approximation). For 95% confidence use $z_{0.025}$ which equals 1.96 (or approximately 2.0). If the degrees of freedom exceed about 30 the corresponding percentage points of the t-distribution are also 1.7 and 2.0, correct to at least one decimal place.

Independent random samples from two populations

A $(1 - \alpha) \times 100\%$ confidence interval for $\mu_A - \mu_B$ is given by

$$\bar{x}_A - \bar{x}_B \pm t_{v,\alpha/2}\sqrt{(s_A^2/n_A + s_B^2/n_B)}$$

where v exceeds the smaller of the two sample sizes less 1, and is given to a better approximation in section 6.4.1. If the population variances are assumed equal, this common variance is estimated by

$$s_p^2 = ((n_A - 1)s_A^2 + (n_B - 1)s_B^2)/(n_A + n_B - 2))$$

and the confidence interval becomes

$$\bar{x}_A - \bar{x}_B \pm t_{n_A + n_B - 2, \alpha/2}s_p\sqrt{(1/n_A + 1/n_B)}.$$

A rather approximate $(1 - \alpha) \times 100\%$ confidence interval for $\sigma_A - \sigma_B$ is given by

$$s_A - s_B \pm z_{\alpha/2}\sqrt{(s_A^2/(2n_A) + s_B^2/(2n_B))}$$

This is again sensitive to the assumption that the populations are approximately normal.

A $(1 - \alpha) \times 100\%$ confidence interval for $p_A - p_B$ is

$$(x/n_A - y/n_B) \pm z_{\alpha/2}\sqrt{[(x/n_A)(1 - x/n_A)/n_A + (y/n_B)(1 - y/n_B)/n_B]}$$

Paired comparisons

Calculate the paired difference for each pair, d_i, $i = 1, \ldots, n$. Then a $(1 - \alpha) \times 100\%$ confidence interval for μ_D is

$$\bar{d} \pm t_{n-1,\alpha/2}s_d/\sqrt{n}$$

McNemar's test is the corresponding procedure for a difference in proportions. Construct a confidence interval for the proportion changing from A to B in the population of people who do change.

Sample-size calculations

You have to choose a width for a specified confidence interval. Then assume some value(s) for the population standard deviation(s), based on experience or a pilot sample, and solve for the sample size(s). The answers will only be a guide, but it is essential to obtain some idea of the likely precision of your experimental results. If the sample sizes are too small you will not be able to draw any reliable conclusions. Excessively large samples are a waste of resources.

Exercises

6.1 The mean amount of coffee powder dispensed by a drinks machine, μ, can be adjusted. Once the mean has been set, the standard deviation of the mass dispensed is 0.12 g. The catering manager aimed for a setting of 2.00 g. The mean of a random sample of 25 cups is 2.047 g. Construct 95% and 90% confidence intervals for the population mean. Would you advise the manager to readjust the setting?

6.2 In an air pollution study, the following amounts for suspended benzene soluble organic matter (in micrograms per cubic metre) were obtained for eight random 1 m³ samples of air:

2.2 1.8 3.1 2.0 2.4 2.0 2.1 1.2

Construct 95% confidence intervals for the mean and standard deviation in the corresponding population, stating any assumptions you make.

6.3 As insurance company obtained quotations for repairing a car wing from a random sample of 80 body shops. The mean ($\bar{x}$) and standard deviation (s) were £542 and £70, respectively.

(i) With what confidence can you assert that the average repair cost in the corresponding population is within £10 of £542?

(ii) Estimate the sample size needed for the probability that the sample mean will be within £10 of the population mean to be 0.99. Why is your estimate only an approximation?

6.4 In a random sample of 300 industrial accidents, it was found that 173 were due, at least partially, to unsafe working conditions. Construct a 95% confidence interval for the proportion in the corresponding population.

6.5 The results of Vickers hardness tests (in megapascals) on random samples of test specimens made from one of two magnesium alloys, A and B, are summarized below:

Alloy A: $n = 9$ $\bar{x}_A = 825.8$ $s_A = 28.9$
Alloy B: $n = 7$ $\bar{x}_B = 893.8$ $s_B = 21.5$

Assume the corresponding populations are near-normal.

(i) Construct a very approximate 90% confidence interval for the difference between the standard deviations in the corresponding populations.

(ii) Construct a 90% confidence interval for the difference in the population means.

(iii) Assume the populations' standard deviations are equal and construct a 90% confidence interval for the difference in the population means.

6.6 The following experiment was designed to investigate any systematic difference in cement content of pavers reported by two laboratories, A and B. Eight pavers were obtained from different sources. Each paver was halved; one half was chosen at random and sent to laboratory A, the other half was sent to laboratory B. The results, expressed as percentages, are given below:

Paver number	Laboratory A (%)	Laboratory B (%)
1	20.3	21.4
2	18.8	19.2
3	17.9	17.4
4	21.5	22.7
5	20.5	21.3
6	19.3	19.1
7	19.8	19.7
8	18.2	19.2

Construct a 90% confidence interval for the mean difference in cement content measurements made by the two laboratories. How many pavers would you recommend testing for the width of a 90% confidence interval for the difference in cement content to be approximately 0.5%?

6.7 The specification for a contactless displacement probe states that the standard deviation of repeat measurements made on a test piece will be less than 0.50 mm. An inspector makes 40 measurements with a standard deviation of 0.67 mm. Construct an approximate 90% confidence interval for the standard deviation. Is there any evidence that the specification has not been met?

6.8 The table below shows the number of defective and satisfactory articles in two independent samples, one taken before and one taken after the introduction of a modification to the process of manufacture.

	Before modification	After modification
Satisfactory	108	128
Defective	18	8

Construct a 95% confidence interval for the difference in proportions of defective articles in the corresponding populations. Has the modification been a success?

6.9 200 motorists were asked for their preference between a type A car and a type B car before an advertising campaign promoting the latter. At the end of the campaign the same motorists were again asked for their preferences. Do you think the campaign had any effect?

		After	
		Prefer A	Prefer B
Before	Prefer A	110	42
	Prefer B	28	20

6.10 Suppose X and Y are Poisson variables with means μ and v respectively. If μ and v exceed at least 5 we can approximate the Poisson distributions by normal distributions, so $X \sim N(\mu, \mu)$ and $Y \sim N(v, v)$. An approximate 95% confidence interval for $\mu - v$ is

$$x - y \pm 2\sqrt{(\mu + v)}$$

and the usual device of replacing μ and v by their estimates x and y gives a practical construction for a 95% confidence interval as

$$x - y \pm 2\sqrt{x + y}$$

(i) The numbers of electrons emitted from two cathode ray tubes, A and B, over 1 hour under test conditions, are 17 and 32, respectively. Construct a 95% confidence interval for the difference in mean emission rates.

(ii) A mini-roundabout was installed at a crossroads two years ago. In the previous ten years there had been 37 accidents. There have been three accidents since the change. Assume there have been no other significant changes to the junction and construct a 90% confidence interval for the reduction in yearly accident rates.

I suggest you start by letting X represent the number of accidents in the ten-year period before the change and λ_B represent the underlying yearly rate. Then if you assume accidents occur as a Poisson process

$$X \sim N(10\lambda_B, 10\lambda_B)$$

and

$$X/10 \sim N(\lambda_B, \lambda_B/10)$$

6.11 (a) Assume we have a sample of n from a normal distribution and intend estimating the lower 1% quantile of this population with the estimator

$$\bar{X} - 2.33S$$

Assume $\bar{X}$ and S are independent and write down the approximate sampling variance of this statistic.

(b) Refer to Example 6.5, and assume the natural logarithm of lifetimes is normally distributed.

(i) Calculate a 90% confidence interval for the lower 1% quantile of the distribution of the natural logarithm of lifetimes of ball-bearings made from steel A. Deduce a 90% confidence interval for the lower 1% quantile of the distribution of lifetimes.

(ii) Construct a 90% confidence interval for the difference in lower 1% quantiles of the distributions of natural logarithms of lifetimes of ball-bearings made from steel A and steel B.

6.12 The folded normal distribution

A steel surface is ground until it appears smooth, but under a powerful microscope it is seen as pitted. The depths of pits are approximately normally distributed, with a mean of δ below the surface and a standard deviation σ. The surface is now highly polished and a layer of material of depth δ is removed. The remaining pits have depths Y which have a folded normal distribution with pdf

$$f(y) = \frac{2}{\sigma\sqrt{2\pi}} e^{-y^2/2\sigma^2} \qquad \text{for } 0 \leqslant y$$

Show that Y has a mean value of

$$\sigma\sqrt{(2/\pi)}$$

and a standard deviation of

$$\sqrt{(1 - 2/\pi)}\sigma$$

6.13 **Approximate sampling distribution of S**

(a) Suppose X has a $N(\mu, \sigma^2)$ distribution and that we have a random sample of n such X_i. Define

$$Y_i = X_i - \mu$$

Refer to Theorem 3 of Appendix A2, and write down the moment generating function of Y. Deduce that

$$E[Y^2] = \sigma^2 \quad \text{and} \quad E[Y^4] = 3\sigma^4$$

Use the result of Exercise 5.11(ii) to obtain

$$\text{var}(Y) = 2\sigma^4$$

(b) Now, define

$$V = \sum Y_i^2$$

Show that

$$E[V] = n\sigma^2 \qquad \text{var}(V) = 2n\sigma^4$$

Use a Taylor series approximation to deduce the approximations

$$E[\sqrt{(V/n)}] = \sigma \qquad \text{var}(\sqrt{(V/n)}) \simeq \sigma^2/2n$$

Hence deduce the approximate results for S. The approximate normal distribution for S relies on both a large sample size and the population being approximately normal.

(c) Assume you have a sample from a normal distribution. Starting from the approximate result that

$$\frac{S - \sigma}{\sigma/\sqrt{2n}} \sim N(0, 1)$$

find a slightly more accurate formula for a $(1 - \alpha) \times 100\%$ confidence interval for σ than

$$s \pm z_{\alpha/2} s/\sqrt{(2n)}$$

7

Asset management plan

7.1 Background

In 1989 the water authorities in England and Wales were taken out of public ownership and sold on the London stock market under the terms of the Water Act. There are now ten water companies, and a Director-General of Water Services (DG) has been appointed to regulate the water supply and sewerage industry. The DG has been given considerable powers and, in particular, requires annual reports to be submitted covering investment, levels of service to the customer and condition of assets. The DG has also instructed the companies to submit, in 1994, asset management plans (AMPs) describing their strategy for the next 20 years. The primary objective of the AMPs is to estimate the expenditure required to improve and maintain assets so that specified standards of performance and levels of service can be achieved. The 1994 AMPs are the second exercise of this kind: the water authorities, which preceded the present water companies, had to prepare AMPs before privatization. It is likely that AMPs will be required every five years. The original AMPs have proved very useful to the companies for data management and planning. They also represent a move away from 'current cost accounting' (which relies on notional lives for assets) towards 'renewals accounting' (which assumes the assets will be maintained indefinitely). This latter is far more realistic for items such as sewers, which have relatively short sections that are repaired or replaced when necessary. The AMPs are also needed to allay public concern about the possibility of companies taking substantial profits while letting the assets slowly deteriorate.

If the AMPs are to be effective they must give a fair reflection of the cost of work needed to preserve standards and an indication of the accuracy of the estimate. A water company can use its AMP to make a case for increasing charges by more than the Retail Price Index, but the DG will need reassurance that an AMP is fair and reasonably accurate before allowing this. If a water company needs to raise extra capital for works that were not included in the AMP it will probably be prohibited from raising charges any further and have to take money out of profit. This could lead to cutbacks in other areas, leading to deterioration in assets, and perhaps take-overs by less scrupulous organizations. It is in both the public's and the companies' interests that they be well managed. The AMP is an essential part of this management process, and involves a considerable investment in personnel time and other resources from the company. It is subject to careful scrutiny by independent assessors, known as cross-certifiers, who are consulting engineers with considerable experience in the water industry. The statistician is responsible for ensuring that the resources are used in such a way as to minimize the overall uncertainty in the AMP, and for estimating this uncertainty.

7.2 Statistical issues

The assets of a typical water company include reservoirs, water treatment works, pumping stations, sewage treatment facilities, thousands of kilometres of water mains and sewers, together with office buildings. A complete survey of all these is inappropriate so sampling is used, and the sampling error is often rather small compared with other sources of uncertainty. I shall not give all the statistical details of a typical AMP here, but describe one possible approach for a part of the water distribution network. There is no unique 'right way', but the method I describe includes two statistical techniques, **stratification** and **ratio estimation**, which have wide-ranging applications. It is long rather than technically difficult, the same statistical technique being used in several places, and Fig. 7.1 may help clarify the argument. The notation is summarized in Table 7.1. Financial details of AMPs are commercially confidential but more details of statistical methods can be found in a paper by O'Hagan *et al.* (1992) and two of my articles (Metcalfe, 1991a; 1991b). The following is loosely based on work I did for the former Northumbrian Water Authority and was a result of helpful discussions with staff there and with the independent consultants. It also incorporates some ideas that arose from more recent discussions with engineers at Northumbria Water Ltd and Yorkshire Water Services Ltd.

The water supply network was split into the raw water system, general divisional resources, dams and strategic supply networks (which were all investigated) and the local distribution network (which was sampled). The local distribution network was divided into 146 zones which were defined so that they were approximately self-contained, inasmuch as work in one would not significantly affect its neighbours. They are typically spurs off the trunk main, which is part of the strategic supply network. A sample of zones was surveyed in detail and schemes that would need to be carried out within the following 20-year period were identified and costed. Some of the potential sources of error are listed below.

1. Uncertainty over the specification of individual schemes.
2. A tendency consistently to overestimate or underestimate the work needed to complete schemes – that is, a bias.

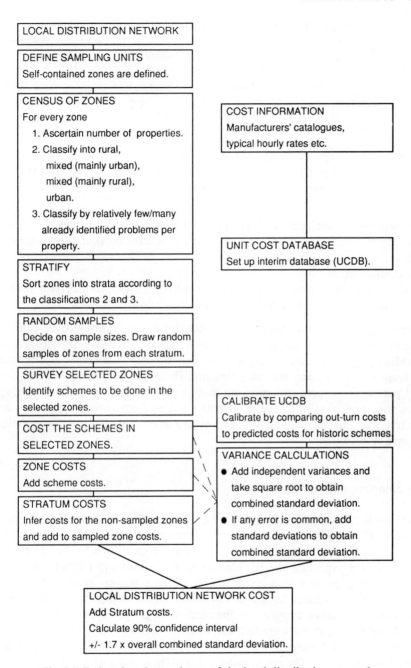

Fig. 7.1 Estimating the total cost of the local distribution network.

Table 7.1 Notation for AMP costs and uncertainties

Level of aggregation	Cost	Source of uncertainty		
		Bias	Individual schemes	Sampling error
Scheme	c_i	$se(r)\%$	$sd(r_i)\%$	
Zone (surveyed)	z_j	$se(z_j)$	$sd(z_j)$	
(other)	$\hat{z}_k$	$se(z_k)$	$sd(z_k)$	$sam(z_k)$
Stratum	$STRATCOST_l$	$STRATSE_l$	$STRATSD_l$	$STRATSAM_l$
Total	$TOTCOST$	$TOTSE$	$TOTSD$	$TOTSAM$

3. Uncertainty over costs for identified work on individual schemes.
4. Incorrect 'unit' costs being used to arrive at scheme costs, another source of bias.
5. Sampling error.
6. Missing schemes that will need to be completed within the 20-year period.
7. Changes in engineering techniques and legislation.

Uncertainty about individual schemes includes variation about their expected size, and variations in unit costs about their average. However, this uncertainty is reduced, in percentage terms, when a large number of schemes are summed. In contrast to this, any bias in estimating the amount of work or unit costs will persist, in percentage terms, no matter how many schemes are summed. This was, and usually is, the dominant source of error. Sampling error was relatively unimportant and is straightforward to estimate if a random sampling scheme is used.

7.3 Sampling scheme

Two statistical devices were used to try to increase the precision of the sampling procedure for a given amount of work. One was to make use of the fact that costs for zones are likely to be approximately proportional to their sizes. The other involved splitting the population of zones into sub-populations called **strata**, so that zones within a stratum would be relatively similar in terms of cost/size, whereas zones from different strata would be dissimilar. The sizes of the zones were taken as given by the number of properties, this being known for all zones. Ten strata were originally used. The zones were first divided into five environmental categories: rural; mixed (mainly rural); mixed (mainly urban); suburban; inner city. The zones within each of these categories were then sorted into two groups: relatively few and relatively many problems already identified per property. As the numbers of zones in two of the strata were small, they were combined with larger strata to reduce the number of strata to eight.

Advantage was taken of the likely relationship between cost and size by taking simple random samples from each stratum and weighting the results (ratio estimators). A total sample size of 33 zones was considered reasonable, given the available resources and the timetable for the AMP. There was not enough time for a formal pilot study, but surveys of several small areas of the sewerage system were available and gave some indication of the possible variance of zone costs. If all zones were of the same size, an optimum allocation in terms of the smallest standard error for the estimate of the total local distribution network AMP cost would be to make the number of zones sampled from each stratum: proportional to the number of zones in the stratum; proportional to the standard deviation within the stratum; and inversely proportional to the square root of the expense of surveying a zone within the stratum (see Barnett, 1974). This could only be used as a very general guide because standard deviations of zone costs within strata were not known, and survey expenses could only be roughly estimated. The differing zone sizes would also need to be taken into consideration for a rigorous optimum allocation. The actual allocation of the sample over the stratified zones is given in Table 7.2. At least three zones were taken from each stratum, with the one obvious exception, so that a reasonable estimate of the stratum variance could be made. The allocation for the larger

Table 7.2 The eight strata and allocation of sample over stratified zones

	Fewer prior identified problems/property	More prior identified problems/property
Rural		2 zones of sizes 1184 and 1400 properties. Both in sample.
Mixed (mainly rural)	13 zones in stratum. Smallest 179 properties, largest 1886 properties. Sample of 3 drawn.	12 zones in stratum. Smallest 212 properties, largest 3539 properties. Sample of 3 drawn.
Mixed (mainly urban)	28 zones in stratum. Smallest 370 properties, largest 14 627 properties. Sample of 5 drawn.	
Suburban	51 zones in stratum. Smallest 104 properties, largest 14 404 properties. Sample of 9 drawn.	19 zones in stratum. Smallest 908 properties, largest 12 900 properties. Sample of 5 drawn.
Inner City	12 zones in stratum. Smallest 727 properties, largest 3075 properties. Sample of 3 drawn.	9 zones in stratum. Smallest 1740 properties, largest 10 000 properties. Sample of 3 drawn.

strata was based on a proportional allocation, with more emphasis on the suburban stratum with more identified problems per household, because it was expected to have a larger variance.

The work to be undertaken in the 33 sample zones was assessed, allocated into schemes, and costed using unit cost formulae which were developed as a separate investigation within the AMP.

7.4 Unit cost formulae

The first four sources of error can all be quantified by 'calibrating' the interim unit cost database against ratios of out-turn costs for past schemes to costs estimated using current AMP methods and the interim database. This will not enable us to separate engineering bias from unit cost errors but there is no need to do so for the AMP. If we did need to do so we would consider both the ratio of out-turn work to estimated work, and the ratio of out-turn cost to estimated cost of out-turn work using the interim data base. We defined ratios

$$r_i = \frac{y_i}{x_i}$$

$$= \frac{\text{out-turn cost}}{\text{retrospective predicted cost using interim unit cost database and AMP methods}}$$

The interim unit cost database is constructed from known costs of materials and typical hourly rates for people working in the industry. Engineers then cost retrospectively selected schemes using only the information that would be available if they had been identified during an AMP survey. Most of Northumbrian Water's engineering work is done by contractors and the out-turn cost can, in principle, be obtained from the invoice. In fact this is not as easy as it may sound because work is often done on a term contract basis, local councils may take a long time to send invoices for reinstating the road or pavement, and so on. The ratios are considered a random sample in time so all schemes within a time period extending up to the time at which the exercise is started must be listed, and either all costed or a random sample costed. In particular, schemes must not be excluded because they turned out to be 'anomalous'! The DG provides indices for bringing all out-turn costs to an April 1994 baseline. The statistical analysis provides a ratio estimate (r) of the calibration factor

$$r = \frac{\sum y_i}{\sum x_i}$$

This is a weighted average of the individual ratios (r_i) with weights equal to the predicted costs (x_i), and seems appropriate when our intention is to estimate the overall cost. There was no evidence of any systematic relationship between the ratios and the predicted costs. The variance of the ratio estimator is, approximately, given by

$$\widehat{\text{var}}(r) = \frac{\sum(y_i - rx_i)^2}{\bar{x}^2 n(n-1)}$$

The standard deviation of this estimator, its standard error $se(r)$, is the square root of the variance. The approximate formula for the variance is derived in Appendix A4, but it is perhaps worth checking that it is at least plausible by putting all the x_i equal to 1. The standard deviation of ratios for individual schemes about the average, $sd(r_i)$ is estimated from

$$sd(r_i) = \sqrt{\left[\frac{\sum(r_i - r)^2}{n-1}\right]}$$

This ignores the nicety that the standard deviation of ratios may decrease as schemes become larger, but it is quite adequate for AMP purposes.

Example 7.1

An interim unit cost database has been set up for work in water quality zones and will be calibrated from the ratios

$$\frac{\text{out-turn cost}}{\text{retrospective predicted cost using interim unit cost database and AMP methods}}$$

for eight schemes. The data are given in Table 7.3. Calculations using the formulae lead to

$$r = 1.29$$
$$se(r) = 0.12$$
$$sd(r_i) = 0.34$$

Table 7.3 Predicted and out-turn costs (monetary units) for eight historic water distribution schemes

Predicted cost	Out-turn cost	Ratio
81	94	1.16
20	18	0.90
98	107	1.09
79	75	0.95
69	52	0.75
144	216	1.50
119	176	1.48
124	207	1.67

Notice that $se(r)$ is approximately equal to $sd(r_i)/\sqrt{8}$. The interim unit cost data base is now calibrated by multiplying all entries by 1.29, and will now be referred to as the (calibrated) unit cost database. It will also be much more convenient to express $se(r)$ and $sd(r_i)$ in percentage terms. So we define

$$se_\%(r) = se(r) \times 100/r = 9.3$$

and

$$sd_\%(r_i) = sd(r_i) \times 100/r = 26$$

7.5 Zone costs

We now consider a typical stratum of 12 zones, and imagine that three have been sampled.

7.5.1 Costing sampled zones

Suppose we identify all the schemes within a typical zone. The schemes are costed using the calibrated unit cost database to obtain costs c_i. Then the estimated zone cost (z_j) is given by

$$z_j = \sum_{\text{schemes}} c_i$$

It is assumed that the variations of individual schemes about the average are independent, so the standard deviation which allows for this, $sd(z_j)$, is calculated by adding the variances and then taking the square root. That is,

$$sd(z_j) = \sum_{\text{schemes}} (c_i \times sd_\%(r_i)/100)^2$$

Adding more schemes reduces this in percentage terms. However, any error in the calibration factor persists in percentage terms, no matter how many schemes are added, because this error is the same for all schemes. So denoting this standard deviation by $se(z_j)$ we have

$$se(z_j) = \sum c_i \times se_\%(r)/100 = z_j \times se_\%(r)/100$$

7.5.2 Costing other zones

We start by estimating the average cost per property in the stratum from the sampled zones. Let the size of a zone be w properties. We know both the sizes (w_j) and costs (z_j) for the three sampled zones and can use the ratio estimator formula, with x and y replaced by w and z, to find an average cost per property p and its associated standard error, $se(p)$.

We can now estimate the costs for the other nine zones:

$$\hat{z}_k = pw_k$$

The 'hat' emphasizes the fact that these are estimates based on the results (z_j) for the sampled zones. It is convenient to calculate an associated sampling standard deviation

$$sam(z_k) = se(p) \times w_k$$

The standard deviation which accounts for uncertainty in the calibration factor is approximately

$$se(z_k) = \hat{z}_k \times se_\% (r)/100$$

The standard deviation which accounts for variation in individual schemes can be estimated, roughly, from the average standard deviation for the sampled zones, $\overline{sd}(z_k)$. That is

$$sd(z_k) = \overline{sd}(z_j) \times \sqrt{(w_k/\bar{w}_j)}$$

where $\bar{w}_j$ is the average size of the sampled zones.

7.6 The stratum cost

All that remains is to sum the zone costs and combine the standard deviations in the appropriate way. If sources of error are considered independent, we square the standard deviations, add these variances, and then take the square root. If any error persists for all zones we add the standard deviations. The subscript l will denote a typical stratum, summations with subscript j are over sampled zones, and summations with subscript k are over the other zones. We were beginning to run out of letters so the following mnemonics were agreed.

$STRATCOST$ is the estimated cost for the stratum.
$STRATSD$ is the standard deviation which accounts for variation of individual schemes about the average (items (1) and (3) in section 7.2).
$STRATSAM$ is the standard deviation which accounts for sampling error (item 5).
$STRATSE$ is the standard deviation which accounts for uncertainty in the calibration factor (items 2 and 4).

In this notation

$$STRATCOST_l = \sum z_j + \sum \hat{z}_k$$
$$STRATSD_l = \sqrt{[\sum (sd(z_j))^2 + \sum (sd(z_k))^2]}$$
$$STRATSAM_l = \sum sam(z_k) = se(p) \sum w_k$$
$$STRATSE_l = \sum se(z_j) + \sum se(z_k) = STRATCOST_l \times se_\%(r)/100$$

Notice that the finite population correction is taken into account, implicitly, because the sampling error is only applied to the non-sampled zones. This seemed the most convenient way to set out the procedure for programming and is equivalent to using the finite population correction in the usual way, which would give

$$STRATSAM_1 = se(p)(\sum w_j + \sum w_k)(1 - \sum w_j/(\sum w_j + \sum w_k))$$

Example 7.1 (continued)

The calibrated unit cost database has already been set up and a random sample of three zones from a stratum of 12 has been selected and costed. Fictitious data are given in Table 7.4. The formulae for the ratio estimator give

$$p = 643 \text{ monetary units (MUs) per property}$$
$$se(p) = 104 \text{ MUs per property}$$

The estimate of the total cost is

$$STRATCOST = 1\,3702\,273$$

and the standard deviation of this estimate due to sampling is estimated as

$$STRATSAM = 1\,788\,931.$$

This estimate is unreliable because of the small sample size, and we will return to this point later. At this stage we have allowed for sampling error but have not made any allowance for possible errors in the estimated costs for the sampled zones. The standard deviation to allow for uncertainty in the calibration factor

Table 7.4 Zone sizes (number of properties) for a stratum and costs (monetary units) for a random sample of 3

Zone size (w)	Cost of surveyed zones (z_j)
212	
629	675 524
712	
858	351 308
1068	
1541	
2011	
2316	
2582	1 588 297
2714	
3138	
3539	

for unit costs is

$$STRATSE = 13\,702\,273 \times 0.093$$
$$= 1\,274\,311$$

The standard deviation to allow for variation in individual schemes about their expected values will depend on the number of schemes and their costs. Because schemes are assumed to be independent it is likely to be relatively small, unless a few very large schemes dominate. In the continuation of this example we will assume a rather high value of $STRATSD$,

$$STRATSD = 2\,000\,000$$

7.7 Total cost of local distribution network

The same principles are applied again.

$$TOTCOST = \sum_{strata} STRATCOST_i$$

$$TOTSD = \sqrt{\sum_{strata} STRATSD_i^2}$$

$$TOTSAM = \sqrt{\sum_{strata} STRATSAM_i^2}$$

$$TOTSE = \sum_{strata} STRATSE_i = TOTCOST \times se_\%(r)/100$$

Notice that any error in the calibration factor persists in percentage terms. The final move is to combine the three independent components of uncertainty. They are quite reasonably assumed independent, so we square the standard deviations, sum the variances and take the square root.

$$SDTOT = \sqrt{(TOTSD^2 + TOTSAM^2 + TOTSE^2)}$$

This is Pythagoras' theorem in three dimensions, and a geometric interpretation is given in Fig. 7.2. The total cost is the sum of a large number of items, and provided none completely dominates we can assume the final estimate of total cost is normally distributed about the hypothetical exact value. An approximate 90% confidence interval for this exact value, expressed as a plus or minus percentage, is

$$TOTCOST \pm 1.7 \times (SDTOT/TOTCOST) \times 100\%$$

The 1.7 is an approximation to the upper 5% point of the standard normal distribution. While individual stratum sampling standard deviations are based on few degrees of freedom, eight such estimates have been combined and the

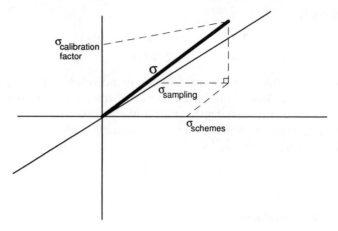

Fig. 7.2 Overall standard deviation (σ) can be obtained from its independent components by Pythagoras' theorem.

net effect is relatively unimportant when compared with uncertainty in the calibration factor. The standard error of the calibration factor was based on eight ratios so there is a good argument for using the upper 5% point from a t-distribution with seven degrees of freedom (which is 1.9). This is not usually done in practice because the main interest is in the overall AMP total which involves several independent unit cost databases and more than eight ratios are usually available to calibrate the more important ones.

Example 7.1 (concluded)

Suppose we have eight strata with similar overall costs and standard deviations to the example stratum. The overall cost would be about 100 million monetary units. The sampling standard deviation which was about 13% in the one stratum will be reduced by a factor of $\sqrt{8}$ to less than 5%. The standard deviation to account for variation in individual schemes will be reduced to slightly more than 5%. The standard deviation which allows for uncertainty in the calibration factor remains at 9.3%. Combining these three standard deviations would give

$$\sqrt{(0.05^2 + 0.05^2 + 0.093^2)} \times 100 = 11.7\%$$

which is much less than their sum. If all the zones had been costed the sampling standard deviation would be zero and the combined value would reduce to 10.6%. It would hardly be worth increasing the sample sizes from the strata to decrease the uncertainty in the AMP total – better to reduce uncertainty about the calibration factor by retrospectively costing more schemes.

Table 7.5 Estimated average costs per property for local distribution network work in monetary units

	Fewer prior-identified problems per property	More prior-identified problems per property
Rural	⎰ 581	2105
Mixed (mainly rural)	⎱	741
Mixed (mainly urban)	537	
Suburban	97	204
Inner city	113	192

7.8 Discussion

7.8.1 Review of stratification procedure

The estimated average costs in monetary units per property for items in the local distribution network of Northumbrian Water Authority are given for the eight sampled strata in Table 7.5. Although there is considerable uncertainty about the average costs per property and hence total costs for individual strata, summing the total costs for eight independent strata considerably reduces the percentage error. While a large difference was expected between rural and urban, the success of the division into high and low density of already indentified problems was, perhaps, more surprising, and the stratification was clearly worthwhile.

7.8.2 Components of uncertainty which have been allowed for

A 90% limit of prediction for the total cost, water and sewerage, for the Authority was

point estimate $\pm\, x\%$

which allows for uncertainty in the calibration factor, error due to sampling of the zones, and variation of individual schemes. The exact value of x is confidential, although it was lower than had been expected at the start of the study, but the proportions of x reported below are those obtained.

If uncertainty in the calibration factor is excluded, the 90% limits of prediction are reduced to

$\pm\, 0.45x\%$

If error due to sampling the zones is also excluded, the 90% limits of prediction are reduced to

$\pm\, 0.35x\%$

If only error due to sampling the zones is included, the 90% limits of predictions are

$$\pm 0.25x\%$$

One of the reasons why the error due to sampling zones is of relatively minor importance is that a considerable proportion of the work, in cost terms, was costed without sampling.

7.8.3 Uncertainty not allowed for

No allowance was made for possible changes in engineering techniques or legislation. The brief description I gave of the calibration procedure glossed over the fact that for certain types of work – bringing bathing beaches up to European Community directives, for example – there were no directly relevant retrospective schemes. A consistent over- or underspecification of such schemes would result in a bias. A less important point is that no statistical allowance for variability between surveyors was made, but this was minimized by agreeing procedures centrally and auditing by the cross-certifiers. You may be able to think of more omissions!

7.8.4 Conclusion

In such a complex study as an asset management plan a statistical analysis can only be a rough guide to the level of uncertainty, but it has two valuable consequences. First, it emphasizes that cost predictions are wide intervals rather than single values, although some people with an interest in financial planning prefer point estimates to intervals. Perhaps statisticians should be more vociferous in pointing out that, at least when forecasting, a single figure suggests a lack of any quantitative assessment of accuracy rather than high precision. Second, it helps identify the weakest points of the study. In this case, loss of accuracy through sampling the local distribution network is less important than the uncertainty over the average unit costs. The only disadvantage of sampling is that a company cannot be sure where the zones in the worst condition are, and will not know where remedial work is most needed. In practice, this is less of a drawback than might at first appear because water companies usually know the 'worst' zones, and these will have been separately surveyed and excluded from the sampled population.

7.9 Summary

Stratification

The population is all local distribution zones. The main objective is to estimate the average cost per property to maintain, and where necessary improve, the network. We expect this cost to be higher in rural zones because dwellings are generally further apart, and we can classify all the zones as urban or rural. We therefore divide the population into sub-populations called strata and take simple random samples from each stratum. This has two advantages.

First, the sample will be representative of the population in terms of the important urban–rural distinction. If the sample sizes are not proportional to the number of properties in each zone appropriate weighting must be used, but the advantage remains. Results can also be given for the different strata, and these may be of interest in themselves.

Second, if costs per property are relatively similar within strata, compared with comparisons between different strata, the final estimate will be more precise than one based on a simple random sample of the same size from the whole population.

A reasonable question to ask is why the stratification has to be done in advance. Post-stratification can be useful, but it is preferable to stratify before drawing the sample for the following reasons.

First, we can ensure that some minimum number of zones is taken from every stratum. In the case study, a simple random sample of zones from the whole population is quite likely to miss some strata altogether. We also need at least three zones from a stratum if we are to make a useful estimate of the within-stratum variance. It is possible to make additional assumptions and thereby avoid having to estimate directly all the within-strata variances, but it is better to avoid this sort of strategy if it is practical to do so.

Second, we can ensure an efficient distribution of the sample over the strata. If we can make reasonable guesses about within-strata variation and sampling costs there is an optimum allocation which will give maximum precision for a given cost. In practice, proportional allocation is often used. Although simple random samples from the whole population would on average give a proportional allocation, the one we happen to get could be far from it!

Third, we avoid accusations that our reason for post-stratifying is to modify the final result in our favour.

Ratio estimator

$$r = \sum y_i / \sum x_i$$
$$\widehat{\mathrm{var}}(r) = \sum (y_i - rx_i)^2 / (\bar{x}^2 n(n-1))$$

Combining sources of uncertainty

The standard deviation of the sum of errors entirely due to a common source is the sum of their standard deviations.

The standard deviation of the sum of independent errors is the square root of the sum of their variances.

8

Making predictions from one variable

8.1 Linear regression

8.1.1 Introduction

Tungsten steel erosion shields are fitted to the low-pressure blading in steam turbines. The most important feature of an erosion shield is its resistance to wear. This is difficult, expensive and destructive to measure directly in terms of abrasion loss, but it is known to be associated with the hardness of steel. For everyday quality control purposes it would be far more convenient to use the hardness measurements, provided they give reasonably reliable estimates of the abrasion losses. Measuring hardness at points on less critical areas of the shields will not affect their performance.

An engineer decided to investigate the feasibility of this strategy by measuring both the abrasion losses and the hardnesses of 25 erosion shields. Each shield was taken from a different batch and they were assumed to be a random sample

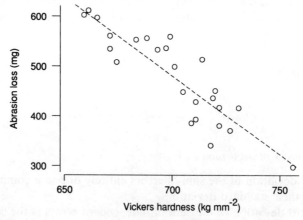

Fig. 8.1 Abrasion loss plotted against Vickers hardness for 25 tungsten steel erosion shields.

Table 8.1 Vickers hardness and abrasion loss measurements on 25 erosion shields

Vickers hardness $(x, \mathrm{kg\,mm}^{-2})$	Abrasion loss (y, mg)
665	597
719	436
659	602
756	297
711	393
671	561
709	385
722	380
718	340
714	513
701	499
671	535
720	450
727	370
711	428
731	416
699	559
661	611
722	417
674	508
705	448
688	556
693	533
697	536
683	553

'in time' from all possible batches. The data pairs are listed in Table 8.1 and plotted in Fig. 8.1. There is a definite tendency for the abrasion loss to decrease as the hardness increases, but as the points do not lie precisely on a straight line we propose a model which includes unexplained variation:

$$abrasion\ loss = \alpha + \beta \times hardness + error$$

The errors are not necessarily measurement errors, and simply represent deviations from the hypothetical line.

8.1.2 The model

Let Y be the variable we wish to predict, the **dependent variable**, and x be some **explanatory variable**. A plot of the available data indicates that a linear relationship between x and y is a sensible approximation. We shall model these observations (n of them) by

$$Y_i = \alpha + \beta x_i + E_i \qquad \text{for } i = 1, 2, \ldots, n$$

and make several assumptions about the errors (E_i).

Even if the abrasion losses and hardnesses of the erosion shields could be measured exactly there would be deviations about the line, because many other factors, unmeasured or unknown, have some influence on the abrasion loss. We will refer to this as **inherent variability** in the population in order to distinguish it from **measurement error**. The errors in the model can represent this inherent variability or errors in measuring Y (or both). However, apart from in sections 8.5 and 8.6, we assume that x is measured precisely. Model assumptions are never satisfied exactly, but it is important to check that the model provides a reasonable approximation to reality. Inherent variability in the population is likely to be the dominant feature of the errors for the erosion shields. Errors in measuring hardness can be minimized by taking an average of several point measurements on each shield. The other assumptions about the errors follow.

A1   The average value of the error is zero, formally $E[E_i] = 0$

There is no way we can check this from our data pairs because any other value would be indistinguishable from the intercept α. It would have to be checked by a separate calibration exercise.

A2   The errors are independent of the values of the explanatory variable.

There is no way we can detect any linear association from our data because it would be incorporated into the slope. Unless we are prepared to carry out a separate calibration exercise we have to accept the assumption as reasonable. Most reputable companies take considerable trouble to ensure that measuring equipment – a balance for weighing in our example – is properly calibrated.

A3   The errors are independent of each other.

It is possible for observations taken over time to have correlated errors, but obvious correlations can be detected at the end of the analysis. It is a common feature of economic time series, and economists allow for it by modifying the analysis.

A4   The errors all have the same variance, which we will denote by σ^2.

Obvious discrepancies should be apparent from the plot and the variance does sometimes change with the magnitude of x. For example, variation is sometimes more constant in percentage terms than absolute terms so if Y tends to increase with x so will the variance. The standard analysis can be modified to take account of this.

A5   The errors are normally distributed.

We can assess whether this is a reasonable approximation at the end of the analysis.

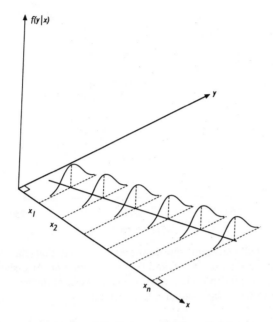

Fig. 8.2 The linear regression model.

To summarize, we are modelling the conditional distribution of Y given a particular value of x, x_p say, as $N(\alpha + \beta x_p, \sigma^2)$. The unknown parameters which have to be estimated from the data are α, β and σ^2. The line on which the mean values of Y lie,

$$y = \alpha + \beta x$$

is known as the **regression** line of y on x (Fig. 8.2).

8.1.3 Fitting the model

Fitting the regression line
From the definition of the model, the sum of squared errors is

$$\sum E_i^2 = \sum (Y_i - (\alpha + \beta x_i))^2$$

We have n data pairs (x_i, y_i) and estimate the parameters α and β by the values, $\hat{\alpha}$ and $\hat{\beta}$, which minimize the sum of squared errors (ψ)

$$\psi = \sum (y_i - (\alpha + \beta x_i))^2$$

This is an application of the **principle of least squares**. The first explicit account was published by A.M. Legendre (1805) in his work on the estimation of comet orbits. Legendre acknowledged Euler's contributions to the method, which had

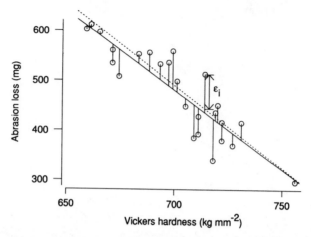

Fig. 8.3 The estimated regression line (unbroken) is such that the sum of squared distances parallel to the y-axis (e.g. r_i and $|r_j|$) is a minimum. Also shown is a typical unmeasurable error (ε_i), its absolute magnitude is the vertical distance to the unknown hypothetical line (dotted). The residual r_i is an estimate of ε_i.

also been used by Karl Friedrich Gauss, working independently, from about 1795 (Plackett, 1972). The values $\hat{\alpha}$ and $\hat{\beta}$ are known as **least-squares estimates**.

In geometric terms we are finding the line

$$y = \hat{\alpha} + \hat{\beta}x$$

such that the sum of squared 'vertical' (measured parallel to the y-axis) distances is a minimum (Fig. 8.3). Necessary conditions for ψ to have a minimum are that

$$\frac{\partial \psi}{\partial \alpha} = 0 \quad \text{and} \quad \frac{\partial \psi}{\partial \beta} = 0$$

The partial differentiation is easy and the result is two linear simultaneous equations in two unknowns.

$$-2\sum(y_i - (\hat{\alpha} + \hat{\beta}x_i)) = 0$$
$$-2\sum x_i(y_i - (\hat{\alpha} + \hat{\beta}x_i)) = 0$$

Rearrangement gives

$$n\hat{\alpha} + \hat{\beta}\sum x_i = \sum y_i$$
$$\hat{\alpha}\sum x_i + \hat{\beta}\sum x_i^2 = \sum x_i y_i$$

From the first equation

$$\hat{\alpha} = \bar{y} - \hat{\beta}\bar{x}$$

Substitution into the second gives

$$\hat{\beta} = \frac{\sum x_i y_i - \bar{y} \sum x_i}{\sum x_i^2 - \bar{x} \sum x_i}$$

but these forms of the numerator and denominator are very prone to rounding errors. For programming purposes the equivalent expressions

$$\sum (x_i - \bar{x})(y_i - \bar{y}) \quad \text{and} \quad \sum (x_i - \bar{x})^2$$

must be used. The equivalence is easily demonstrated. For the first expression,

$$\sum (x_i - \bar{x})(y_i - \bar{y}) = \sum x_i (y_i - \bar{y}) - \bar{x} \sum (y_i - \bar{y})$$
$$= \sum x_i y_i - \bar{y} \sum x_i - \bar{x} \times 0$$

The result for the second is identical with the y replaced by x. The fitted regression line is,

$$y = \hat{\alpha} + \hat{\beta} x$$

or equivalently

$$y = \bar{y} + \hat{\beta}(x - \bar{x})$$

The latter form emphasizes that the line passes through the centroid of the data, $(\bar{x}, \bar{y})$. It follows that the estimator of the slope is independent of $\bar{Y}$.

Example 8.1

For the tungsten shield data,

$$\bar{x} = 701.08 \qquad\qquad \bar{y} = 476.9$$
$$\sum (x_i - \bar{x})^2 = 14\,672 \quad \sum (x_i - \bar{x})(y_i - \bar{y}) = -46\,373$$

The estimated regression line is

$$y = 476.9 - 3.161(x - 701.08)$$

or equivalently

$$y = 2693.1 - 3.161x$$

and is shown in Fig. 8.1.

Residuals
The residuals (r_i) are our best estimates of the values taken by the errors. They are defined as the differences between the y_i and their estimated values $(\hat{y}_i)$, known as **fitted values**, using the estimated regression line. That is

$$r_i = y_i - \hat{y}_i$$

where

$$\hat{y}_i = \hat{\alpha} + \hat{\beta} x_i$$

In geometric terms, the residuals are the absolute values of the vertical distances from the points to the estimated regression line (Fig. 8.3). For any set of data,

$$\sum r_i = 0 \quad \text{and} \quad \sum (x_i - \bar{x})r_i = 0$$

The proof of the first result is as follows.

$$\sum r_i = \sum (y_i - (\hat{\alpha} + \hat{\beta}x_i))$$

But $\hat{\alpha} = \bar{y} - \hat{\beta}\bar{x}$, so the sum of residuals can be written as

$$= \sum (y_i - (\bar{y} - \hat{\beta}\bar{x} + \hat{\beta}x_i))$$
$$= \sum (y_i - \bar{y}) - \hat{\beta}\sum (x_i - \bar{x})$$
$$= 0 - \hat{\beta} \times 0 = 0$$

The proof of the second is left as an exercise. These results reinforce the earlier remarks that errors with non-zero mean, and linear association between errors and x, cannot be detected from the data. We estimate the variance of the errors (σ^2) by the **residual mean square** (also known as the **error mean square**)

$$s^2 = \sum r_i^2 /(n - 2)$$

We divide by $n - 2$ rather than n because we have lost two degrees of freedom by using two parameters estimated from the data, $\hat{\alpha}$ and $\hat{\beta}$. That is, given any $n - 2$ residuals, the remaining two are determined by the constraints

$$\sum r_i = 0 \quad \text{and} \quad \sum (x_i - \bar{x})r_i = 0$$

With this definition we would have no estimate of the variance if we only had two points, and this is appropriate. However, a more substantial reason for adopting this definition is that it is an unbiased estimator of σ^2, in other words, if s^2 is imagined to be averaged over repeated analyses it would equal σ^2. This turns out to be very useful for comparing different models when we introduce more explanatory variables.

Example 8.1 (continued)

For the tungsten shield data s^2 is 1759, so s is approximately 42.

8.1.4 Properties of the estimators

These results are a special case of those for multiple regression. They are proved, for the general multiple regression model, in Appendix A5. The proofs only need some elementary matrix theory and are worth reading. First

$$\hat{\beta} \sim N(\beta, \sigma^2 / \sum (x_i - \bar{x})^2)$$

(Fig. 8.4). This seems reasonable enough: on average $\hat{\beta}$ would equal the hypothetical β, and its variance decreases as the sample size increases because there will be more terms in $\sum (x_i - \bar{x})^2$. By the usual argument, a $(1 - \varepsilon) \times 100\%$

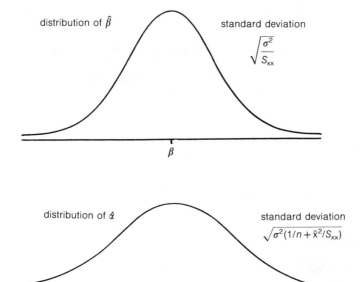

Fig. 8.4 Normal distributions of estimators of (a) slope and (b) intercept.

confidence interval for β is given by

$$\hat{\beta} \pm t_{n-2,\varepsilon/2}\sqrt{(s^2/\sum(x_i - \bar{x})^2)}$$

where we use ε instead of the more usual α to avoid confusion with the intercept. Given any value of x, x_p, the mean value of Y is $\alpha + \beta x_p$. This is estimated by $\hat{\alpha} + \hat{\beta}x_p$, which, for our present purposes, is more conveniently written as

$$\bar{Y} + \hat{\beta}(x_p - \bar{x})$$

The variance of this quantity is

$$\text{var}(\bar{Y}) + (x_p - \bar{x})^2 \,\text{var}(\hat{\beta})$$

since $\bar{Y}$ and $\hat{\beta}$ are independent. Now

$$\bar{Y} = \sum Y_i/n$$

and the variance of the Y_i – remember they are conditional on x_i – is σ^2. Furthermore, the errors are assumed to be independent, so

$$\text{var}(\bar{Y}) = n\sigma^2/n^2 = \sigma^2/n$$

To summarize,

$$\hat{\alpha} + \hat{\beta}x_p \sim N(\alpha + \beta x_p, \sigma^2(1/n + (x_p - \bar{x})^2/\sum(x_i - \bar{x})^2))$$

and a $(1 - \varepsilon) \times 100\%$ confidence interval for the mean value of Y given that x equals x_p is

$$\hat{\alpha} + \hat{\beta}x_p \pm t_{n-2,\varepsilon/2}s\sqrt{(1/n + (x_p - \bar{x})^2/\sum(x_i - \bar{x})^2)}$$

The result for the intercept is obtained by putting x_p equal to 0, but it is not always of practical interest.

Example 8.1 (continued)

A 95% confidence interval for the slope of the regression of abrasion loss on hardness is

$$-3.161 \pm 2.069\sqrt{(1759/14\,672)}$$

which gives

$$[-3.88, -2.44]$$

A 95% confidence interval for the mean value of Y when x equals its mean of 701.08 is

$$476.9 \pm 2.069\sqrt{(1759/25)}$$

which gives

$$[460, 494]$$

The unbiased properties of the estimators depend on assumptions A1 and A2. The construction of the confidence intervals depends on A1–A4. If errors are positively correlated the calculated confidence intervals will be too narrow. The assumption of normality is not very critical because we can rely on the central limit theorem. From the plot of the data the assumptions of linearity and constant variance seem reasonable for the tungsten shields. If the data could be sensibly ordered, chronologically for example, the assumed independence of the errors (A3) could be tested by plotting r_{i+1} against r_i and calculating the correlation between r_i and r_{i+1}. No ordering is known for the tungsten shields data, so this test cannot be done, but there is no reason to suspect correlated errors.

8.1.5 Predictions

Confidence intervals for the mean value of Y given x, and prediction intervals for a single value of Y given x, rely on the assumption that the model remains valid. For this reason they should be restricted to values of x_p within the range of x-values used in fitting the regression unless it is known that the assumption of a linear relationship remains reasonable beyond these limits. For example, it could be disastrous to assume that an empirical linear relationship established between extension and load of a steel cable will continue to hold for higher

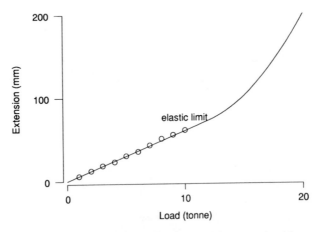

Fig. 8.5 Extension plotted against load for a steel cable.

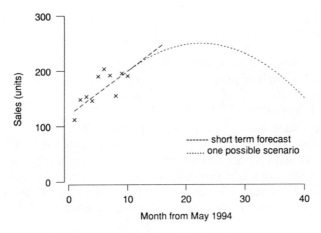

Fig. 8.6 Sales plotted against time.

loads than those used in the experiment (Fig. 8.5). In contrast to this, short-term forecasts often have to be made by assuming that present linear trends will continue, because there is no other information. This is reasonable if underlying trends change 'slowly' compared with the 'short term' (Fig. 8.6). The data themselves cannot justify assuming the linear relationship remains realistic beyond the limits of available data, and if extrapolation cannot be avoided, the assumption is subjective. The following intervals, wide though they may be, only allow for uncertainty in the estimates of the parameters α, β and σ^2 and depend on the underlying assumption of a linear model. Prediction intervals

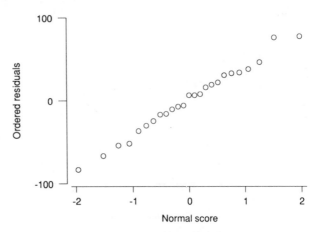

Fig. 8.7 Normal score plot for residuals from regression of abrasion loss on Vickers hardness.

are sensitive to the assumption of normality and this should be tested by a normal score plot of the residuals (Fig. 8.7).

Prediction interval for a single observation given x
A $(1 - \varepsilon) \times 100\%$ limit of prediction for a single observation Y when $x = x_p$ is:

$$(\hat{\alpha} + \hat{\beta} x_p) \pm t_{n-2, \varepsilon/2} s \sqrt{1 + 1/n + (x_p - \bar{x})^2 / \sum (x_i - \bar{x})^2}$$

This is the confidence interval for the mean value with the addition of 1 under the square-root sign to allow for the variance of a single value of Y about its mean. We add this variance to the variance which accounts for uncertainty in the estimates of α and β, and then take the square root, because variation about the mean is independent of errors in estimating the mean. Notice that even if the parameters were known exactly a 95% prediction interval for Y would be

$$\alpha + \beta x_p \pm 1.96\sigma$$

Example 8.1 (continued)

A 95% confidence interval for the mean abrasion loss of erosion shields with hardness 670 is

$$476.9 - 3.161(670 - 701.08) \pm 2.069 \times 41.94 \sqrt{(1/25 + (670 - 701.08)^2/14672)}$$

which gives

$$575 \pm 28$$

A 95% prediction interval for the abrasion loss of a single erosion shield with

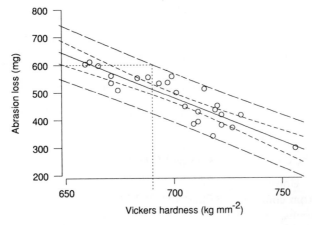

Fig. 8.8 Confidence limits for the mean, and limits of prediction, for the erosion shields.

a hardness measurement of 670 is

$$575 \pm 91$$

The width of a confidence interval for the mean value of Y, given x equal x_p, becomes wider as x_p moves further from $\bar{x}$. This is because of the $(x_p - \bar{x})^2$ term which reflects the increasing influence of the uncertainty in the estimate of the slope. There is a similar, but less marked, effect for limits of prediction. If the limits of prediction and confidence intervals are calculated for a few values of x they can be superimposed on the plot of the data and joined with smooth curves. This has been done for 95% intervals with the erosion shield data in Fig. 8.8.

8.1.6 MINITAB regression command

Read the Vickers hardness and abrasion loss pairs into C1 and C2, respectively, and name the variables 'hardness' and 'abrloss'. First plot the data (Fig. 8.1).

PLOT   C2   C1

All the calculations so far are covered by the following command and sub-command:

REGRESS   C2   1   EXPLAN   C1   STDRES   C98   FITS   C99
RESIDUALS   C5;
PREDICT   670.

The regression procedure in Release 9 of MINITAB has been considerably enhanced. While you can type the above commands in the session window, using the mouse allows you to see more of the options available.

Stat ▶   Regression ▶   Regression
   Response: C2
   Predictors: C1
   Storage ☐ Residuals ☐ Standard. resids ☐ Fits
   Options
   Prediction intervals for new observations: 670

The residuals, standardized residuals, and fits end up in columns

   RESI1, SRES1, FITS1

They could be copied to C5, C98 and C99 by

   Manip ▶ Copy
     Copy from columns: RESI1 SRES1 FITS1
     To columns: C5   C98   C99

Notice you must specify that there is only one explanatory variable, but storing the standardized residuals (see below) and fitted values are options. The output is shown in Fig. 8.9. The 'Coef' column contains $\hat{\alpha}$ and $\hat{\beta}$ and the 'Stdev' column

---

MTB   > regress c2 1 explanatory var c1 stdres c98 fits c99;
SUBC > residuals c5;
SUBC > predict when x is 670.

The regression equation is
abrloss = 2693 − 3.16 hardness

Predictor	Coef	Stdev	t-ratio	p
Constant	2692.8	242.9	11.09	0.000
hardness	−3.1607	0.3462	−9.13	0.000

s = 41.94   R-sq = 78.4%   R-sq(adj) = 77.4%

Analysis of Variance

SOURCE	DF	SS	MS	F	p
Regression	1	146569	146569	83.34	0.000
Error	23	40451	1759		
Total	24	187020			

Unusual Observations

Obs	hardness	abrloss	Fit	Stdev.Fit	Residual	St.Resid
4	756	297.00	303.34	20.78	− 6.34	−0.17 X
9	718	340.00	423.44	10.23	−83.44	−2.05R

R denotes an obs. with a large st. resid.
X denotes an obs. whose X value gives it large influence.

Fit	Stdev.Fit	95% C.I.	95% P.I.
575.15	13.64	(546.92, 603.38)	(483.90, 666.40)

Fig. 8.9 MINITAB regression analysis of (hardness, abrasion loss) pairs from 25 erosion shields.

contains their estimated standard deviations. The t-ratio is the ratio of the coefficient to its standard deviation. The 'p' column contains the probability of such a large absolute value of the t-ratio if the value of the parameter in the hypothetical population were to be zero. This is known as the **p-value**. In this example we know α is positive and have good physical reasons for expecting β to be negative, so the p-value is more an indication of the precision of our estimate. A 95% confidence interval for β is given by

$$-3.1607 \pm 0.3462 t_{0.025,23}$$

In general, if the p-value exceeds 0.05 the 95% confidence interval for β will include 0. 'R-sq' and the 'Analysis of Variance' table are explained in the next section. Two 'Unusual Observations' are flagged. The pair (756, 297) has a relatively large influence on the estimate of the slope because the hardness value of 756 is furthest from the mean of the hardness values. The **standardized residual** corresponding to the pair (718, 340) is, approximately, the residual divided by the estimated standard deviation of the errors (s). If the errors are normally distributed we would expect an average of 1 in 20 standardized residuals to have absolute values greater than about 2, so there is nothing remarkable about the value of -2.05. An exact description of standardized residuals (optionally put into C98) is given in the next chapter. You should check that the 95% confidence interval and 95% prediction interval correspond to the confidence interval for the mean value of abrasion loss, given a hardness of 670, and the prediction interval for a single value which we have just calculated. Release 9 of MINITAB can plot the data, fitted regression line, the envelope curves for confidence and prediction intervals, and save the graph in a file. You need to type the following in the session window:

```
% FITLINE   C2   C1;
  GSAVE 'filename'.
```

Example 8.1 (continued)

The specification for abrasion losses of the erosion shields is that only 1 in 1000 should exceed 700 mg. In the past, random samples of 20 shields have been taken from each batch of 1600 and the abrasion losses $\{y\}$ have been measured. It has been assumed that abrasion losses within a batch are normally distributed, with mean μ_Y and standard deviation σ_Y, in which case an average of 99.9% shields per batch have losses less than

$$\mu_Y + 3.09 \sigma_Y$$

A batch is acceptable if this upper 0.1% point is less than 700. It has been estimated by

$$\bar{y} + 3.09 s_y$$

where $\bar{y}$ is the sample mean and s_y is the sample estimate of the within-batch standard deviation. This statistic has a variance (Exercise 6.11) of

$$\sigma_Y^2/20 + 3.09^2\sigma_Y^2/(2 \times 20)$$

The rule was to accept a batch if the inspector was 90% confident that the upper 0.1% point was less than 700, a requirement that

$$\bar{y} + 3.09s_y + t_{0.1,19}s_y\sqrt{(1/20 + 9.55/40)} < 700$$

which is equivalent to

$$\bar{y} + 3.8s_y < 700$$

Typical values for $\bar{y}$ and s_y are 460 and 50, but both the mean and standard deviation vary significantly from batch to batch.

The objective of fitting the regression analysis was to try to replace direct measurement of abrasion loss by Vickers hardness measurements $\{x\}$. If we do this we can predict the abrasion losses from our regression line to obtain a sample of estimated abrasion losses $\{\hat{y}\}$, given by

$$\hat{y}_i = 2693.1 - 3.161x_i$$

We can then proceed as before, using the $\hat{y}_i$ instead of directly measured y_i. The snag is that the standard deviation of the estimates $(s_{\hat{y}})$ is not a reliable estimate of σ_Y. The estimates $\hat{y}$ all lie exactly on the regression line, whereas individual values of abrasion loss are scattered about their expected values with a standard deviation of σ. This variation is independent of the variation in the $\hat{y}$, so we can estimate σ_Y^2 by

$$s_y^2 = s_{\hat{y}}^2 + s^2$$

We have established that there is a useful relationship between hardness and abrasion loss. It would now be a good idea to improve our estimate of the line by measuring both abrasion losses and hardnesses of at least another 50 shields. This would only have to be done once, and occasionally checked in case the process changed, and is not an unreasonable requirement when compared with measuring abrasion losses of 20 shields from every batch. As it is much easier to measure hardness than abrasion loss, and the test is not destructive, we could also afford to increase this sample size considerably.

We could put our regression analysis to a different use. For some critical application – an aircraft engine with heavy use for long-haul flights, perhaps – we might prefer that all the erosion shields have losses of less than 600 mg. If we draw a line on Fig. 8.8 through 600 parallel to the x-axis, we see it intersects the upper curve of the 95% prediction interval where the corresponding value of Vickers hardness is 690. We could measure the hardness of all the shields and select those with values greater than $690 \, \text{kg mm}^{-2}$. We would then expect 97.5% of the shields to have losses less than the preferred upper limit.

8.1.7 Summarizing the algebra

The analysis of variance (ANOVA) table is a convenient summary of the main results. The original corrected sum of squares of the y can be split into the sum of squared residuals, known as the **residual sum of squares** or **error sum of squares** (RSS), and a sum of squares that is attributed to the regression.

$$\sum(y_i - \bar{y})^2 = \sum((y_i - \hat{y}_i) + (\hat{y}_i - \bar{y}))^2$$
$$= \sum(y_i - \hat{y}_i)^2 + \sum(\hat{y}_i - \bar{y})^2 + 2\sum(y_i - \hat{y}_i)(\hat{y}_i - \bar{y})$$

The cross product term is equal to zero because it can be written

$$2\sum r_i(\hat{\beta}(x_i - \bar{x})) = 2\hat{\beta}\sum r_i(x_i - \bar{x}) = 0$$

and it follows that

$$\sum(y_i - \bar{y})^2 = \sum(\hat{y}_i - \bar{y})^2 + \sum r_i^2$$
$$= \text{regression sum of squares} + \text{residual sum of squares}$$

In the following ANOVA table the mean square is the 'sum of squares' divided by the 'degrees of freedom'. One degree of freedom is allocated to the regression sum of squares because there was one explanatory variable. Note that the $E[\text{mean square}]$ column includes the unknown parameters of the model and tells us, for example, that on average s^2 would equal σ^2 (Appendix A5). It is therefore algebraic rather than numerical and is not usually given by computer packages. The proof of the expected value of the regression mean square is straightforward (Exercise 5.11(ii)). Start from

$$\text{var}(\hat{\beta}) = E[(\hat{\beta} - \beta)^2]$$
$$= E[\hat{\beta}^2 - 2\hat{\beta}\beta + \beta^2]$$

but $E[\hat{\beta}] = \beta$ so

$$\text{var}(\hat{\beta}) = E[\hat{\beta}^2] - \beta^2$$

Now

$$\text{var}(\hat{\beta}) = \sigma^2 / \sum(x_i - \bar{x})^2$$

The denominator is a constant and the result follows.

Source of variation	Corrected sum of squares	Degrees of freedom	Mean square	$E[\text{mean square}]$
Regression	$\sum(\hat{y}_i - \bar{y})^2$	1	$\hat{\beta}^2\sum(x_i - \bar{x})^2$	$\sigma^2 + \beta^2\sum(x_i - \bar{x})^2$
Residual	$\sum r_i^2$	$n - 2$	s^2	σ^2
Total	$\sum(y_i - \bar{y})^2$	$n - 1$		

Example 8.1 (continued)

The MINITAB analysis of variance for the erosion shields is shown in Fig. 8.9. The residual row is referred to as 'Error'. The 'F' column contains the F-ratio, the ratio of the regression mean square to the error mean square, and the 'p' is the probability of obtaining such a large value if we hypothesise that the slope $\beta = 0$, in which case the expected values of the numerator and denominator would be equal. MINITAB also gives the value of R^2, which is the proportion of the corrected sum of squares of the y accounted for by the regression (78.4%).

8.2 Intrinsically linear models

Tests on steel specimens subject to fluctuating loading have indicated that the number of cycles to failure (N) is approximately inversely proportional to the stress range (S, equal to maximum stress less minimum stress) raised to some power between 3 and 4. That is

$$N = kS^{-m}$$

Appropriate values of k and m depend on the type and thickness of steel, and are obtained experimentally. The proposed nonlinear relationship can be transformed to a straight line by taking logarithms. Then

$$\ln N = \ln k - m \ln S$$

and if x and y are defined as $\ln S$ and $\ln N$, respectively, standard regression techniques can be used. If it is assumed that

$$\ln N_i = \ln k - m \ln S_i + E_i$$

where the E_i satisfy the usual assumptions, then

$$N_i = kS_i^{-m} H_i$$

where H_i has a lognormal distribution with a median value of 1. If there is good reason to suppose that a model

$$N_i = kS_i^{-m} + E_i$$

is more appropriate, and the errors are significant, it can be fitted by numerical minimization of the least-squares function. The statistical theory needed to obtain estimates of the standard deviations of such estimators is more complicated than for the linear regression covered in this chapter, but many statistical packages offer routines to carry out the calculations. Such nonlinear least-squares or nonlinear regression routines are usually used to estimate parameters in relationships which cannot be reduced to straight lines. An example is the logistic curve

$$y = \frac{a}{1 + br^t} - c \qquad \text{for } 0 < t$$

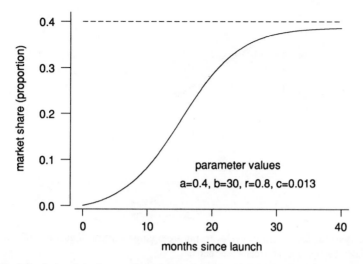

Fig. 8.10 A logistic curve for market share of a new design of self-tuning digital radio.

The parameters a, b, and r must all be positive, r must be less than 1, and c would usually be chosen so the curve starts at the origin. The general shape is indicated in Fig. 8.10. It is sometimes used to model market share of a new product over time. Unfortunately, while it is easy to fit retrospectively, estimates

Table 8.2 Amplitude of fluctuating stress and cycles to failure for 15 steel specimens

Amplitude of stress ($N m^{-2}$)	Cycles to failure ($\times 1000$)
500	20
450	19
400	19
350	40
300	48
250	112
200	183
150	496
100	1 883
90	2 750
80	2 181
70	3 111
60	9 158
50	15 520
40	47 188

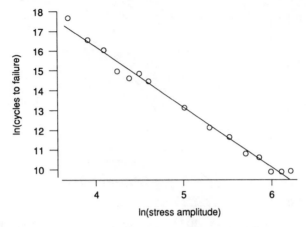

Fig. 8.11 Natural logarithm of cycles to failure against natural logarithm of amplitude of fluctuating stress for 15 steel specimens.

of the parameters made when it is in the initial or rapidly increasing phases are rather unreliable!

Example 8.2

Fifteen steel specimens with the same geometry were prepared and allocated to fluctuating loads of different amplitudes. The number of cycles to failure were recorded. The data are given in Table 8.2 and a plot of $\ln N$ against $\ln S$ is shown in Fig. 8.11. A linear relationship between the log variables is plausible and the fitted regression line is

$$\ln N = 28.4474 - 3.0582 \ln S$$

The estimates of k and m, for this steel in this geometry, are 2.26×10^{12} and 3.06, respectively.

8.3 Conditional distributions

We start by looking at bivariate distributions, which were introduced in Chapter 5. The concepts generalize in a natural way to higher-order multivariate distributions but the straightforward geometric interpretations are then lost. As their name suggests, conditional distributions are defined in terms of conditional probability.

8.3.1 Conditional distributions for discrete variables

Assume that the discrete random variables X and Y have a bivariate probability distribution $P(x, y)$. From the definition of conditional probability,

$$\Pr(Y = y | X = x) = \Pr(X = x \text{ and } Y = y)/\Pr(X = x)$$

It only remains to introduce the notation for the conditional distribution defined by this equation:

$$P(y|x) = P(x, y)/P_X(x)$$

Example 8.3

The following discrete distribution is a simple model for rainfall events in a catchment. The events are classified by duration (x), which can be short, medium and long, and by the amount of water (y), which can be small (S), medium (M) and large (L). The random variables X and Y and their bivariate distribution are given in Table 8.3. To find the conditional distribution of Y when $x = 3$ we first need to calculate $P_X(3)$, which equals 0.7. Then

$$P(y|3) = P(3, y)/0.7 \qquad \text{for } y = 1, 2, 3$$

For example, $P(1|3)$ equals 0.143. This is less than the value of 0.3 which can be calculated for $P_Y(1)$, because small amounts of rain are less likely to occur in long-duration rainfall events.

8.3.2 Conditional distributions for continuous variables

Assume that the continuous random variables X and Y have a bivariate probability density function $f(x, y)$. The conditional distribution of Y given x is defined by

$$f(y|x) = f(x, y)/f_X(x)$$

Table 8.3 Distribution of rainfall events

			Amount (y)		
			S 1	M 2	L 3
Duration (x)	S	1	0.1		
	M	2	0.1	0.1	
	L	3	0.1	0.2	0.4

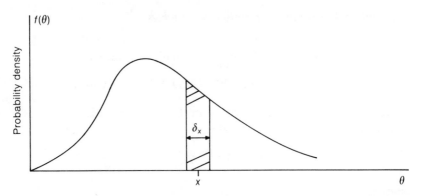

Fig. 8.12 Probability that X lies within a line segment of length δx centred on x.

This definition can be justified in terms of conditional probability.

$$\Pr(Y \text{ lies within } \delta y \text{ of } y | X \text{ lies within } \delta x \text{ of } x)$$

$$= \frac{\Pr(Y \text{ lies within } \delta y \text{ of } y \text{ and } X \text{ lies within } \delta x \text{ of } x)}{\Pr(X \text{ within } \delta x \text{ of } x)}$$

It can be seen in Fig. 8.12 that

$$\Pr(X \text{ lies within } \delta x \text{ of } x) \simeq f_X(x)\delta x$$

Similar results hold for the bivariate and conditional distributions, and the conditititional probability statement can be written as:

$$f(y|X \text{ lies within } \delta x \text{ of } x)\delta y \approx \frac{f(x, y)\delta x\, \delta y}{f_X(x)\delta x}$$

Letting δx and δy tend to zero, in which case the approximation becomes exact, and noting the cancellation gives

$$f(y|x) = f(x, y)/f_X(x)$$

The conditional distribution when $x = x_p$ can be interpreted geometrically as a section of the joint pdf through x_p, parallel to the (y, z) plane, and scaled by $1/f_X(x_p)$ so that the area equals 1. This is shown in Fig. 8.13.

Example 8.4

A petrol station sells quantities X of leaded fuel and Y of unleaded fuel each month. X and Y, both measured in millions of litres, have the joint pdf,

$$f(x, y) = 2(x + 2y)/3 \qquad \text{for } 0 < x, y < 1$$

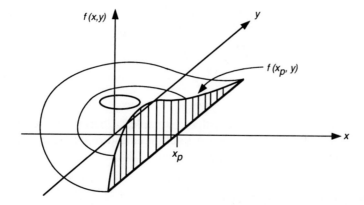

Fig. 8.13 Section through a bivariate pdf cut parallel to y–z plane.

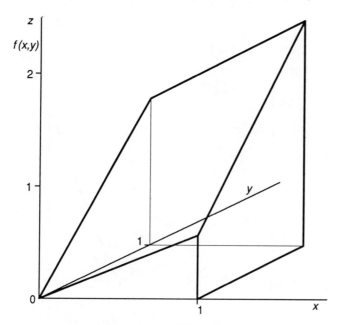

Fig. 8.14 Bivariate distribution $f(x, y) = 2(x + 2y)/3$ defined over unit square.

which is shown in Fig. 8.14. The conditional distribution of y given x is

$$f(y|x) = [2(x + 2y)/3]/[2(x + 1)/3]$$
$$= (x + 2y)/(x + 1) \qquad \text{for } 0 < y < 1$$

To check that it is a pdf, we can integrate to obtain

$$\int_0^1 \frac{x + 2y}{x + 1} \, dy = \frac{[xy + y^2]_0^1}{x + 1} = 1$$

An example calculation is the probability that the unleaded fuel sales are less than 0.5 if the sales of leaded fuel are 0.8. This is given by the integral

$$\int_0^{0.5} \frac{0.8 + 2y}{0.8 + 1} \, dy = 0.361$$

Example 8.5

This is slightly more complicated than Example 8.4 because the domain of y depends on x. Let X and Y represent total manufacturing time and inspection time for a certain article, and suppose X and Y have the joint distribution

$$f(x, y) = 8xy \qquad \text{for } 0 < x < 1, 0 < y < x$$

which is shown in Fig. 8.15. Then

$$f(x) = \int_0^x 8xy \, dy = 4x^3$$

$$f(y) = \int_y^1 8xy \, dx = 4y(1 - y^2)$$

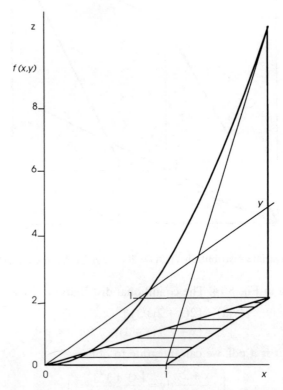

Fig. 8.15 Bivariate distribution $f(x, y) = 8xy$ defined over triangle.

The conditional distribution of y on x is

$$f(y|x) = 2y/x^2 \qquad \text{for } 0 < y < x$$

Notice that X and Y are not independent.

8.3.3 The bivariate normal distribution

The bivariate normal distribution has the rather formidable looking pdf

$$f(x, y) = \frac{1}{2\pi\sigma_X\sigma_Y\sqrt{(1-\rho^2)}}$$

$$\times \exp\left\{-\frac{1}{2(1-\rho^2)}\left[\left(\frac{x-\mu_X}{\sigma_X}\right)^2 - 2\rho\left(\frac{x-\mu_X}{\sigma_X}\right)\left(\frac{y-\mu_Y}{\sigma_Y}\right) + \left(\frac{y-\mu_Y}{\sigma_Y}\right)^2\right]\right\}$$

The parameters μ_X, μ_Y, σ_X and σ_Y are the means and standard deviations of the marginal distributions of X and Y. The parameter ρ is the correlation between X and Y. The easiest way to find the conditional distribution of y on x is to use the standardized distribution which has a mean of 0 and a standard deviation of 1. The general result then follows from a simple scaling argument.

Suppose that (X, Y) has a bivariate normal distribution. Then if W and Z are defined by

$$W = (X - \mu_X)/\sigma_X \quad \text{and} \quad Z = (Y - \mu_Y)/\sigma_Y$$

(W, Z) has a standardized bivariate normal distribution and the pdf is

$$f(w, z) = \frac{1}{2\pi\sqrt{(1-\rho^2)}}\exp\left\{\frac{-1}{2(1-\rho^2)}(w^2 - 2\rho wz + z^2)\right\}$$

The marginal distributions are standard normal. For example, write $(w^2 - 2\rho wz + z^2)$ as $[(z - \rho w)^2 + (1 - \rho^2)w^2]$ then substitute θ for $(z - \rho w)$ and integrate to obtain

$$f_W(w) = \frac{1}{\sqrt{2\pi}}\exp(-\tfrac{1}{2}w^2)$$

Some straightforward algebra leads to the conditional distribution,

$$f(z|w) = \frac{1}{\sqrt{2\pi}\sqrt{(1-\rho^2)}}\exp\left\{\frac{-1}{2(1-\rho^2)}(z - \rho w)^2\right\}$$

This is a normal distribution with mean, $E[Z|w]$, equal to ρw and a variance of $(1 - \rho^2)$. The regression line of z on w is

$$z = \rho w$$

This result can be rescaled so it is explicitly in terms of Y and x. First, use the relationship between Z and Y to write

$$E\left[\frac{Y - \mu_Y}{\sigma_Y}\middle| W = w_p\right] = \rho w_p$$

If W is w_p then X equals $\mu_X + \sigma_X w_p$, and we define x_p by $\mu_X + \sigma_X w_p$. Then

$$E\left[\left.\frac{Y - \mu_Y}{\sigma_Y}\right| X = x_p\right] = \rho \frac{x_p - \mu_X}{\sigma_X}$$

and finally

$$E[Y|X = x_p] = \mu_Y + \rho \frac{\sigma_Y}{\sigma_X}(x_p - \mu_X)$$

Furthermore, if the conditional distribution of Z given w has a variance of $(1 - \rho^2)$, the conditional distribution of Y given x has a variance $(1 - \rho^2)\sigma_Y^2$. Notice this variance does not depend on the value of x. The regression line of Y on x is

$$y = \mu_Y + \rho \frac{\sigma_Y}{\sigma_X}(x - \mu_X)$$

An identical argument leads to the regression line of X on y which is

$$x = \mu_X + \rho \frac{\sigma_X}{\sigma_Y}(y - \mu_Y)$$

The two regression lines are not the same. Since $|\rho| < 1$, the regression line of y on x is less steep than the major axis of the elliptical contours of the bivariate distribution shown in Fig. 8.16, whereas the regression line of x on y is steeper.

The following argument explains why the regression of y on x crosses contours at the points where their tangents are parallel to the y-axis. To begin with, $f(y|x_p)$ is a scaled section of the joint pdf $f(x, y)$ cut through x_p parallel to the (y, z) plane. It is also a normal distribution so its mean coincides with its highest point. The highest point above the line through x_p parallel to the y-axis is where the line is tangent to a contour. The regression line consists of all these highest points. Although the diagram was drawn with a positive ρ, similar arguments hold for negative values.

To conclude, if x is σ_X above μ_X the mean value of Y conditional on x is closer to μ_Y than σ_Y, except in the degenerate case when $|\rho| = 1$. This is known as 'regression towards the mean'. It is a geometric property of the bivariate normal distribution, and could have been anticipated from the following common-sense observation. If $|\rho| \neq 1$ the value of Y is not exactly related to the given x and other things will tend to bring its average value closer to the mean of all the Y-values. Suppose a car is driven at two standard deviations above the mean on a given stretch of road (Fig. 8.16). The square root of stopping distance will not, on average, be as extreme as two standard deviations above the mean for all cars. This is because other factors such as driver reaction time, standard of maintenance and potential efficiency of brakes are unlikely to be as far from their means as two standard deviations.

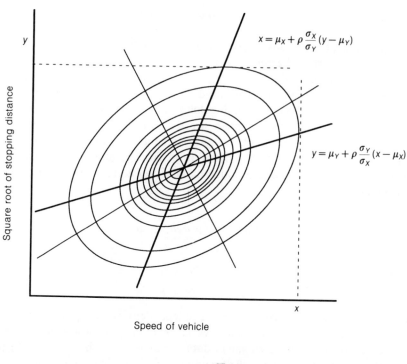

$$x = \mu_X + \rho \frac{\sigma_X}{\sigma_Y}(y - \mu_Y)$$

$$y = \mu_Y + \rho \frac{\sigma_Y}{\sigma_X}(x - \mu_X)$$

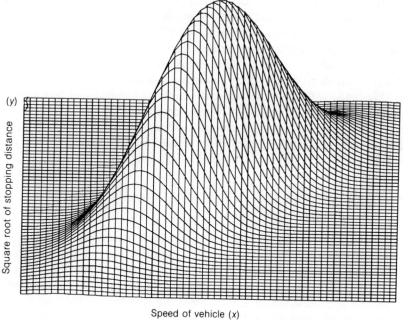

Fig. 8.16 Contours of bivariate normal pdf and regression lines and perspective plot.

8.4 Relationship between correlation and regression

The measured abrasion loss of shields for turbine blades would vary even if they were all the same hardness. This variation would be partly due to other differences in the test pieces (that is, inherent variation in the population being sampled), but measurement error could also make a noticeable contribution. The experiment in which fluctuating loads were applied to steel specimens involved measurement of the applied stresses and the times to fracture. These are relatively easy measurements to make and the measurement errors should be negligible compared with inherent variation in the population. The distinction between measurement error and inherent variation is crucial for the remainder of this chapter.

8.4.1 Values of x are assumed to be measured without error and can be preselected

The model for a linear regression of Y on x is

$$Y_i = \alpha + \beta x_i + E_i$$

with E_i having a zero mean and being independent of the x_i and each other. The errors, E_i, may represent inherent variation in the population, measurement error, or both. If the tungsten shield data are assumed to be a sample from a bivariate normal distribution, the assumptions of the linear regression model are all satisfied. However, the model merely describes the distribution of Y for fixed values of x and there is no reason why these values of x cannot be chosen in advance by the experimenter. In fact there are considerable advantages to be gained by doing so. The x-values should be chosen to cover the range of values over which predictions may be required, and an even spacing of x-values over this range will allow us to assess whether a straight-line relationship is plausible. Although the values of x may be chosen by the investigator, the experimental material must be randomly allocated to these values. Also, if the tests are made consecutively their order should be randomized to help justify the first three assumptions A1–A3 made about the E_i.

8.4.2 The data pairs are assumed to be a random sample from a bivariate normal distribution

Assume that (X, Y) have a bivariate normal distribution and are measured without error. If a random sample from this distribution is available, either the regression of Y on x or the regression of X on y can be estimated. The choice depends on which variable the investigator wishes to predict. You should note that the assumptions of a random sample are not compatible with choosing the values of x in advance. If the values of x are chosen in advance only the regression of Y on x, discussed in section 8.4.1, is appropriate.

An alternative to either of the regressions is a correlation analysis. This is appropriate if a measure of the association between the variables is required rather than an equation for making predictions. In a correlation analysis both X and Y are treated as random variables, whereas the regression of Y on x estimates the conditional distribution of Y for given x, specific values of which happen to have arisen at random. The following approximate distribution of the sample correlation (r), due to R.A. Fisher (1921), can be used to construct confidence intervals for ρ:

$$\operatorname{arctanh}(r) \sim N(\operatorname{arctanh}(\rho), 1/(n-3))$$

Example 8.6

The following data are measurements (in parts per million) of chromium and nickel in a random sample of six pieces of shale from a particular site.

Identifier of piece	A	B	C	D	E	F
Chromium content (x)	400	490	380	440	460	470
Nickel content (y)	130	165	95	140	135	145

They are plotted in Fig. 8.17, and the calculated value of r is 0.901. If the data are assumed to be a random sample from a bivariate normal distribution, a 95% confidence interval for $\operatorname{arctanh}(\rho)$ is

$$\operatorname{arctanh}(0.901) \pm 1.96\sqrt{(1/3)}$$

which becomes

$$1.478 \pm 1.132$$

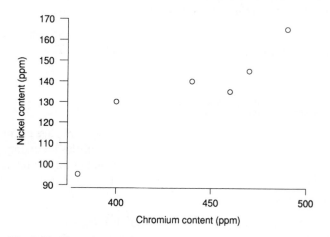

Fig. 8.17 Chromium and nickel contents of six pieces of shale.

and this gives

[0.346, 2.609]

If we are 95% confident that arctanh(ρ) is between 0.346 and 2.609, we are equally confident that ρ is between tanh(0.346) and tanh(2.609). A 95% confidence interval for ρ is therefore

[0.33, 0.99]

8.5 Fitting straight lines when both variables are subject to error

A project has been designed to investigate any systematic difference between land and aerial survey methods. Let X_i and Y_i represent n measurements to be made by land and aerial survey, respectively. It is assumed that they are subject to independent errors, E_i and H_i, respectively, of zero mean and constant variances σ_E^2 and σ_H^2. Let u_i and v_i represent the unobservable error-free measurements. It is supposed that

$$v_i = \alpha + \beta u_i$$

for some α and β, and one of the main aims of the project is to see whether there is any evidence that these parameters differ from 0 and 1, respectively.

Substituting

$$u_i = X_i - E_i \quad \text{and} \quad v_i = Y_i - H_i$$

into the model relating v_i and u_i gives

$$Y_i = \alpha + \beta X_i + (H_i - \beta E_i)$$

Despite first appearances, this is not the standard regression model because the assumption that the errors are independent of the predictor variable is not satisfied:

$$\text{cov}(X_i, (H_i - \beta E_i)) = E[(X_i - u_i)(H_i - \beta E_i)]$$
$$= E[(E_i(H_i - \beta E_i)] = -\beta\sigma_E^2$$

8.5.1 Maximum likelihood solution

The usual approach to this problem is to assume the errors are normally distributed and that the ratio of σ_H^2 to σ_E^2, denoted by λ in the following, is known. In practice, some reasonable value has to be postulated for λ, preferably based on information from replicate measurements. The (maximum likelihood) estimates of α and β are:

$$\hat{\beta} = \{(S_{yy} - \lambda S_{xx}) + [(S_{yy} - \lambda S_{xx})^2 + 4\lambda S_{xy}^2]^{1/2}\}/(2S_{xy})$$
$$\hat{\alpha} = \bar{y} - \hat{\beta}\bar{x}$$

where S_{xy} is a shorthand for $\sum(x - \bar{x})(y - \bar{y})$, etc. (These follow from the likelihood function L (that is, the joint pdf treated as a function of the unknown parameters) which is proportional to

$$\sigma_E^{-1}\sigma_H^{-1}\exp\left\{-\tfrac{1}{2}\sigma_E^{-2}\sum_{i=1}^{n}(x_i - u_i)^2 - \tfrac{1}{2}\sigma_H^{-2}\sum_{i=1}^{n}(y_i - (\alpha + \beta u_i))^2\right\}$$

if α, β, σ_E, and u_i are treated as the unknown parameters.)

Estimates of σ_E^2 and σ_H^2 are given by

$$\hat{\sigma}_H^2 = \lambda\hat{\sigma}_E^2 = (S_{yy} - \hat{\beta}S_{xy})/(n - 2)$$

If $\lambda = 1$ the values of $\hat{\alpha}$ and $\hat{\beta}$ are the slope and intercept of the line such that the sum of squared perpendicular distances from the plotted points to it is a minimum.

Upper and lower points of the $(1 - \alpha) \times 100\%$ confidence interval for β are given by

$$\lambda^{1/2}\tan(\arctan(\hat{\beta}\lambda^{-1/2}) \pm \tfrac{1}{2}\arcsin(2t_{\alpha/2}\theta))$$

where

$$\theta^2 = \frac{\lambda(S_{xx}S_{yy} - S_{xy}^2)}{(n - 2)((S_{yy} - \lambda S_{xx})^2 + 4\lambda S_{xy}^2)}$$

and $t_{\alpha/2}$ is the upper $(\alpha/2) \times 100\%$ point of the t-distribution with $n - 2$ degrees of freedom. An approximation to the standard deviation of $\hat{\beta}$ is given by one-quarter of the width of this interval. Since $\hat{\beta}$ is independent of $\bar{x}$

$$\mathrm{var}(\hat{\alpha}) \simeq \hat{\sigma}_H^2/n + \bar{x}^2\,\mathrm{var}(\hat{\beta}) + \hat{\beta}^2\hat{\sigma}_E^2/n$$

The use of these formulae is demonstrated in the following example.

Example 8.7

The data pairs in Table 8.4 are the measured heights (in metres) above sea level of 25 points from a land survey (x) and an aerial survey (y). The points were equally spaced over a hilly 10 km by 10 km area. Replicate measurements of the height of a single point suggest that errors in the aerial survey measurements have a standard deviation three times that of errors in the land survey. Calculations give

$$\bar{x} = 780.6 \qquad \bar{y} = 793.0$$
$$S_{xx} = 177\,970 \qquad S_{xy} = 179\,559 \qquad S_{yy} = 181\,280$$

λ is assumed to be 9. The formulae of this section lead to the following estimates

$$\hat{\alpha} = 5.379 \qquad \hat{\beta} = 1.008\,99$$
$$\hat{\sigma}_E = 0.716 \qquad \hat{\sigma}_H = 2.147$$

Table 8.4 Heights of 25 points determined by land and aerial surveys

Land survey estimate (x, m)	Aerial survey estimate (x, m)
720.2	732.9
789.5	804.9
749.7	760.5
701.5	712.3
689.2	702.0
800.5	812.8
891.2	902.7
812.8	820.0
780.6	793.6
710.5	720.2
810.4	825.6
995.0	1008.6
890.5	902.4
829.4	845.1
808.7	820.3
781.7	796.1
868.7	885.2
904.2	920.1
780.7	790.2
649.6	660.0
732.1	741.2
770.4	781.2
733.7	745.6
694.9	707.3
620.0	633.4

95% confidence intervals for α and β are

  $[-7.6, 18.3]$ and $[0.993, 1.025]$

respectively. There is no evidence of any systematic difference between the results of the two surveys because the confidence intervals for α and β include 0 and 1 respectively. (If a standard regression analysis of Y on x is used, which corresponds to λ tending to infinity, the parameter estimates would be:

  5.368,   1.008 93,   0   and   2.254

respectively. 95% confidence intervals for α and β would be

  $[-3.3, 14.0]$ and $[0.998, 1.020]$

respectively.)

8.6 Calibration lines

In the UK, a 'breathalyser' test may be administered to motorists who are suspected of having drunk more than the legal limit of alcohol. If this test is positive they are then required to take a clinical test. The results of the clinical test may be used as evidence in court, and refusal to take it is an offence in itself.

A 'breathalyser' instrument is to be calibrated against blood tests for alcohol. A random sample from a pool of male volunteers was asked to drink a specified volume, v, of alcohol, wait a specified time, T_1, take the 'breathalyser' test (y), wait a further specified time, T_2, and then provide a blood sample. The alcohol content of the blood, x, was measured (in milligrams of alcohol per 100 ml of blood). The values for v were chosen in advance to span a range from 0 to 9 units (equivalent to a bottle of ordinary-strength table wine) and volunteers were randomly assigned to values of v. T_1 and T_2 were independent random numbers drawn from realistic distributions. Although the calibration line will be used to predict x from the roadside breathalyser reading, y, it is not necessarily best to use the estimated regression line of x on y. The blood test gives a very accurate measure of the blood alcohol level, but the breathalyser reading is subject to considerable measurement error as well as variation in waiting times and the relationship between x and y for different men. This has been demonstrated by administering several breathalyser tests in quick succession to the same person. The assumption of independence of the errors and the explanatory variable, explained in section 8.5, is more plausible if we fit the regression of Y on x. This will give unbiased estimators of α and β in the model,

$$Y_i = \alpha + \beta x_i + E_i$$

where the E_i have zero mean and are independent of the x_i and each other. If the breathalyser gives a reading of y_p, an approximately unbiased estimate of the corresponding x is

$$x_p = (y_p - \hat\alpha)/\hat\beta$$

This follows from the argument that y_p is an unbiased estimator of $E[Y_p]$ and

$$E[Y_p] = \alpha + \beta x_p$$

It is not exactly unbiased because although $\hat\beta$ is unbiased for β, $1/\hat\beta$ is not unbiased for $1/\beta$. If the sample is large this is of little practical significance. A construction for what can loosely be called 'limits of prediction' is shown in the following example.

Example 8.8

Nineteen male volunteers were randomly allocated 19 drinks containing from 0 to 9 units of alcohol, in increments of half a unit. After waiting a randomly assigned length of time they took the breathalyser test (y) and after a further randomly assigned length of time, the alcohol content of their blood (x) was measured (milligrams of alcohol per 100 ml of blood). The results are given in Table 8.5 and the data are plotted in Fig. 8.18.

The estimated regression line is

$$y = 28.425 + 0.3023x$$

and the estimated standard deviation of the errors is 10.21. The line is drawn on Fig. 8.18 together with curves giving the 90% limits of prediction for Y given x. The construction of limits for x given y is shown for a case when the breathalyser reading is 80. The blood alcohol level would be estimated as 171 and a 90% prediction interval would be between 131 and 200. The formula for predicting x from y is

$$x = (y - 28.425)/0.3023$$

(This is substantially different from the regression of x on y which would give $x = -57.5 + 2.73y$.)

This motorist would certainly be required to take a clinical test as the legal maximum in the UK (1983) is 80 mg/100 ml blood.

Table 8.5 Blood alcohol levels measured from blood samples and breathalyser readings for 19 adult male volunteers

Blood alcohol (x, mg alcohol per 100 ml blood)	Breathalyser reading (y)
0	40
14	21
20	35
46	34
71	59
74	44
86	47
79	51
96	76
114	70
112	52
152	73
150	71
165	88
190	67
190	90
202	103
226	88
229	101

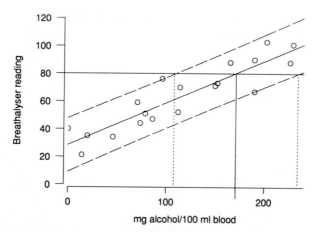

Fig. 8.18 Breathalyser reading plotted against blood alcohol determined by laboratory test for 19 adult males.

8.7 Summary

Regression model

The standard regression model is

$$Y_i = \alpha + \beta x_i + E_i$$

where the $E_i \sim N(0, \sigma^2)$ and are independent of the x_i and each other.

Estimating the parameters

$$S_{xy} = \sum (x_i - \bar{x})(y_i - \bar{y}), \text{ and } S_{xx}, S_{yy} \text{ are defined similarly.}$$
$$\hat{\beta} = S_{xy}/S_{xx} \qquad \hat{\alpha} = \bar{y} - \hat{\beta}\bar{x}$$
$$s^2 = (S_{yy} - S_{xy}^2/S_{xx})/(n-2)$$

The estimated regression line is

$$y = \hat{\alpha} + \hat{\beta}x = \bar{y} + \hat{\beta}(x - \bar{x})$$

Residuals

The residuals r_i are the differences between the y_i and their fitted values $\hat{y}_i$. That is:

$$\hat{y}_i = \hat{\alpha} + \hat{\beta}x_i \quad \text{and} \quad r_i = y_i - \hat{y}_i$$

They are estimates of the values taken by the errors, and are often referred to as 'errors' in computer output (MINITAB). The computational formula for s^2 is algebraically equivalent to $\sum r_i^2/(n-2)$.

Confidence interval for slope

A $(1-\alpha) \times 100\%$ confidence interval for β is given by

$$\hat{\beta} \pm t_{n-2,\alpha/2}\sqrt{(s^2/S_{xx})}$$

Confidence interval for mean of Y given x

If $x = x_p$, a $(1-\varepsilon) \times 100\%$ confidence interval for the mean value of Y is

$$\hat{\alpha} + \hat{\beta}x_p \pm t_{n-2,\varepsilon/2}s\sqrt{(1/n + (x_p - \bar{x})^2/S_{xx})}$$

Limits of prediction for a single value of Y given x

If $x = x_p$, $(1-\varepsilon) \times 100\%$ limits of prediction for a single value of Y are

$$\hat{\alpha} + \hat{\beta}x_p \pm t_{n-2,\varepsilon/2}s\sqrt{(1 + 1/n + (x_p - \bar{x})^2/S_{xx})}$$

Bivariate normal distribution

A regression analysis assumes that the errors in measuring the x are negligible. Apart from that restriction, the x can be selected in advance or arise as a result of sampling some bivariate distribution. The bivariate normal distribution is relatively easy to work with, and, even if it is not directly appropriate, it is often plausible for some transformation of variables such as the natural logarithm. If we have a random sample from a bivariate normal distribution we can regress Y on x, regress X on y, or calculate the correlation coefficient r as an estimate of the population correlation ρ. A $(1-\alpha) \times 100\%$ confidence interval for ρ is given by

$$\tanh[\operatorname{arctanh}(r) \pm z_{\alpha/2}/\sqrt{(n-3)}]$$

MINITAB

Suppose we have data pairs (x_i, y_i) and the x- and y-values are in C1 and C2, respectively. A typical regression analysis of y on x might proceed

```
PLOT       C2  C1
REGRESS    C2   1   C1;
  RESID    C5.
NSCO       C5  C6
PLOT       C5  C6
```

Predictions can be obtained using the subcommand **PREDICT**. Using the mouse:

Graph ▶ Plot...
Stat   ▶ Regression ▶ Regression...
Calc ▶ Mathematical Expressions...
Graph ▶ Plot...

Exercises

8.1 The following data are the sizes (x_i, thousand properties) and costs of meeting European Community requirements (y_i, coded monetary units) for eight water supply zones.

Size	1.0	2.3	4.5	5.1	6.7	6.8	7.2	9.3
Estimated cost	11	4	41	36	45	87	80	81

(a) (i) Plot the data.
   (ii) Fit the regression line of y on x, making the usual assumptions.
   (iii) Draw the line on your graph.
   (iv) Construct 90% limits of prediction for the cost for a zone of 8400 properties.
   (v) Do you think the usual assumptions are realistic in this context?

(b) As for (a) except take y as ln(estimated cost).
(c) Find the least-squares estimator of the slope, b, of a line constrained to pass through the origin:

$$Y_i = bx_i + E_i$$

where the E_i satisfy the usual assumptions. Fit this model to the data, and draw the line on the same graph as (a). Predict the cost for a zone of 8400 properties.

(d) Which of the three analyses would you use in a report for the water company?

8.2 The following pairs are baseflow (x, minimum flow in river before storm) and peakflow (y) of the River Browney for eight storms during 1983, both measured in cubic metres per second.

x	2.03	2.35	4.14	1.27	2.52	0.74	0.46	0.31
y	23.90	31.55	17.27	30.97	38.82	3.21	1.42	1.58

(a) (i) Plot the data on graph paper.
   (ii) Calculate a 90% confidence interval for the correlation between X and Y in the corresponding population.

(iii) Fit a regression line of y on x and construct a 90% confidence interval for the slope.

(b) Repeat using the increase in flow $(y - x)$ in place of y.

8.3 At a certain stage of their training apprentices take three practical and two theoretical exams. Let X be the number of practical exams passed and Y be the number of theoretical exams passed. The probability distribution $P(x, y)$, based on extensive records, is given below.

x	0	1	2	3
y				
0	0.1	0.1		
1	0.1	0.1	0.1	
2	0.1	0.1	0.1	0.2

Find the conditional probability distribution of X given that $y = 2$.

8.4 The number of defects per metre, X, for a certain fabric has a Poisson distribution with mean λ. However, λ is itself a random variable with pdf given by

$$f(\lambda) = e^{-\lambda} \qquad \text{for } 0 \leqslant \lambda$$

Find the unconditional probability function for X.

8.5 Let X denote the weight in tonnes of a powder stocked by a supplier at the beginning of the week. Assume X has a uniform distribution on $[0, 1]$. Let Y denote the weight of powder sold during the week and suppose Y is uniformly distributed on the interval $[0, x]$ where x is a specific value of X.

(i) Find the joint pdf of X and Y.
(ii) If the supplier stocks half a tonne what is the probability of selling more than a quarter of a tonne?
(iii) If a quarter of a tonne was sold, what is the probability that more than half a tonne was stocked?

8.6 Bayesian confidence interval

Assume we have a random sample of size n from a distribution with mean μ and variance σ^2. We wish to construct a confidence interval for μ. The sample mean $\bar{X}$ has, at least approximately, a normal distribution with mean μ and variance σ^2/n. Denote its pdf by

$$f(\bar{x}|\mu) = (2\pi\phi)^{-1/2} e^{-(\bar{x} - \mu)^2/2\phi}$$

where we have written ϕ in place of σ^2/n to make the subsequent algebra slightly simpler. Now suppose we have some idea about possible values for μ and express these by a prior distribution, which we will assume normal.

We write

$$\mu \sim N(\theta, \psi)$$

where θ is our best guess for μ, and ψ is the variance of the distribution which represents our uncertainty about the value of μ. The choice of a normal distribution enables us to find a general formula in terms of standard functions. The result is not sensitive to this assumption if ψ is relatively large.

In the Bayesian approach the unknown mean μ is explicitly treated as a variable, whereas in the frequentist approach of Chapter 6 it was simply an unknown constant. Now, by Bayes' theorem,

$$
\begin{aligned}
f(\mu|\bar{x}) &= f(\mu, \bar{x})/f(\bar{x}) \\
&= f(\mu)f(\bar{x}|\mu)/f(\bar{x}) \\
&\propto f(\mu)f(\bar{x}|\mu)
\end{aligned}
$$

The constant of proportionality is found by using the fact that the area under any pdf is 1, rather than by attempting to find the unconditional distribution of $\bar{x}$.

(i) Show that it follows from the prior distribution

$$\mu \sim N(\theta, \psi)$$

and the sampling distribution

$$\bar{x}|\mu \sim N(\mu, \phi)$$

that the posterior distribution of μ is

$$\mu|\bar{x} \sim N(\eta, \lambda)$$

where

$$\lambda^{-1} = \psi^{-1} + \phi^{-1}$$

and

$$\eta = \theta(\psi^{-1}/\lambda^{-1}) + \bar{x}(\phi^{-1}/\lambda^{-1})$$

(ii) Demonstrate that this tends towards the frequentist confidence interval for μ, with σ known, as ψ tends to infinity.

(iii) The result is quite general, inasmuch as $\bar{x}$ can be replaced by any normal variable x with mean μ and variance ϕ. Lee (1989) gives an example of dating rocks. In the 1960s the age of Ennerdale granophyre was estimated as 370 million years, with a standard deviation of 20 million years, by observing the relative proportions of potassium-40 and argon-40 in the rock. Later, independent estimates were made from relative proportions of rubidium-87 and strontium-87. The quoted estimate was 421 million years, with a standard deviation of 8 million years. Use the result of (i) to combine these two estimates. Would it make any difference if the Rb/Sr estimate had preceded that based on the K/Ar ratio?

8.7 This application has been contributed by Gary Edwards of Northumbrian Water Ltd. The data in Table D.7 are the costs (in monetary units adjusted to current prices) of schemes to prevent flooding during extreme storms. The number of properties that are affected is also given. More schemes are planned, but the average number of properties protected by each of these prospective schemes is less. A fictitious distribution is given below.

Number of properties at risk	Number of schemes planned
1	28
2	5
3	4
5	2
6	1
8	1

Write a short report giving your estimate of the total cost (current prices) of these schemes and associated 90% limits of prediction (making any approximations you need to).

8.8 Figure 8.19 represents a junction box in an oil pipeline. The oil enters at A and leaves at either B or C, and there is no leakage. There are three meters 1, 2 and 3 which give readings of flow which are subject to independent measurement errors with zero mean and the same variance σ^2. Let the flows be θ_1, θ_2, and θ_3. Because there is no leakage θ_1 equals the sum of θ_2 and θ_3 and we can arbitrarily choose any two of the flows to be the unknown parameters. The measurements of the flows are Y_1, Y_2, and Y_3. Choosing

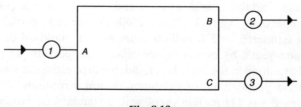

Fig. 8.19

θ_2 and θ_3 as the unknown parameters leads to an error sum of squares,

$$\psi = (Y_1 - (\theta_2 + \theta_3))^2 + (Y_2 - \theta_2)^2 + (Y_3 - \theta_3)^2$$

Necessary conditions for a minimum are that the partial derivatives equal zero.

(a) Find the least-squares estimators of θ_1, θ_2 and θ_3.
(b) What is the variance of these estimators?

9
Making predictions from several explanatory variables

9.1 Regression on two explanatory variables

9.1.1 An introductory example

A small retailing company sells personal computers through ten shops. The business has been successful and the director would like to expand it by purchasing more shops. She has recorded the gross weekly sales, floor area, and average number of pedestrians passing the entrance during opening hours, for each shop, and these are given in Table 9.1 (courtesy of J. Turcan).

If the floor area and pedestrian traffic were ascertained for any shops offered for sale, a formula relating sales to these variables could help the director to decide whether to bid for them. We assume that any such shops have not been selling computer equipment, so relevant sales figures are not available. A reasonable statistical model might be

$$\text{sales} = \beta_0 + \beta_1 \text{ (pedestrian traffic)} + \beta_2 \text{ (floor area)} + \text{error}$$

The deterministic part of this model can be represented in three dimensions by a plane, shown in Fig. 9.1. There are obvious limitations to this procedure. First, the relationship is just an empirical approximation over the range of values of floor area and pedestrian traffic of the ten shops owned by the company. It should not be relied on for any predictions outside this range. Second, it assumes that any shops offered for sale are randomly drawn from the same population as the ten used for fitting the model. Assumptions of this nature are inevitable in this sort of situation, but if there is some obvious reason why it is implausible, such as a shop being located in a different type of area from the original ten, we should be very wary of any prediction. If the ten shops are in a variety of localities, it may be possible to include some 'sociological' variable in the regression model.

The model relating sales to pedestrian traffic and floor area is an example of a multiple regression with two explanatory variables. The principles remain the same if there are more than two explanatory variables, though the geometrical interpretation is lost. We let Y, x_1 and x_2 represent sales, traffic and floor area,

Table 9.1 Sales, size and pedestrian traffic for ten shops

Average traffic (pedestrians/hour)	Floor area (m²)	Average weekly gross sales (× £1000/week)
564	130	4.90
1072	140	5.80
326	90	4.00
1172	100	5.65
798	110	5.20
584	130	5.00
280	135	3.70
970	150	6.25
802	125	5.40
605	100	4.38

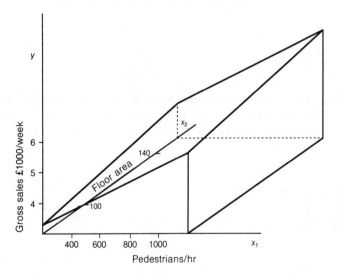

Fig. 9.1 Deterministic planar relationship between sales, pedestrian traffic and floor area.

respectively. Then the multiple regression model is

$$Y_i = \beta_0 + \beta_1 x_{1i} + \beta_2 x_{2i} + E_i$$

The variable on the left-hand side is usually called the dependent variable and I shall refer to the x-variables as explanatory variables. The E_i are the random components of the model, conveniently known as 'errors', and are assumed to have a zero mean and to be independent of the explanatory variables. The fitting procedure is simplified if the E_i are also assumed to have a constant variance σ^2 and to be independent of each other. The method of least squares is used to fit the model.

If a random sample of size n is to be taken from the population it can be modelled by

$$Y_i = \beta_0 + \beta_1 x_{1i} + \beta_2 x_{2i} + E_i \qquad \text{for } i = 1, \ldots, n$$

It is worth noting that the same model will be used for a designed experiment, for which the x have been preselected and the runs carried out in a random order, in Chapter 10.

The assumptions about the errors are exactly the same as for the regression on one explanatory variable covered in Chapter 8. The principle of least squares is again used to fit the model and the details are given in Appendix A5. The least-squares estimates are the values of β_0, β_1 and β_2 which minimize

$$\psi = \sum (y_i - (\beta_0 + \beta_1 x_{1i} + \beta_2 x_{2i}))^2$$

The residuals r_i are the differences between y_i and the fitted values $\hat{y}_i$. That is,

$$r_i = y_i - \hat{y}_i$$

where

$$\hat{y}_i = \hat{\beta}_0 + \hat{\beta}_1 x_{1i} + \hat{\beta}_2 x_{2i}$$

The residuals can be thought of as estimates of the errors, and they are shown in Fig. 9.2. The estimated variance of the errors, s^2, is found by dividing the sum of squared residuals (RSS) by their degrees of freedom, which is $n - 3$ if

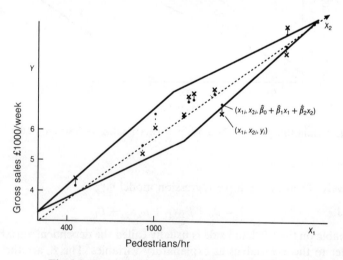

Fig. 9.2 The fitted regression plane and residuals.

three parameters (β_0, β_1 and β_2) estimated from the data:

$$s^2 = \sum r_i^2 / (n-3)$$

This is shown to be an unbiased estimator in Appendix A5.

It should always be remembered that addition of further covariates can only reduce RSS. In Fig. 9.3(a) the sales (y_i) are plotted against pedestrian traffic (x_{1i}). The ten points are marked with an L, M or H depending on whether the corresponding area (x_{2i}) is among the lowest three, middle four or highest three, respectively. The fitted regression line is also shown. The points are shown in three dimensions in Fig. 9.3(b), where they have been translated parallel to the x_2-axis according to their area value. A plane which includes the original regression line and is parallel to the x_2-axis is shown. The distances, parallel to the y-axis, from the points in Fig. 9.3(b) to the plane, are identical to those from the points in Fig. 9.3(a) to the line. In general, it is almost certain that there will be some other plane which will result in a decrease in RSS – even if x_{2i} were random numbers! The fitted regression plane which minimizes RSS is shown in Fig. 9.3(c). An 'algebraic' explanation is that if $\hat{\beta}_2$ is set at zero, and $\hat{\beta}_1$ is the slope of the regression line, then the RSS for the regression plane will equal that for the line. However, it would be extraordinary if no better combination of estimates could be found when $\hat{\beta}_2$ is allowed to take any value.

The results of the regression analysis will indicate whether or not the area is worth including as an explanatory variable. The fitted plane is

$$y = 2.01 + 0.002\,33x_1 + 0.0111x_2$$

and

$$s = 0.2875$$

The standard deviations of the coefficients of traffic and area are given below.

Predictor	Coefficient	Standard deviation	t-ratio
constant	2.0114	0.6047	3.33
traffic	0.002 327 4	0.000 325 2	7.16
area	0.011 133	0.004 914	2.27

A 90% confidence interval for the coefficient of area is

$$0.0111 \pm t_{7,0.05} 0.004\,914$$

which gives

$$0.0111 \pm 0.009$$

We have not estimated the coefficient very precisely but we are reasonably confident it is at least positive. Common sense suggests it would be positive because more floor space allows room for more customers and more stock can be displayed, but it is possible that customers like the 'boutique' effect of smaller shops. For comparison, the regression of sales on traffic only is

$$y = 3.26 + 0.00246x_1$$

with $s = 0.354$. Notice that the coefficient of x_1 is different; this will always be the case unless x_1 and x_2 are uncorrelated.

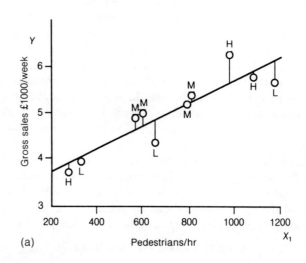

(a)

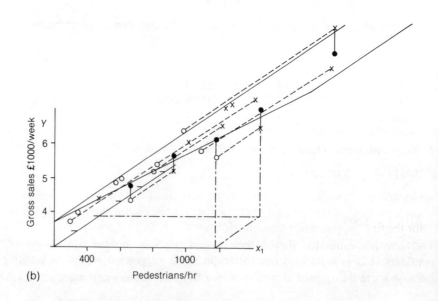

(b)

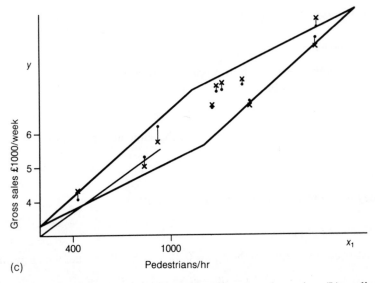

Fig. 9.3 Sales of PC. (a) Sales plotted against traffic, areas shown low (L) medium (M) high (H). (b) Points projected parallel to x_2-axis from $\odot$ to X. The plane includes the regression line for sales on x_1 and is parallel to the x_2-axis. (c) The fitted regression plane and residuals.

Example 9.1

The director is thinking about buying one of two shops which are up for sale. She does not have funds to buy both. The first has traffic of 475 pedestrians per hour and a floor area of $200\,m^2$. The prediction for sales is 5.38 (£1000/wk) and 90% limits for this are ± 1.24. The second has traffic of 880 pedestrians per hour and a floor area of $110\,m^2$. The prediction for sales is 5.27 and 90% limits for this are ± 0.76. The limits of prediction for the first shop are very wide because its floor area is a long way from the mean value for the ten for which the regression plane was fitted. The limits allow for uncertainty about the parameter estimates, but they assume the form of the model remains correct for floor areas up to $200\,m^2$. This is a long way beyond the data used to fit the model, so there is no empirical evidence to support this assumption. It may be that increases in floor area are beneficial until there is space to display all the varieties of goods, and detrimental thereafter. If the purchase price and operating costs (taxes, electricity and staff) are similar for the two shops (reasonable as council taxes are higher in busy streets), I would recommend buying the second. The prediction for the first shop is much less reliable than the, already large, limits of prediction indicate. It is useful because it demonstrates that, even if the plane can be extrapolated, the predicted sales for the first shop barely exceed those for the second.

9.2 Multiple regression model

9.2.1 The model

Let Y be the variable we wish to predict, the dependent variable, and $x_1, \ldots, x_k$ be k explanatory variables. We model n observations by

$$Y_i = \beta_0 + \beta_1 x_{1i} + \cdots + \beta_k x_{ki} + E_i \qquad \text{for } i = 1, 2, \ldots, n$$

The 'usual' assumptions about the E_i are the same as for regression on a single explanatory variable. That is

$$E_i \sim N(0, \sigma^2)$$

and they are independent of the x_i and each other.

9.2.2 Estimates of parameters

The least-square estimates of the β_j, $j = 0, 1, \ldots, k$, are the values which minimize

$$\psi = \sum (y_i - (\beta_0 + \beta_1 x_{1i} + \cdots + \beta_k x_{ki}))^2$$

They are denoted by $\hat{\beta}_j$. The residuals are defined by

$$r_i = y_i - (\hat{\beta}_0 + \hat{\beta}_1 x_{1i} + \cdots + \hat{\beta}_k x_{ki})$$

The least-squares estimate of σ^2 is written as s^2 and defined by

$$s^2 = \sum r_i^2 / (n - k - 1)$$

The smaller s, the more the variation in the dependent variable is accounted for by the model. Another indicator is the statistic R^2, usually pronounced as it is written but formally known as the **coefficient of determination**. It can be thought of as the proportion of the variance of the dependent variable accounted for by the model and is defined by:

$$R^2 = (S_{yy} - \sum r_i^2)/S_{yy}$$
$$= 1 - \sum r_i^2 / S_{yy}$$

It is easily misunderstood because it can only increase as more explanatory variables are added. A slight modification is the 'adjusted' R^2, written R^2_{adj}, and defined by

$$R^2_{adj} = 1 - [\sum r_i^2 / (n - k - 1)] / [S_{yy}/(n - 1)]$$
$$= 1 - s^2 / [S_{yy}/(n - 1)]$$

It increases or decreases as s decreases or increases.

9.2.3 Geometrical interpretations

We have already interpreted

$$y = \beta_0 + \beta_1 x_1 + \beta_2 x_2$$

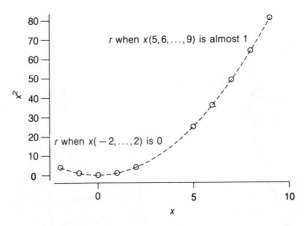

Fig. 9.4 Correlations of sets of points clustered around a parabola depend on their distance from the vertex.

as a plane in three dimensions. However a special case is when x_2 is equal to x_1^2. Then we have part of a parabola

$$y = \beta_0 + \beta_1 x + \beta_2 x^2$$

There is no requirement that the explanatory variables be uncorrelated, but high correlations lead to numerical inaccuracy and make interpretation more difficult. The correlation between x and x^2 will be large if x has a small coefficient of variation, because

$$(x + \delta x)^2 \simeq x^2 + 2x\delta x$$

when δx is small compared with x. Such undesirable correlations can be dramatically reduced by redefining x in terms of deviations from its mean (see Fig. 9.4, Appendix A5 and Exercise 9.1).
   A general quadratic surface in three dimensions is given by

$$y = \beta_0 + \beta_1 x_1 + \beta_2 x_2 + \beta_3 x_1^2 + \beta_4 x_2^2 + \beta_5 x_1 x_2$$

and examples are shown in Fig. 9.5. It is, again, quite often advisable to redefine x_1 and x_2 in terms of deviations from their means. The quadratic surface is still referred to as a **linear model** by statisticians because it is linear in the unknown parameters.

Example 9.2

A hydrological model for the design of storage reservoirs requires annual run-offs for all the watersheds in the region. Run-off is measured by stream flow, but only some of the rivers are gauged. Catchment area and annual precipitation are available for all rivers, so we would like to estimate average run-off from

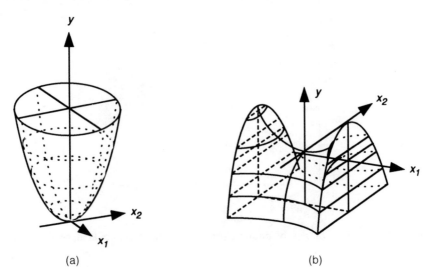

Fig. 9.5 Two examples of quadratic surfaces: (a) paraboloid; (b) saddle.

these variables if it is not measured directly. We will use 13 small gauged rivers to estimate a regression model for the small rivers in the region. If we had been asked to choose which 13 rivers would be gauged we should have taken a stratified random sample. The stratification would be used to ensure that the range of sizes, within the small category, would be represented when fitting the model. However, we often have to make do with available data and rely on hydrologists' judgement that the rivers are fairly typical. When we construct a regression model for the larger rivers we will include other variables, such as whether the valley is steep- or gentle-sided, and this will help to reduce any bias caused by a non-random selection of rivers. The regression models will be used by engineers to predict run-off for various randomly generated rainfall scenarios.

The data in Table 9.2 are last years' precipitation (x_1, in millimetres), area (x_2, in square kilometres) and average run-off (y, in thousands of litres per second) for 13 small rivers in Kentucky. We start with a regression of y on the product of x_1 and x_2, which can be interpreted as a volume of water, and compare this with a full quadratic model. The MINITAB output is shown in Fig. 9.6. The regression of run-off on volume, $s = 50.7$ compared with a standard deviation of unconditional run-offs of 56.4, leaves a great deal of unexplained variation. You may notice the non-zero intercept, but this can be attributed to either a nonlinear relationship between run-off and volume or, perhaps, residual flow from infiltrated water. It would be helpful to have data from another year which might increase the range of rainfall values.

The full quadratic surface is no improvement. Although s has become very slightly smaller (it is now 49.4), prediction intervals will usually be wider because

Table 9.2 Precipitation, area and run-off for 13 small rivers in Kentucky

Precipitation (mm in year)	Area (km²)	Run-off ($\times 1000\,\mathrm{l\,s^{-1}}$)
1109	5.66	506
1102	6.45	425
1031	14.39	450
1137	3.94	428
1152	13.18	535
1228	5.48	495
1101	13.67	530
1217	19.12	551
1111	5.38	406
1193	9.93	542
1209	1.72	502
1225	2.18	509
1176	4.40	383

of uncertainty in the estimates of the five parameters. A formal interpretation of the analysis of variance table suggests this more complicated model has little predictive value. If we suspend disbelief and imagine that area and precipitation do not affect run-offs, the regression mean square is another estimate of variance of errors, independent of the residual (error in MINITAB output) mean square (the RSS divided by the degrees of freedom). The ratio of these estimates is 1.72 and the probability of a ratio greater than this is 0.248.

The next stage was to drop the quadratic terms and fit precipitation, area and volume (Fig. 9.7). I think volume should be retained because of its physical significance and I would choose to use a regression of run-off on precipitation and volume, for which $s = 44.6$. However, the regression plane of run-off on precipitation and area gives an equally good fit, and very similar predictions within the range of values used to fit the regression. A normal score plot for the residuals from my preferred model is shown, and it seems reasonable to assume normality when constructing prediction intervals.

Example 9.3

The data in Table 9.3 are from a process which produces plastic sheet by 'bubble-blowing' molten plastic. The tensile strength (y) of the sheet and the rate of stretching during processing (x) are given for 35 runs. A high tensile strength is desirable and we have been asked to advise the process engineer. The data have been coded for commercial reasons. Edited output from a MINITAB

```
MTB > name c1 'precip' c2 'area' c3 'runoff'
MTB > let c11=c1*c1
MTB > let c12=c1*c2
MTB > let c22=c2*c2
MTB > name c11 'precsqu' c12 'volume' c22 'areasqu'
MTB > describe c1-c3
```

	N	MEAN	MEDIAN	TRMEAN	STDEV	SEMEAN
precip	13	1153.2	1152.0	1157.5	60.8	16.9
area	13	8.12	5.66	7.70	5.41	1.50
runoff	13	481.7	502.0	484.4	56.4	15.6
	MIN	MAX	Q1	Q3		
precip	1031.0	1228.0	1105.5	1213.0		
area	1.72	19.12	4.17	13.43		
runoff	383.0	551.0	426.5	532.5		

```
MTB > gstd # In Release 9 set to std graphics for character plots
          Use the GPRO command to enable Professional Graphics.
MTB > plot c3 c12
```

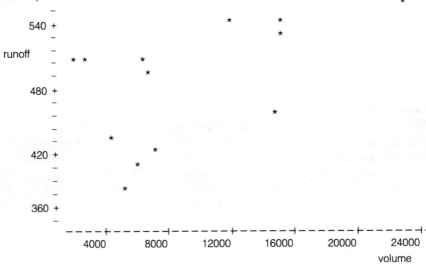

```
MTB > regress c3 on 1 explan c12

The regression equation is
runoff = 439 + 0.00458 volume
```

Predictor	Coef	Stdev	t-ratio	p
Constant	439.15	25.91	16.95	0.000
volume	0.004583	0.002344	1.95	0.077

```
s=50.71              R-sq=25.8%              R-sq(adj)=19.0%
```

Analysis of Variance

SOURCE	DF	SS	MS	F	p
Regression	1	9826	9826	3.82	0.077
Error	11	28287	2572		
Total	12	38113			

Unusual Observations

Ob.	volume	runoff	Fit	Stdev.Fit	Residual	St. Resid
8	23269	551.0	545.8	35.7	5.2	0.14 X

X denotes an obs. whose X value gives it large influence.
MTB > regress c3 on 5 explan c1 c2 c11 c12 c22
* NOTE * precip is highly correlated with other predictor variables
* NOTE *   area is highly correlated with other predictor variables
* NOTE * precsqu is highly correlated with other predictor variables
* NOTE * volume is highly correlated with other predictor variables

The regression equation is
runoff= − 5209 + 8.7 precip + 78.5 area − 0.00330 precsqu
              − 0.0592 volume − 0.101 areasqu

Predictor	Coef	Stdev	t-ratio	p
Constant	−5209	8745	−0.60	0.570
precip	8.69	14.85	0.58	0.577
area	78.49	76.35	1.03	0.338
precsqu	−0.003303	0.006296	−0.52	0.616
volume	−0.05917	0.06748	−0.88	0.410
areasqu	−0.1013	0.7501	−0.14	0.896

s=49.41   R-sq=55.2%   R-sq(adj)=23.1%

Analysis of Variance

SOURCE	DF	SS	MS	F	p
Regression	5	21021	4204	1.72	0.248
Error	7	17092	2442		
Total	12	38113			

Unusual Observations

Obs.	precip	runoff	Fit	Stdev.Fit	Residual	St.Resid
8	1217	551.0	556.2	49.2	−5.2	−1.28 X
13	1176	383.0	474.6	19.7	−91.6	−2.02 R

R denotes an obs. with a large st. resid.
X denotes an obs. whose X value gives it large influence.

Fig. 9.6 Preliminary regression models for small Kentucky rivers.

session is shown in Fig. 9.8. A regression of y on both x and x^2 gives a somewhat smaller estimated standard deviation of the errors (s) than a regression of y on x only. Although the difference, 23.8 compared with 26.4, is not great the implications of fitting the parabolic curve rather than the line are important. The former suggests that there may not be much to be gained, in terms of strength, by attempting to increase the rate of stretching any further. The high correlation between x and x^2 makes it difficult to decide how much notice we should take of the squared term. Subtracting the mean from x before squaring gives some improvement, but because of the uneven distribution of x-values a significant correlation between x and x^2 remains. A short 'search' shows that $(x − 339)^2$ is uncorrelated with $(x − 339)$. The 95% confidence intervals for the

MTB > regress c3 3 c1 c2 c12
* NOTE *   area is highly correlated with other predictor variables
* NOTE * volume is highly correlated with other predictor variables

The regression equation is
runoff= −544+ 0.838 precip + 45.4 area − 0.0332 volume

Predictor	Coef	Stdev	t-ratio	p
Constant	−544.4	564.7	−0.96	0.360
precip	0.8380	0.4789	1.75	0.114
area	45.37	48.62	0.93	0.375
volume	−0.03323	0.04127	−0.81	0.442

s=44.94   R-sq=52.3%   R-sq(adj)=36.4%

Analysis of Variance

SOURCE	DF	SS	MS	F	p
Regression	3	19934	6645	3.29	0.072
Error	9	18178	2020		
Total	12	38113			

Breakdown of the Regression SS

SOURCE	DF	SEQ SS
precip	1	5662
area	1	12963
volume	1	1309

Unusual Observations

Obs.	precip	runoff	Fit	Stdev.Fit	Residual	St.Resid
13	1176	383.0	468.8	15.6	− 85.8	−2.04R

R denotes an obs. with a large st. resid.

MTB   > regress c3 2 c1 c12;
SUBC > resid c5.

The regression equation is
runoff = −73 + 0.439 precip + 0.00523 volume

Predictor	Coef	Stdev	t-ratio	p
Constant	−73.3	251.5	−0.29	0.777
precip	0.4392	0.2146	2.05	0.068
volume	0.005232	0.002089	2.51	0.031

s = 44.65   R-sq = 47.7%   R-sq(adj) = 37.2%

Analysis of Variance

SOURCE	DF	SS	MS	F	p
Regression	2	18176	9088	4.56	0.039
Error	10	19937	1994		
Total	12	38113			

Breakdown of the Regression SS

SOURCE	DF	SEQ SS
precip	1	5662
volume	1	12514

```
Unusual Observations
Obs.          precip        runoff         Fit        Stdev.Fit      Residual       St.Resid
 13            1176          383.0        470.2          15.4          -87.2         -2.08R
R denotes an obs. with a large st. resid.

MTB > regress c3 2 c1 c2

The regression equation is
runoff = -141 + 0.496 precip + 6.27 area

Predictor       Coef        Stdev        t-ratio           p
Constant      -140.8        255.5        -0.55          0.594
precip         0.4957       0.2164        2.29          0.045
area           6.274        2.432         2.58          0.027

s = 44.14            R-sq = 48.9%              R-sq(adj) = 38.6%

Analysis of Variance
SOURCE          DF           SS           MS            F            p
Regression       2          18625        9313          4.78         0.035
Error           10          19488        1949
Total           12          38113

Unusual Observations
Obs.          precip        runoff         Fit        Stdev.Fit      Residual      St. Resid
 13            1176          383.0        469.7          15.3          -86.7        -2.09R
R denotes an obs. with a large st. resid.

MTB > nsco c5 c6
MTB > name c5 'resids' c6 'normsco'
MTB > plot c5 c6
```

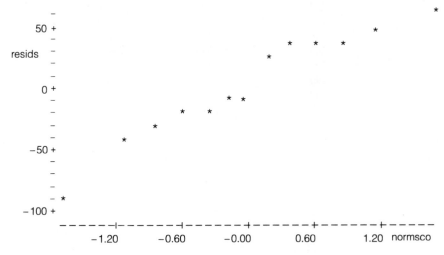

Fig. 9.7 Preferred regression models for small Kentucky rivers.

Table 9.3 Rate of stretching during processing and tensile strength for plastic sheet. The units are scaled for commercial reasons

Rate of strength (x)	Tensile strength (y)
40	173
65	179
75	171
75	151
85	192
95	217
105	186
115	211
120	187
130	183
140	189
145	203
145	181
160	241
165	187
170	254
178	235
179	197
180	203
190	263
228	222
229	197
236	217
245	233
255	246
290	254
343	330
380	284
385	321
415	333
498	321
500	329
510	290
520	316
750	337

coefficient of $(x - 339)^2$ is

$$-0.000\,36 \pm 0.000\,25$$

and we are quite confident that addition of this term is an improvement over $(x - 339)$, or equivalently x, on its own. However, the fitted parabola is sensitive to the result from a single run at a high stretch rate and I would recommend some more runs with high stretch rates.

9.2.4 Multicollinearity

There is no requirement that the explanatory variables be uncorrelated, but if there is an exact linear relationship between x_1 and x_2, it will not be possible to obtain unique estimates of β_1 and β_2. Suppose that

$$x_2 = ax_1 + b$$

Then

$$\alpha + \beta_1 x_1 + \beta_2 x_2 = \alpha + \beta_1 x_1 + \beta_2 (ax_1 + b) = (\alpha + \beta_2 b) + (\beta_1 + a\beta_2)x_1$$

We can estimate $\alpha + \beta_2 b$ and $\beta_1 + a\beta_2$ but not α, β_1 and β_2 separately. In practice, a near-linear relationship would lead to numerical instability and either x_1 or x_2 should be dropped from the model (unless it is a scaling issue; see Exercise 9.1).

9.3 Categorical variables

Categorical variables can easily be handled with multiple regression. A freight ferry, the *Toucan*, sails from port A to port B and from there to port C before returning to port A. Each voyage therefore consists of three 'legs' AB, BC and CA. The *Toucan* always takes on fuel at port A. A fuel additive, which the manufacturers claim will improve fuel consumption through a catalytic effect on the oxidation of the fuel, was included in the fuel line to the engine on a random selection of 180 from 360 voyages. For each leg of each voyage the passage time, fuel consumed, work done, draught and weather were noted. The work done was measured with an integrating power meter which was a standard fitting on the *Toucan*. The weather was classified as strong following wind, slight following wind, calm, slight head wind and strong head wind. The main purpose of fitting regression models was to investigate whether the additive had any effect.

The leg of the voyage was coded with two indicator variables, x_3 and x_4. They were set as:

	x_3	x_4
leg AB	0	0
leg BC	1	0
leg AC	0	1

```
MTB > name cl 'extrat' c2 'strength'
MTB > plot c2 c1
```

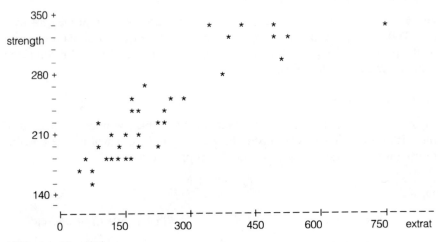

```
MTB > let c11=c1*c1
MTB > let c12=(c1-mean(cl))**2
MTB > let c3  =(c1-339)
MTB > let c13=(c1-339)**2
MTB > name c3 'scextr'
MTB > name c11 'extrsq'
MTB > name c12 'mcxtrsq'
MTB > name c13 'scexsq'
MTB > regress c2 1 c1
```

The regression equation is
strength=163 + 0.303 extrat

Predictor	Coef	Stdev	t-ratio	p
Constant	163.020	7.927	20.56	0.000
extrat	0.30300	0.02751	11.02	0.000

s=26.37   R-sq=78.6%   R-sq(adj)=78.0%

Analysis of Variance

SOURCE	DF	SS	MS	F	p
Regression	1	84413	84413	121.35	0.000
Error	33	22955	696		
Total	34	107368			

Unusual Observations

Obs.	extrat	strength	Fit	Stdev.Fit	Residual	St.Resid
27	343	330.00	266.95	5.31	63.05	2.44R
35	750	337.00	390.27	14.76	−53.27	−2.44RX

R denotes an obs. with a large st. resid.
X denotes an obs. whose X value gives it large influence.

```
 MTB > regress c2 2 c1 c11
```

The regression equation is
strength=135 + 0.549 extrat −0.000364 extrsq

Predictor	Coef	Stdev	t-ratio	p
Constant	134.52	12.13	11.09	0.000
extrat	0.54934	0.08818	6.23	0.000
extrsq	−0.0003636	0.0001249	−2.91	0.007

s = 23.81   R-sq = 83.1%   R-sq(adj) = 82.0%

Analysis of Variance

SOURCE	DF	SS	MS	F	p
Regression	2	89221	44610	78.66	0.000
Error	32	18148	567		
Total	34	107368			

Breakdown of Regression SS

SOURCE	DF	SEQ SS
extrat	1	84413
extrsq	1	4807

Unusual Observations

Obs.	extrat	strength	Fit	Stdev.Fit	Residual	St.Resid
27	343	330.00	280.16	6.60	49.84	2.18R
35	750	337.00	341.98	21.28	−4.98	−0.47 X

R denotes an obs. with a large st. resid.
X denotes an obs. whose X value gives it large influence.

MTB > correlation c1 c3 c11 c12 c13

	extrat	scextr	extrsq	mcxtrsq
scextr	1.000			
extrsq	0.960	0.960		
mcxtrsq	0.711	0.711	0.880	
scexsq	−0.003	−0.003	0.279	0.702

MTB > regress c2 2 c3 c13

The regression equation is
strength = 279 + 0.303 scextr −0.000364 scexsq

Predictor	Coef	Stdev	t-ratio	p
Constant	278.955	6.563	42.50	0.000
scextr	0.30280	0.02484	12.19	0.000
scexsq	−0.0003636	0.0001249	−2.91	0.007

s = 23.81   R-sq = 83.1%   R-sq(adj) = 82.0%

Analysis of Variance

SOURCE	DF	SS	MS	F	p
Regression	2	89221	44610	78.66	0.000
Error	32	18148	567		
Total	34	107368			

Breakdown of Regression SS

SOURCE	DF	SEQ SS
scextr	1	84413
scexsq	1	4807

Unusual Observations

Obs.	scextr	strength	Fit	Stdev.Fit	Residual	St.Resid
27	4	330.00	280.16	6.60	49.84	2.18R
35	411	337.00	341.98	21.28	−4.98	−0.47 X

R denotes an obs.with a large st. resid.
X denotes an obs. whose X value gives it large influence.

Fig. 9.8 Regressions for strength of plastic film against stretch rate during processing.

It follows that the coefficients β_3 and β_4 represent the differences in fuel consumption, relative to leg AB, for legs BC and AC, respectively. The covariate x_5 was used to code legs for which the additive was in use, when it was put at 1 rather than 0. Because the additive was included in the fuel line, rather than mixed in the fuel tank, there was no reason to allow for a 'carry-over' effect when investigating any catalytic action. There is no physical reason why draught or weather should affect the relationship between fuel consumption and shaft work, and if they were included in the regression model there was no improvement in fit. The following regression model was eventually selected:

$$\text{Fuel consumed on leg} = \beta_0 + \beta_1 \times (\text{work done on leg})$$
$$+ \beta_2 \times (\text{work done on leg/time of sailing on leg})$$
$$+ \beta_3 x_3 + \beta_4 x_4 + \beta_5 x_5 + \text{error}$$

Regression models based on an assumed linear relationship between fuel rate and power were also tried but had rather smaller R^2 values. A plot of fuel used against work done is shown in Fig. 9.9, and it is clear that any other effects are relatively small. A 95% confidence interval for β_5 is $[-1.17, 0.30]$. In fact the additive is quite expensive and a value of β_5 equal to -1.00 would be needed for it to be economical, so trials were discontinued. There was some statistical evidence that the leg had an effect on fuel consumed. A possible explanation

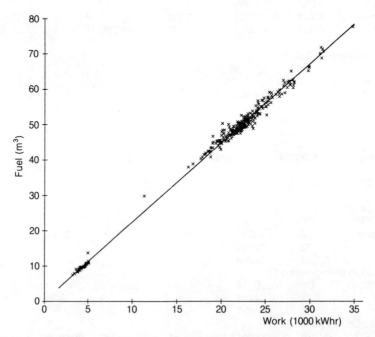

Fig. 9.9 Fuel against work done for a ferry.

for this finding is that the speed might be held more constant on one leg, thereby giving a slight improvement in fuel consumption.

At an anecdotal level, the chief engineer reported that use of the additive left the engine much cleaner. If the additive does have a cleaning effect, this might also be a mechanism for saving fuel during the period between engine overhauls. The analysis described would not detect any such effect. One approach would be to compare trends in fuel consumption for the *Toucan*, using the additive over the period, with another ship which was not using the additive. Alternatively, we might compare trends in fuel consumption between overhauls for the *Toucan* when the additive was used, with trends between overhauls when it was not used.

9.4 Chrome plating of rods for hydraulic machinery

9.4.1 Background

A company frequently chrome-plates batches of rods for use in hydraulic machinery. The main steps in the process are described below.

- The rods are preground.
- The surface is prepared for plating.
- The rods are fitted with 'thieves' at both ends to reduce the high build-up of chrome at the edges.
- Batches of nine, or occasionally six, rods are loaded into a plating bath (Fig. 9.10) and plated.
- The rods are prepared for grinding to size
- The rods are ground to size, and finally linished (a polishing process).

Ideally the thickness of chrome would be constant along the rod but it is a feature of the electroplating process that the depth of chrome tends to increase

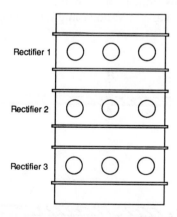

Fig. 9.10 Plan view of plating bath.

towards the ends of the rods (Fig. 9.11). The thieves are designed to minimize this effect and three types, which we will denote A, B and C, were in use at the time. The aim of the investigation was to determine the effect of the thief type, and other variables, on the relative increase in chrome at the ends. The variables could then be specified to minimize this difference.

9.4.2 Routinely collected data

The following data are routinely recorded for each run of the plating bath.

- The length and type of rods to be plated.
- The position of each rod in the bath.
- Location of long and short anodes in the bath.
- Types of thieves used (A, B or C).
- Number of rods loaded for plating (six or nine).
- Day of the week.
- Plating time.
- Current.
- The electrolyte concentration is monitored but, provided all goes well, it is maintained at a specified value by a feedback control loop.

In addition the diameters of the finished rods, i.e. after grinding to size and linishing, are monitored.

9.4.3 Retrospective analysis

At the time of the investigation an order for 39 rods of the same length and type was already in progress. It was possible to measure chrome depths on these rods before they were ground to size but not to preselect all the values of the variables. There are many pitfalls to be wary of when analysing such 'happenstance' (Box et al., 1978) data. Salameh and Metcalfe (1992) describe the tentative conclusions which could reasonably be drawn from such an analysis, together with some reservations, and the implications for any prospective designed experiment.

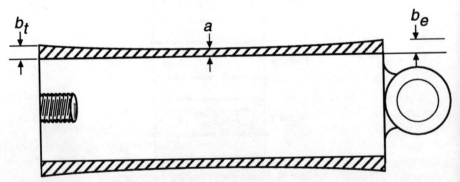

Fig. 9.11 Section through chromes rod; *reldiff* is $(\bar{b} - a)/a$, where $\bar{b}$ is $(b_t + b_e)/2$.

The rods were tapped and threaded at one end (the thread end) and had an eye which was not plated at the other end (the eye end). Measurements of chrome depth at the middle and ends of each rod were made at $10°$ spacings using a permoscope. There was little variation among the 36 measurements and the depth was recorded as their average. The dependent variable in the following regressions, denoted *reldiff* for relative difference (see Fig. 9.11), is

$$reldiff = \frac{\text{average of chrome depth at ends} - \text{chrome depth at middle}}{\text{chrome depth at middle}}$$

The anode length does not appear as an explanatory variable because long anodes were always used with rectifier 2 and short anodes with rectifiers 1 and 3, and any effect of anode length was therefore indistinguishable from any rectifier effect. The time and current combinations used gave a similar number of ampere-hours spread over longer or shorter time intervals. The day of the week was coded from 1 for Monday to 5 for Friday; a linear progression was thought reasonable because the anodes are cleaned on Mondays. The qualitative variables are represented by *rec2*, *rec3*, *thief B*, *thief C*, and *capacity* and *position* which are coded as 0 or 1 according to the following scheme:

rec2   1 if rod in rectifier 2, 0 otherwise
rec3   1 if rod in rectifier 3, 0 otherwise
thief B   1 if thief B used, 0 otherwise
thief C   1 if thief C used, 0 otherwise
capacity   1 if rod is one of 6 rods loaded, 0 if one of 9 loaded
position   1 if rod in middle of rectifier, 0 if rod at either end.

Note that the coefficient of *thief B* measures the difference between thief B and thief A, and so on.

A first multiple regression analysis produced the following model (1).

$$reldiff = 0.264 - 0.0031\,rec2 - 0.0102\,rec3 - 0.000\,0021\,time \times current$$
$$- 0.0107\,thief\,B - 0.00207\,thief\,C + 0.0317\,capacity$$
$$+ 0.00284\,position + 0.000\,175\,time - 0.009\,11\,day + error$$

The estimated standard deviation of the errors is 0.0070, R^2_{adj} is 80%, and t-ratios (estimated cofficient/estimated standard deviation of coefficient) are

4.71, -0.81, -2.27, -3.18, -2.16, -0.37, 4.14, 1.19, 0.54, -2.16

for the constant and coefficients of *rec2*, *rec3*,..., *day*, respectively.

If the absolute value of a t-ratio exceeds about 1.7 there is some empirical evidence that the corresponding variable has an effect on the relative difference – that is to say, the 90% confidence interval for the coefficient does not include zero. If the t-ratio is less than 1.7 in absolute value we will feel happier about dismissing an apparent difference as due to chance. These conclusions are crucially dependent on the assumption that the errors are independent of the values of the explanatory variables and each other. One of the hazards of analysing happenstance data is that the errors may not be independent. In a designed experiment, randomization is a justification for treating the errors as independent. In model 1 the interpretation of some of the more significant explanatory variables is difficult. The negative sign for the product of time and current is hard to explain physically. A plausible explanation is that a longer plating time reduces the difference, but this is offset by the positive coefficient for time! The negative day effect, which implies that cleaner anodes result in a larger relative difference, is also surprising. Clean anodes might well result in a greater depth of chrome, but there is no reason why this should be associated with an increase in the relative difference. The significant effect of rectifier 3 has no obvious explanation. Since all the explanatory variables are quite highly correlated, it is possible that other combinations of variables which have more sensible physical interpretations provide regression models which are nearly as good.

If we assume that rectifier 3 is no different from rectifier 1 and drop it from the equation, together with the non-significant position variable, we still obtain a negative day effect but it can reasonably be attributed to chance. This model 2 is:

$$reldiff = 0.20 + 0.003\,60\,rec2 - 0.000\,0016\,time \times current - 0.0034\,thief\,B$$
$$- 0.0078\,thief\,C + 0.0204\,capacity - 0.000\,151\,time$$
$$- 0.003\,19\,day + error$$

The estimated standard deviation of the errors is 0.0075, R^2_{adj} is 77.4%, and the t-ratios are

$$3.80, \quad 1.38, \quad -2.14, \quad -0.84, \quad -1.51, \quad 3.34, \quad -0.49, \quad -0.89$$

for the constant and coefficients of $rec2, \ldots, day$, respectively. The most significant effect is still the capacity.

If day is also dropped we have a regression model 3:

$$reldiff = 0.162 + 0.0038\,rec2 - 0.000\,0014\,time \times current - 0.00024\,thief\,B$$
$$- 0.0117\,thief\,C + 0.0157\,capacity - 0.000\,272\,time + error$$

The estimated standard deviation of the errors is 0.0075, R^2_{adj} is 77.5%, and the t-ratios are

$$5.13, \quad 1.46, \quad -2.32, \quad -0.62, \quad -3.96, \quad 5.06, \quad -0.99$$

for the constant and coefficients of $rec2, \ldots, time$, respectively. A neat way of writing this model for a report is:

$$reldiff = 0.162 + 0.0038rec\,2 - 0.000\,0014\,time \times current$$

$$- \begin{cases} 0 & thief\,A \\ 0.0024 & thief\,B \\ 0.0117 & thief\,C \end{cases} + 0.0157\,capacity - 0.000\,272\,time + error$$

This model is almost as good (a simple definition of 'good' is that the smaller the standard deviation of the errors – or equivalently the larger R^2_{adj} – is, the better the model) as model 1 and it does make physical sense. The t-ratios suggest that *thief C* and *capacity* have an effect on the relative difference. The negative sign of the coefficient for *thief C* implies that the *reldiff* is reduced.

If *rec2* and *thief B* are dropped a model 4 is obtained:

$$reldiff = 0.175 - 0.000\,001\,3\,time \times current - 0.0116\,thief\,C$$
$$+ 0.0159\,capacity - 0.000\,332\,time + error$$

The estimated standard deviation of the errors is 0.0075, R^2_{adj} is 77%, and the t-ratios are

$$5.67, \quad -2.39, \quad -4.43, \quad 5.08, \quad -1.23$$

for the constant and coefficients of *time* $\times$ *current*, ..., *time*, respectively. All the explanatory variables are now statistically significant except *time*, and its sign is consistent with a hypothesis that the longer the time the less the relative difference. It is retained because it also appears in an interaction term. However, it makes no difference to the conclusion about *thief C* if it is dropped.

The end result of these analyses is that if it is assumed that rectifier 3 is no different from rectifier 1 (at least in the context of relative difference) and that cleaner anodes do not increase the relative difference, together with the 'usual' assumptions for multiple regression, then there is some evidence that thief type C is the best for reducing the relative difference. There is rather more substantial evidence that only loading six rods, not using the middle rectifier, is associated with larger relative differences. It is recommended that these interim conclusions should be followed up by a designed experiment. Then an effect of, for example, day of the week could be allowed for by treating days as blocks (section 10.3.2).

9.4.4 Hazards of retrospective analysis

Box *et al.* (1978) discuss seven problems that may occur in the analysis of happenstance data.

1. Inconsistent data.
2. Range of explanatory variables limited.
3. Confounding, or near-confounding, of effects.
4. Nonsense correlations due to lurking variables.
5. Errors correlated over time.
6. Dynamic relationships.
7. Feedback.

In this application the main problems come under categories 3 or 4. Any effect of using long anodes was indistinguishable from any difference between rectifier 2 and the other two rectifiers. Moreover, the correlations between the other explanatory variables made interpretation of the regression equations equivocal. These correlations could be avoided in a designed experiment, when the values of the explanatory variables are chosen in advance. A more serious limitation is the lack of randomization when collecting retrospective data. It is possible that the operator systematically makes some other slight changes to the procedure when using thieves of type C, which function in a different manner from types A and B. To the best of our knowledge this was not the case, but there could be some less obvious feature of the process associated with thieves of type C. In a designed experiment it is essential to explain to the operator that no variations to the experimental protocol must be made, and randomization must be used throughout to provide a proper justification for the statistical analysis.

There was no reason to suppose the data were inconsistent (category 1), as no changes to the procedures or source of electrolyte were made during the period. If batch numbers were included in the regression models there was no evidence of an effect and the possibility of a correlation of errors over time (category 5) did not seem a major problem. The fact that the range of explanatory variables was limited (category 2) restricts the conclusion about thief type to present operating ranges. There were no dynamic relationships (category 6) to affect the data collected. The feedback loops (category 7) are only a problem if they relate the dependent variable to an explanatory variable. There were no such feedback loops for this process.

9.4.5 Conclusion

We do not claim that a regression analysis of data extracted from past records is a substitute for properly designed experimental programmes. However, such experimental programmes should be ongoing and a preliminary analysis of past records may be a useful starting point. Possible advantages of this approach are discussed below.

First, it can form the basis for discussions with plant managers and operators, make them more enthusiastic about carrying out properly designed and randomized experiments, and emphasize the need to adhere strictly to the experimental protocol.

Second, it should help with the choice of treatments to be tested in the designed experiment and the range of values, or number of levels, for other explanatory variables. In the case described, the type C thieves are much easier to attach than the other types. The results of the regression analysis suggest they are also more effective in reducing the build-up of chrome at the edges. An experiment has been designed to seek the optimum plating conditions with the type C thief. It might then be advisable to run a further experiment to compare the type C thief with the type B at 'optimal' conditions, although it is likely that the possible reduction in build-up of chrome would not offset the practical advantages of using the type C thief.

Third, it provides an estimate of the random variation inherent in the process which is necessary to provide guidance about sample sizes for a designed experiment.

Finally, if a process has been unchanged over a long period, a time-series analysis may provide useful information which could not be elicited from a relatively short experiment. The length of record may offset the fact that variables which would be strictly controlled during an experiment have varied over the period. If such variation is 'smooth' it will appear as a trend, or cycle, whereas if it is 'haphazard' it will be incorporated in the random variation.

9.5 Summary

Multiple regression model

$$Y_i = \beta_0 + \beta_1 x_{1i} + \cdots + \beta_k x_{ki} + E_i$$

where the standard assumptions are that the E_i follow the $N(0, \sigma^2)$ distribution and are independent of the x and each other. There is no requirement that the x are uncorrelated, but if they are chosen in advance it is best to arrange that they are. This makes interpretation easier and improves the efficiency of the estimation.

Quadratic surface for two explanatory variables

$$Y_i = \beta_0 + \beta_1 x_1 + \beta_2 x_2 + \beta_3 x_1^2 + \beta_4 x_1 x_2 + \beta_5 x_5^2 + E_i$$

A special case of the multiple regression model. It is linear in the unknown parameters $\{\beta_j\}$.

Residuals

The residuals are defined by

$$r_i = y_i - \hat{y}_i$$

where

$$\hat{y}_i = \hat{\beta}_0 + \hat{\beta}_1 x_{1i} + \cdots + \hat{\beta}_k x_{ki}$$

The residuals can be thought of as estimates of the errors. The residual sum of squares, *RSS*, is defined by

$$RSS = \sum_i r_i^2$$

The unbiased estimate of the variance of the errors (σ^2) is the residual mean square (s^2) given by

$$s^2 = RSS/(n - k - 1)$$

The residuals are often referred to as 'errors' in computer output (MINITAB, for example).

Confidence intervals for coefficients

Statistical packages will provide estimates, $\hat{\beta}_j$, of the parameters, β_j. They also provide estimates of the standard deviations of these estimators $(\widehat{sd}(\hat{\beta}_j))$. A $(1 - \alpha) \times 100\%$ confidence interval for a particular β_j will be given by

$$\hat{\beta}_j \pm t_{n-k-1,\alpha/2} \widehat{sd}(\hat{\beta}_j)$$

How good is the fit?

If we add more explanatory variables a minimum requirement for an improved model is that the estimate of the standard deviation of the errors (s) should be reduced. It may be appropriate to require the 90% confidence intervals for the β_j to exclude zero, which is equivalent to an absolute value of a t-ratio greater than about 1.7, but this is a somewhat arbitrary criterion. I would retain explanatory variables if there is a good physical explanation for their effect, provided the estimated coefficient has the correct sign! I would also include all three product terms when fitting a quadratic surface, even if one of the t-ratios is small. If there are substantial correlations between explanatory variables, apparently different models may give very similar predictions within the range of the explanatory variables used to fit the model.

The proportion of the variance accounted for by the model, R^2, will always increase when extra explanatory variables are added.

$$R^2 = [\text{Sum of squares attributed to regression}/S_{yy}] \times 100\%$$
$$= [1 - \text{RSS}/S_{yy}] \times 100\%$$

where

$$S_{yy} = \sum(y - \bar{y})^2/(n - 1)$$

We define R^2_{adj} by

$$R^2_{adj} = \{1 - [(RSS)/(n - k - 1)]/[S_{yy}/(n - 1)]\} \times 100\%$$
$$= (1 - s^2/s_y^2) \times 100\%$$

where s_y is the standard deviation of the $\{y_i\}$ when no model is fitted. It will increase or decrease as s decreases or increases. Its only advantage is that it allows us to compare models with Y as the dependent variable with those using $\ln Y$ (for example) as the dependent variable. The advantage of s is that it has a straightforward interpretation.

The residuals should be looked at to check that the usual assumptions are plausible. The limits of predictions are most sensitive to discrepancies. First, a plot of residuals against the fitted values of y may show that the variability is related to Y. This could be modelled, and allowed for in limits of prediction.

The precision of the estimates of the coefficients would be improved if generalized least squares (section 11 of Appendix A5) is used for fitting.

Second, plots of the residuals against the explanatory variables, ideally in three dimensions against pairs of them, may indicate that quadratic terms, and interactions if the three dimensional plot is used, should be included.

Third, if the observations can be ordered, usually over time, the residuals can be plotted against time. If the trend is significant, time could be included as an explanatory variable.

The correlation between residuals at time t and $t + 1$ can also be calculated, even if time is already an explanatory variable. If there is a significant correlation generalized least squares should really be used. For example, a negative correlation might arise because a high yield from one run tends to leave a tarry residue in the reactor which has a detrimental effect on the yield of the next run.

Finally, a normal score plot may suggest that the assumption of normal errors is unrealistic. It may be more appropriate to use some transform of Y as the dependent variable.

Predictions

There is no empirical evidence that the model remains realistic outside the range of the explanatory variables used to fit it. If it must be used for extrapolation there should be good physical reasons for supposing the assumed relationship extends beyond the data. Extrapolating linear relationships usually seems safer than extrapolating quadratics, but even so should not be relied on very far. The limits of prediction are based on the usual assumptions and, if extrapolation is involved, assume the form of the model remains valid. A useful guide for 90% limits of prediction, provided the sample is of a reasonable size, is

estimate $\pm 1.7s$

It is too narrow because it does not allow for uncertainty in the parameter estimates. An underestimate of the 95% limits of prediction is

estimate $\pm 2s$

MINITAB (refer to section 9.1.1).

Suppose we have the data triples (x_{1i}, x_{2i}, y_i) in C1, C2 and C3 respectively. The command TPLOT gives a plot in the style of Fig. 9.3(a):

TPLOT   C3   C1   C2

The fitted plane and predictions would be obtained from

```
REGRESS   C3   2   C1   C2   STDRES   C98   FITS   C99;
   RESIDUALS C50;
   PREDICT   475   200;
   PREDICT   880   110.
```

The residual checks described in this summary could be implemented by

(i) PLOT   C50   C99

(ii) TPLOT   C1   C2   C50

(iii) $\begin{cases} \text{LAG} \qquad 1 \quad \text{C50} \quad \text{C51} \\ \text{CORREL} \qquad \text{C50} \quad \text{C51} \end{cases}$

(iv) $\begin{cases} \text{NSCO} \quad \text{C50} \quad \text{C60} \\ \text{PLOT} \quad \text{C50} \quad \text{C60} \end{cases}$

Using the mouse:

    Graph ▶ Plot ▶ Character Graphs ▶ Pseudo 3-D Plot...
    Stat ▶ Regression ▶ Regression...
    Graph ▶ Plot...
    Stat ▶ Time Series ▶ Lag...

Exercises

9.1 Calculate the correlation between x and x^2 if $\{x_i\}$ are

    (i) 101, 102, 103, 104, 105, 106, 107
    (ii) -3, -2, -1,   0,   1,   2,   3.

9.2 The data in Table 5.2 were the drying time y (in hours) for a random sample of 11 dishes of varnish, each of which had a different amount of additive, x, mixed with it.

    (i) Plot the data.
    (ii) Fit the regression of y and x and draw the line on your graph.
    (iii) Fit a regression of y on x and x^2, and draw the parabola on your graph,
    (iv) Predict the amount of additive which will lead to the shortest drying time.

9.3 Show that the formulae for s^2 and limits of prediction in multiple regression, given in Appendix A5, reduce to those for a linear regression of y on one explanatory variable when the number of explanatory variables is one.

9.4 In the manufacture of car tyres it would be useful to be able to predict abrasion loss (y) from measurements of the hardness (x_1) and tensile strength (x_2) of the rubber. Both x_1 and x_2 are relatively easy to measure. An experiment was carried out and 30 data triples (y_i, x_{1i}, x_{2i}) were obtained. The model

$$Y_i = \alpha_0 + \alpha_1 x_{1i} + E_i$$

was fitted to the data by least squares. The residual sum of squares was 102 565 and the total corrected sum of squares S_{yy} (i.e. $\sum (y_i - \bar{y})^2$) was 225 011. The model

$$Y_i = \beta_0 + \beta_1 x_{1i} + \beta_2 x_{2i} + E_i$$

was fitted and the residual sum of squares was reduced to 35957. Calculate the values of s, R^2, and R^2_{adj} for the two models.

The following exercises need access to a multiple regression routine.

9.5 The data in Table D.8 of Appendix D were obtained from a process at ICI Runcorn (Wetherill, 1982). A melt consisted of calcium chloride, sodium chloride, calcium and sodium. The temperature in degrees Celsius (x_1), percentage calcium in the salt phase (x_2), and percentage calcium in the metal phase (y) were measured at the end of 28 separate runs.

(i) Fit a regression plane for y on x_1 and x_2.
(ii) Investigate the residuals for any evidence of curvature.
(iii) Fit a quadratic surface, before and after scaling x_1 and x_2 by subtracting their means.

9.6 The 57 data in Table D.9 of Appendix D are from a study carried out by the Transportation Centre at Northwestern University, Chicago, in 1962. Each datum consists of five variables for a traffic analysis zone. They are in five columns.

column 1: trips per occupied dwelling unit
column 2: average car ownership
column 3: average household size
column 4: socio-economic index
column 5: urbanization index.

(i) Fit the best regression you can find for the number of trips in terms of the variables in columns 3–5 (including their interactions if appropriate).
(ii) Fit the best regression you can find for the average car ownership in terms of the variables in columns 3 to 5.
(iii) Fit the regression of average car ownership in terms of all the other variables.

9.7 A company specializes in setting up computer systems for insurance companies. A manager has kept a record of the time taken (x_1) and the number of snags encountered (y) for the latest 40 contracts, together with the assessment of difficulty she made before work started (x_2). The data, which are genuine but arose in a completely different context, are given in Table D.10.

(i) Regress y on x_1 only.
(ii) Regress y on x_2 only.
(iii) Regress y on x_1 and x_2. Does anything surprise you? Can you explain your finding?
(iv) The company has three systems analysts, including the manager. She is responsible for accepting a contract, but once accepted any one of

the analysts may do the work. How could she allow for analysts having different perceptions of problems, in her regression model?

9.8 Refer to the data in Table 6.3. Use a regression analysis to test for a difference between means in the corresponding populations. How does this compare with a two sample t-test:

(i) if variances are not assumed the same?
(ii) if variances are assumed the same?

9.9 On 25 October 1991 the *Evening Standard*, a London newspaper, printed the headline, 'Are you heading towards a crash?' The article referred to a formula devised by the Transport and Road Research Laboratory, UK, which gives the expected number of accidents in a year, A_c, for various categories of motorist. The formula was reproduced with the caption: 'For the brainy ... how to work out the risks using the Transport and Road Research Laboratory formula.' The formula was later printed, red on yellow, on the front cover of the *Sunday Times* magazine of 22 December, above a headline, '1991: a year of glorious follies.' The formula that attracted journalists' attention is given below:

$$A_c = 0.006\,33 \exp\{s + g\}(1 + 1.6p_d)(p_b + 0.65p_r + 0.88p_m)M^{0.279}$$
$$\times \exp\{b_1/A + b_2/(X + 2.6)\}$$

into which the following substitutions are made:

s = 0 for men; -0.02 for women
g = 0 for social groups A, B, C1; -0.72 for social groups C2, D, E
p_d = the proportion of driving done in the dark
p_b = the proportion of driving done in the town
p_r = the proportion of driving done in the countryside
p_m = the proportion of driving done on motorways
M = annual mileage
b_1 = 13 for social groups A, B, C1; 23 for social groups C2, D, E
b_2 = 3.5 for men; 2.3 for women
A = age
X = years since passing test.

(a) (i) Calculate your expected number of accidents next year.
    (ii) Does it imply that women or men have fewer accidents per mile, other things being equal, or not?
(b) (i) Is it in the form of a standard multiple regression model?
    (ii) Write down the expression that would have been minimized to fit the model by least squares.

10

Design of experiments

There are a few general principles which should be kept in mind when planning experiments. First, decide which variable you wish to predict and thereby maximize, minimize, or keep within a specification. I shall refer to this as a **response variable**.

Second, list the variables which may affect the response. Split them into those which you can set to given values, the **control** variables, and those over which you have no control but can measure in some way, the **concomitant** variables.

Third, allow for known sources of variation by choosing an appropriate experimental design. The relatively simple strategy of **blocking**, a generalization of the paired comparison procedure for comparing two treatments, is often used. You should try, during the course of your experiment, to cover the variation that will be met in normal operating conditions. For example, it would be rash to suggest a shipyard change all its welding practices because of successful trials using 25 mm thickness plates of a single steel type.

Fourth, randomize the experimental materials to runs and the order of runs, subject to the constraints imposed by the experimental design. The reason for randomizing is that it makes the independence assumptions about the errors in the regression model plausible, and hence provides the basis for assessing the accuracy of the results. You should not rely on randomizing to even out known sources of variation. It may not, and there is no point in continuing with an obviously biased experiment because you would obtain the 'right' answer if you were to repeat the experiment, with independent randomizations, millions of times! Even when randomization does turn out much as you would have liked, the estimates, for a given amount of resources, will be less accurate than if an appropriate experimental design had been used. Admittedly, the use of concomitant variables may help to salvage the situation but undue reliance on these is one of the limitations of happenstance data.

Finally, you expect the experiment to provide potentially valuable information about the process. If this potential is to be realized you need to communicate the results clearly and succinctly to the people making the decisions (Greenfield, 1993). A report should start with a summary of the objectives of the experiment, the findings, and recommendations for changes with some attempt at quantifying the benefits (Lindley, 1985). This may be the only part of your report which receives immediate attention, but the technical detail must also be clearly documented. You are unlikely to have covered every aspect of the problem. You may also have been 'unlucky', inasmuch as a confidence interval for some parameter clearly indicates an improvement but does not in fact include the actual value of that parameter (5% of all independent 95% confidence intervals constructed, for example). You should advise that the consequences of the changes be monitored, and you or someone else will need to compare these new findings with the results of your original experiment. This is the essence of evolutionary operation, which is described in the next section.

10.1 Evolutionary operation

The yield of a liquid chemical used in the pharmaceutical industry is known to depend on the pressure and temperature inside the reactor. Both variables can be controlled quite accurately, and the specified settings have been 160 kPa and 190°C for as long as anyone can remember. It is safe to operate the process with changes in temperature and pressure up to at least 10%. A graduate student on a placement proposed experimenting with small changes in these two variables. The plant would continue to produce satisfactory product and the experimental programme should identify optimum settings for pressure and temperature. If these turned out to be different from the current specification, substantial savings would be made. The site manager was interested in the possible benefits but remained sceptical because this was presumably how the specified values had been decided.

The student then suggested that previous experiments might have relied on a one-variable-at-a-time strategy, and explained that this might not have led to the best settings. A possible dependence of yield on pressure and temperature is represented by the contour diagram in Fig. 10.1. This is an example of a **response surface**, and the general shape shown in the diagram is quite common in the chemical industry. If we fix pressure at 150 kPa and vary the temperature we will find a highest yield of 32 kg at 190°C. Now suppose we fix the temperature at 190°C and vary the pressure. The highest yield becomes 34 kg when the pressure is 160 kPa. This is a long way from the optimum of 48 kg. The pressure and temperature are said to interact – that is, the effect of changing one depends on the value of the other. This is just one possible scenario, and alternatives include the possibility that the present settings are already at the highest point or that the response surface is of a quite different shape (multi-peaked or saddle-shaped, for example). The objective of an experimental programme is to infer the nature of the surface, as efficiently as possible, and to use this information to specify optimum operating conditions.

The simplest design which makes changes in both variables is the 2^2 factorial design. Each factor appears as high (coded $+1$) or low (coded -1). This gives four runs, allowing a plane to be fitted with one degree of freedom for error. The results from such an experiment are given in Table 10.1. The runs were

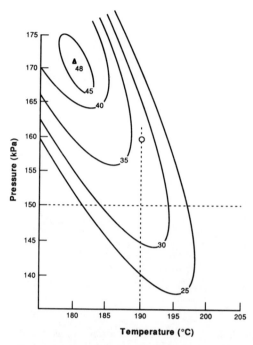

Fig. 10.1 Contours of yield (kg) of a chemical process. The highest point with pressure fixed at 150 occurs when temperature is 190°C. The highest point when temperature is fixed at 190°C is marked ○ and corresponds to pressure of 160 kPa.

Table 10.1 Yield from chemical process: experiment 1

Pressure (5 kPa from 160 kPa)	Temperature (5°C from 190°C)	Yield (kg)
1	1	46.2
−1	−1	43.2
1	−1	44.9
−1	1	43.7

carried out in a random order, which helps justify the assumption of random errors in the regression model. The fitted plane, where y is the yield, x_1 the pressure and x_2 the temperature, is

$$y = 44.50 + 1.05x_1 + 0.450x_2$$

with $s = 0.40$, and $R^2 = 97\%$.

The standard deviation of all the estimated coefficients is 0.20 but, with only one degree of freedom for error, the 90% confidence intervals are inevitably

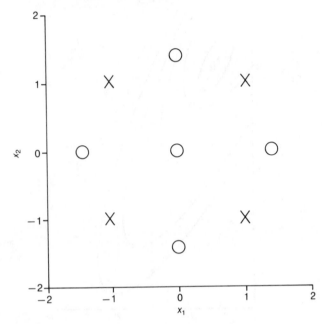

Fig. 10.2 Composite design consists of 2^2 factorial ($\times$) as first stage, followed by star design ($\bigcirc$) as second stage.

wide and include 0. As any movement would be away from the specified settings, we thought it prudent to augment the experiment with another five runs arranged as a **star design** (Fig. 10.2). Two-stage designs like this are called **composite designs**. The runs were carried out in a random order and the data are given in Table 10.2. The plane was fitted to all nine data

$$y = 43.84 + 0.88x_1 + 0.76x_2$$

with $s = 0.90$ and $R^2 = 69\%$.

Table 10.2 Yield from chemical process: experiment 2

Pressure (5 kPa from 160 kPa)	Temperature (5°C from 190°C)	Yield k(g)
0	−1.4	42.1
1.4	0	43.9
−1.4	0	41.9
0	0	43.6
0	1.4	45.1

The standard deviation of the coefficients of x_1 and x_2 has actually increased to 0.32 because the reduction due to additional data has been more than offset by the increase in s. The confidence intervals are, nevertheless, narrower because of the increase in degrees of freedom to six. In fact the 90% confidence intervals for x_1 and x_2 are 0.88 ± 0.62 and 0.76 ± 0.62, respectively.

It is possible to fit a quadratic surface to the data in the composite design, but this results in an increase in s to 1.17, which is not much less than the unconditional standard deviation of the nine yields, 1.40. As a result of all these analyses we felt sufficiently confident to carry out a similar composite design based on a new centre point in the direction of steepest ascent of the plane. To find this direction, suppose we increase x_1 by 1 unit. We need to find the increase in x_2 such that

$$\frac{\text{change in } y}{\text{change in vector } (x_1, x_2)}$$

is a maximum. Let this value of x_2 be d. Then we wish to maximize

$$g = \frac{0.884 + 0.758d}{\sqrt{1 + d^2}}$$

with respect to d. Elementary calculus leads to the result that

$$d = 0.758/0.884 = 0.86$$

An increase in pressure by 1 unit (5 kPa) did seem reasonable. It kept the process well within safe operating conditions, and the corresponding increase in temperature was rounded to 4°C. Another composite design experiment was carried out and the results are given in Table 10.3. A quadratic surface fitted to these data does have a smaller associated estimate of standard deviation of the

Table 10.3 Yield from chemical process: experiment 3

Pressure (5 kPa from 165 kPa)	Temperature (5°C from 194°C)	Yield (kg)
1.0	−1.0	46.7
−1.0	1.0	47.1
−1.4	0.0	45.9
1.4	0.0	43.7
0.0	−1.4	45.9
0.0	0.0	48.0
1.0	1.0	44.2
0.0	1.4	45.1
−1.0	−1.0	47.7

errors (1.16) than a plane (1.26). This surface is

$$y = 47.9 - 0.881x_1 - 0.533x_2 - 1.27x_1^2 - 0.475x_1x_2 - 0.909x_2^2$$

and it is represented by contours, drawn using MINITAB, in Fig. 10.3. But the coefficients have not been determined very accurately, there are only three degrees of freedom for error and s is not strikingly smaller than the unconditional standard deviation of the yields, 1.50. We would not have wished to make any firm recommendations on the basis of this fit, and decided to carry out a further experiment centred on the same operating point. An alternative to the composite design, which still allows the fitting of interaction and quadratic terms and has a slight advantage that these are all uncorrelated, is the 3^2 design, in which each factor now appears at three levels, high, medium or low. The results from this design are given in Table 10.4. An analysis of all 18 results centred on 165 kPa and 194°C gives the equation,

$$y = 48.805 - 0.889x_1 - 0.519x_2 - 1.533x_1^2 - 0.537x_1x_2 - 1.167x_2^2$$

```
MTB  > read 'table10p3.dat' c1 c2 c3
MTB  > name c1 'Pressure' c2 'Temp' c3 'Yield'
MTB  > let c11 = c1*c1
MTB  > let c12 = c1*c2
MTB  > let c22 = c2*c2
MTB  > name c11 'pressqu' c12 'interact' c22 'tempsqu'
MTB  > regress c3 5 c1 c2 c11 c12 c22;
SUBC > coeff c234.
```

The regression equation is
Yield = 47.9 − 0.881 Pressure − 0.533 Temp − 1.27 pressqu
          − 0.475 interact − 0.909 tempsqu

Predictor	Coef	Stdev	t-ratio	p
Constant	47.947	1.158	41.41	0.000
Pressure	−0.8813	0.4116	−2.14	0.122
Temp	−0.5328	0.4116	−1.29	0.286
pressqu	−1.2657	0.6870	−1.84	0.163
interact	−0.4750	0.5791	−0.82	0.472
tempsqu	−0.9086	0.6870	−1.32	0.278

$s = 1.158$ R-sq = 77.6%   R-sq(adj) = 40.2%

```
MTB > note\ We now have to set up a column (c30) which contains
MTB > note\ number of factors (2 in our example)
MTB > note\ number of blocks (0 in our example)
MTB > note\ location of the blocking variable (0 as none)
MTB > note\ location of factors (columns 1 & 2 in this example)
MTB > note\ regression coefficients in the following order:
MTB > note\ constant; linear; quadratic; interactions.
MTB > note\ First copy the coefficients, in c234, to constants.
MTB > copy c234 k100 k1 k2 k11 k12 k22
MTB > note\ Now set up the information for the macro.
MTB > copy 2 0 0 1 2 k100 k1 k2 k11 k22 k12 c30
MTB > %contour 'Yield' c30 'Pressure' 'Temp'
MTB > note\ change order of coefficients for Temp vertical
MTB > copy 2 0 0 2 1 k100 k2 k1 k22 k11 k12 c40
MTB > %contour 'yield' c40 'Temp' 'Pressure'
```

Contour Plot for : Yield

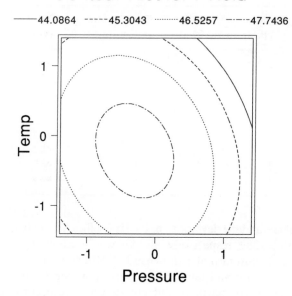

Contour Plot for : Yield

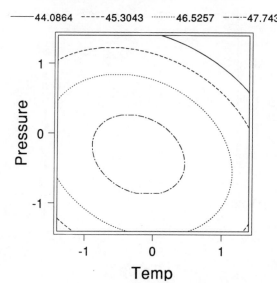

Fig. 10.3 Contour plot for surface fitted to yields from Experiment 3.

Table 10.4 Yield from chemical process: experiment 4

Pressure (5 kPa from 165 kPa)	Temperature (5°C from 194°C)	Yield (kg)
−1	1	47.8
1	1	44.8
−1	0	48.9
0	−1	49.7
1	0	47.1
0	1	47.1
0	0	48.2
1	−1	46.2
−1	−1	46.8

with $s = 0.94$. The standard deviations of the six coefficients are 0.550, 0.252, 0.252, 0.389, 0.332, and 0.389, respectively, and we were reasonably confident about the fitted model. A contour plot is shown in Fig. 10.4 and the estimated optimum conditions are pressure and temperature corresponding to about -0.25 and -0.2, respectively. If you partially differentiate y with respect to x_1 and x_2, and set the derivatives equal to zero you will obtain the exact maximum for the

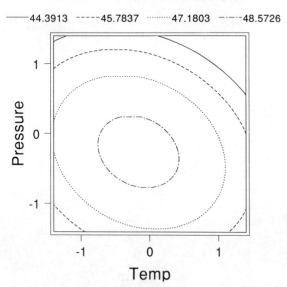

Fig. 10.4 Contour plot for yields from all 18 runs of Experiments 3 and 4.

fitted surface at -0.26 and -0.16. The three-dimensional representation of the response surface (Fig. 10.5) was produced with the experimental design package, DEX, which is particularly easy to use. We recommended that the specified operating conditions be changed to a pressure of 163.7 kPa and a temperature of 193.2°C, that the yields be carefully monitored, and that occasional further experiments with changes in pressure of 2 kPa and temperature of 2°C be carried out. Our 95% prediction limit for the yield at the estimated optimum is

$$48.96 \pm 2.34$$

For further reading on evolutionary operation see the book by Box and Draper (1969).

10.2 More than two factors

We often have more than two control variables for a process. Devore (1982) asks readers to analyse data from a study of desizing process used to remove impurities from cellulose goods. The data themselves are from Mousa (1976). Four factors were tried at low and high levels: enzyme concentration (A, 0.50 and $0.75 \, \mathrm{g \, l^{-1}}$), pH (B, 6.0 and 7.0), temperature (C, 60°C and 70°C), and time (D, 6 hours and 8 hours). The response was percentage starch by weight.

Let the four variables x_1, x_2, x_3 and x_4 be coded -1 or $+1$ for low or high levels of the four factors A, B, C and D. A full factorial experimental design will consist of all 2^4 possible combinations of the four factors at high and low levels. These 16 runs should be carried out in a random order. The convention for naming treatments is that the lower-case letter appears if the corresponding factor is at the upper levels. The 1 corresponds to all factors being at the low

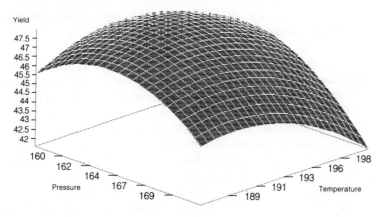

Fig. 10.5 DEX plot of yield against temperature and pressure.

level. Since all variable values, including all possible interactions, are restricted to -1 or $+1$ it is convenient to abbreviate these to $-$ and $+$ respectively.

x_1	x_2	x_3	x_4	Treatment
$-$	$-$	$-$	$-$	1
$-$	$-$	$-$	$+$	d
$-$	$-$	$+$	$-$	c
$-$	$-$	$+$	$+$	cd
$-$	$+$	$-$	$-$	b
$-$	$+$	$-$	$+$	bd
$-$	$+$	$+$	$-$	bc
$-$	$+$	$+$	$+$	bcd
$+$	$-$	$-$	$-$	a
$+$	$-$	$-$	$+$	ad
$+$	$-$	$+$	$-$	ac
$+$	$-$	$+$	$+$	acd
$+$	$+$	$-$	$-$	ab
$+$	$+$	$-$	$+$	abd
$+$	$+$	$+$	$-$	abc
$+$	$+$	$+$	$+$	abcd

The results of two replicates are given in Table 10.5. Since the design is nicely balanced – for example, the low level of A appears with the same eight possible combinations of B, C and D as does the high level – the effect of moving from low A to high A can be estimated by comparing the two means, and so on. We will fit a regression model and the estimated effect of A will correspond to twice the coefficient of x_1, because we have coded low A as -1 and high A as $+1$. A full model will consist of the main effects x_1, x_2, x_3, x_4, all six second-order interactions $x_1 \times x_2$ and so on, the four third-order interactions $x_1 \times x_2 \times x_3$ and so on, the fourth-order interaction $x_1 \times x_2 \times x_3 \times x_4$ and an indicator variable for the replicate. This latter will be denoted by x_{16}, and coded 0 for replicate 1 and 1 for replicate 2. Another consequence of the balanced design is that the correlations between any two of the variables $x_1, \ldots, x_{16}$ are all zero, so that estimates of their coefficients do not depend on other terms in the model. Note, however, that the estimated standard deviations of these coefficients do depend on which other terms are included. None of the t-ratios for the third- and fourth-order interactions is much above 1, so 90% confidence intervals would include zero. I happily drop them from the model, because finding a physical interpretation for high-order interactions is awkward. The t-ratio for the replicate exceeds 2, so I decide to retain it. The model with main effects and

Table 10.5 Percentage starch by weight for desizing experiment

Treatment	1st replicate	2nd replicate
(1)	9.72	13.50
a	9.80	14.04
b	10.13	11.27
ab	11.80	11.30
c	12.70	11.37
ac	11.96	12.05
bc	11.38	9.92
abc	11.80	11.10
d	13.15	13.00
ad	10.60	12.37
bd	10.37	12.00
abd	11.30	11.64
cd	13.05	14.55
acd	11.15	15.00
bcd	12.70	14.10
abcd	13.20	16.12

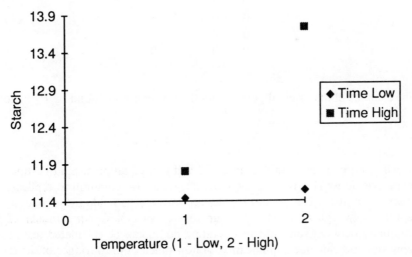

Fig. 10.6 Starch (% weight) by temperature and time.

290 Design of experiments

second-order interactions is

$$y = 11.55 + 0.07x_1 - 0.25x_2 + 0.50x_3 + 0.64x_4 + 0.33x_1x_2 + 0.09x_1x_3$$
$$- 0.17x_1x_4 + 0.15x_2x_3 + 0.16x_2x_4 + 0.46x_3x_4 + 1.16x_{16}$$

with $s = 1.19$.

The standard deviation of all the coefficients is 0.21, except for the constant (0.30) and replicate (0.42). The most significant variables correspond to temperature, time and their interaction. High time and high temperature lead to a particularly high percentage of starch (Fig. 10.6). For some reason the percentage starch appears to have been significantly higher in the second replicate. There is nothing remarkable about the residuals.

10.2.1 Quadratic effects

If we want to fit a surface that includes quadratic effects we will have to introduce more levels of the variables. The most economical way to achieve this would be to use nine additional points. For example,

x_1	x_2	x_3	x_4
$-\alpha$	0	0	0
α	0	0	0
0	$-\alpha$	0	0
0	α	0	0
0	0	$-\alpha$	0
0	0	α	0
0	0	0	$-\alpha$
0	0	0	α
0	0	0	0

The distance of the original points from the centre can be found by Pythagoras' theorem; it is

$$\sqrt{(1^2 + 1^2 + 1^2 + 1^2)} = 2$$

So with four control variables, an α of 2 will give equal precision of estimation in all directions. This is the preferred value. It is also common to replicate the observation at the centre, and the design can even be made orthogonal if it is replicated enough times (12 for four control variables). An estimate of the variability between runs, at this operating point, can be calculated from these replicates and can give an indication of how much of the variance of the errors is attributable to the approximate nature of the regression model. However, such comparisons do assume that the variation between runs at the same operating point is independent of its location. Devore (1982) provides data from a study of the relationship between brightness of finished paper (y) and the control variables, hydrogen peroxide (x_1, percentage by weight), sodium hydroxide (x_2, percentage by weight), silicate (x_3, percentage by weight) and process temperature (x_4). The data themselves (Table 10.6) are from Slove (1964). The

Table 10.6 Brightness of paper

H_2O_2 (x_1)	NaOH conc. (x_2)	Silicate conc. (x_3)	Temperature (x_4)	Brightness (y)
−1	−1	−1	−1	83.9
+1	−1	−1	−1	84.9
−1	+1	−1	−1	83.4
+1	+1	−1	−1	84.2
−1	−1	+1	−1	83.8
+1	−1	+1	−1	84.7
−1	+1	+1	−1	84.0
+1	+1	+1	−1	84.8
−1	−1	−1	+1	84.5
+1	−1	−1	+1	86.0
−1	+1	−1	+1	82.6
+1	+1	−1	+1	85.1
−1	−1	+1	+1	84.5
+1	−1	+1	+1	86.0
−1	+1	+1	+1	84.0
+1	+1	+1	+1	85.4
−2	0	0	0	82.9
+2	0	0	0	85.5
0	−2	0	0	85.2
0	+2	0	0	84.5
0	0	−2	0	84.7
0	0	+2	0	85.0
0	0	0	−2	84.9
0	0	0	+2	84.0
0	0	0	0	84.5
0	0	0	0	84.7
0	0	0	0	84.6
0	0	0	0	84.9
0	0	0	0	84.9
0	0	0	0	84.5
0	0	0	0	84.6

model which includes all the interaction and squared terms is:

$$y = 84.671 + 0.650x_1 - 0.258x_2 + 0.133x_3 + 0.108x_4 + 0.038x_1x_2$$
$$- 0.075x_1x_3 + 0.213x_1x_4 + 0.200x_2x_3 - 0.187x_2x_4 + 0.050x_3x_4$$
$$- 0.135x_1^2 + 0.028x_2^2 + 0.028x_3^2 - 0.072x_4^2$$

with $s = 0.353$.

The t-ratios for the coefficients of the interactions x_1x_4, x_2x_3, x_2x_4 and x_1^2 all exceed 2. If all squared terms are dropped, s increases to 0.369 and if the interactions are also dropped s increases to 0.430. I would use the model which included all interaction and squared terms for planning purposes. The standard deviation of the seven brightness readings at the centre is 0.17, which suggests that either modelling error is the main constituent of the errors or that measurement error increases at operating conditions away from the centre. Another possible explanation would be correlated errors, if the seven centre points were consecutive runs. However, the order of runs should have been randomized making this an extremely unlikely event! Notice that these composite designs require far fewer runs than the 81 needed for a 3^4 design. The slight correlations among the explanatory variables are not usually inconvenient. Admittedly, the 81 runs would lead to more accurate estimation of the parameters of the model, but such a large experimental programme would often be quite impractical.

10.2.2 Fewer runs

Sometimes there may not be sufficient time or experimental material for a full factorial experiment. A fractional factorial design offers a useful compromise. I shall explain the idea in the context of the desizing experiment. A single replicate of the full factorial consists of 16 combinations. These are followed by the sign

Table 10.7 Percentage starch by weight for a half replicate of the desizing experiment

Enzyme concentration (x_1)	pH (x_2)	Temperature (x_3)	Time (x_4)	Treatment	Percentage starch (y)
−1	−1	+1	+1	cd	13.05
+1	−1	+1	−1	ac	11.96
+1	+1	+1	+1	abcd	13.20
+1	−1	−1	+1	ad	10.60
+1	+1	−1	−1	ab	11.80
−1	−1	−1	−1	1	9.72
−1	+1	−1	+1	bd	10.37
−1	+1	+1	−1	bc	11.38

of the fourth-order interaction, and signs of a typical third-order interaction and of a pair of second-order interactions when the fourth-order interaction is positive. Remember that -1 and $+1$ have been abbreviated to $-$ and $+$.

x_1	x_2	x_3	x_4	Treatment	$x_1x_2x_3x_4$	$x_1x_2x_3$	x_1x_2	x_3x_4
$-$	$-$	$-$	$-$	1	$+$	$-$	$+$	$+$
$-$	$-$	$-$	$+$	d	$-$			
$-$	$-$	$+$	$-$	c	$-$			
$-$	$-$	$+$	$+$	cd	$+$	$+$	$+$	$+$
$-$	$+$	$-$	$-$	b	$-$			
$-$	$+$	$-$	$+$	bd	$+$	$+$	$-$	$-$
$-$	$+$	$+$	$-$	bc	$+$	$-$	$-$	$-$
$-$	$+$	$+$	$+$	bcd	$-$			
$+$	$-$	$-$	$-$	a	$-$			
$+$	$-$	$-$	$+$	ad	$+$	$+$	$-$	$-$
$+$	$-$	$+$	$-$	ac	$+$	$-$	$-$	$-$
$+$	$-$	$+$	$+$	acd	$-$			
$+$	$+$	$-$	$-$	ab	$+$	$-$	$+$	$+$
$+$	$+$	$-$	$+$	abd	$-$			
$+$	$+$	$+$	$-$	abc	$-$			
$+$	$+$	$+$	$+$	abcd	$+$	$+$	$+$	$+$

The estimated effect of the fourth-order interaction is found by subtracting the average of the eight results at conditions corresponding to the $-$ signs from the average of the eight results at conditions corresponding to the $+$ signs. If we forgo this comparison we might get away with only eight runs. The drawbacks can be seen from the display of signs. The third-order interaction $x_1x_2x_3$ will be indistinguishable from x_4, and the second-order interactions x_1x_2 and x_3x_4 will be indistinguishable from each other. This is because if $x_1x_2x_3x_4$ equals $+1$ then x_1x_2 and x_3x_4 must both equal $+1$ or -1. Similarly for $x_1x_2x_3$ and x_4 and so on. Indistinguishable effects are known as **aliases**. I shall leave you to convince yourself that the aliases of any effect can be found from the product of that effect with ABCD, subject to a rule that any letter squared equals 1. For example,

$$AB(ABCD) = A^2B^2CD = CD$$

So AB is an alias of CD. If the main effects are dominant the half fraction corresponding to the $+$ signs, or equally the half corresponding to the $-$ signs, may suffice. If the interaction is significant, and we wish to distinguish exactly what is interacting with what, we will have to do the other half of the full factorial. The general idea can be extended to quarter fractions, and so on, if there are a large number of variables, and there often are in the early stages of an experimental programme. A quarter replicate is described in Example 10.2 and further details are given in Box *et al.* (1978), Devore (1982) and Montgomery (1991), among many others. Alternatively an experimental design package such as DEX will do it all for you.

Example 10.1

Suppose that there had only been time for a half replicate of the desizing experiment, and that the fraction corresponding to the positive sign of the fourth-order interaction had been selected by the toss of a coin. The data that would have been obtained are given in Table 10.7. A regression on the main effects gives

$$y = 11.510 + 0.380x_1 + 0.178x_2 + 0.888x_3 + 0.295x_4$$

and the standard deviation of all the coefficients is 0.320. There are only three degrees of freedom for error, and the 90% confidence interval for the most significant coefficient, that of temperature, is 0.89 ± 0.75.

It is also possible that a third-order interaction is contributing towards any apparent temperature effect, but this would inevitably go undetected from this single experiment. We can fit a regression on x_1, x_2, x_3, x_4 with x_1x_2, x_1x_3 and x_1x_4, but this uses up all the degrees of freedom. A normal score plot of the coefficients of the seven variables might show up dominant effects. The principle is that if none of the variables has any effect on y, the estimated coefficients would be a random sample from a normal distribution. The normal score plot for this example is shown in Fig. 10.7, and nothing stands out. However, this certainly does not imply nothing has an effect. If the error is negligible all the variables would affect the percentage starch, yet the normal score plot would look the same. It is also important to remember that any effect attributed to x_1x_4 could equally well be due to x_2x_3 or some combination of both interactions.

Although the results from this half replicate would have hinted that temperature had some effect, the natural variation in the process is too great for the technique to be very useful in this situation. We would probably have followed up with at least the other half replicate. Fractional replication generally becomes more useful as the number of control variables increases. It can be combined with star designs to fit main effects, two-variable interactions and quadratic effects.

Example 10.2

The diagram of the cement kiln shown in Fig. 10.8 and the background information are based on work by Norman and Naveed (1990, for example), but the experiment and data are fictitious.

The meal which is fed into the kiln is a mixture of limestone, clay, sand and

```
MTB > nsco c51 c52
MTB > print c51 c52

              coeff      norscor

   x1        0.3800      0.35147
   x2        0.1775     -0.35147
   x3        0.8875      1.36459
   x4        0.2950      0.00000
  x1x2       0.4325      0.75613
  x1x3      -0.1975     -0.75613
  x1x4      -0.2850     -1.36459

MTB > plot c51 c52
```

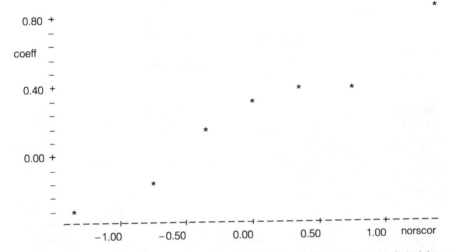

Fig. 10.7 Normal score plot of parameter estimates from half replicate of desizing experiment.

iron ore. The main ingredient is the limestone, and the proportions of the other materials are specified for given grades of cement. The response for a particular grade was the percentage of free lime in the cement product, for which the target range was between 1% and 2%. Six control variables were identified: feed rate (x_1); rotation speed (x_2); proportion of fuel to oxidant (x_3); fuel rate (x_4); fan 1 speed (x_5); and fan 2 speed (x_6). Small changes about the standard operating conditions were made, and they were kept sufficiently small for all combinations to be feasible. If larger changes had been made it would not have been sensible to try a slow rotation speed with fast feed, or low fuel rate with fast feed, and so on. The overall objective was to maximize the ratio of feed rate to fuel rate while remaining safely within the target, for free lime. The kiln was operated

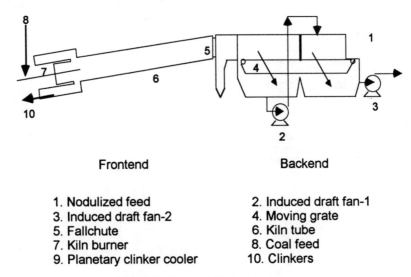

Frontend   Backend

1. Nodulized feed   2. Induced draft fan-1
3. Induced draft fan-2   4. Moving grate
5. Fallchute   6. Kiln tube
7. Kiln burner   8. Coal feed
9. Planetary clinker cooler   10. Clinkers

Fig. 10.8 Cement kiln and feed grate.

continuously, and during the experiment samples of cement were taken for X-ray analysis at 5-minute intervals. The process does not respond immediately to changes in the control variables, so after each change an hour was allowed for the process to reach a steady state before samples were taken. Each run lasted for 4 hours and the response was the average of the 48 analyses. A full factorial experiment, 2^6, would need 64 runs, and a star design to allow for the estimation of quadratic effects would require an extra 13 runs. The whole programme would have taken 16 days. A quarter replicate of the full factorial, 2^{6-2}, with the star design was decided on. It could be extended to a half (2^{6-1}) or full factorial (2^6) with a follow-up programme. The 2^{6-2} design can be constructed by first writing down a full factorial for four factors, and then associating two additional columns with two chosen interactions. Suppose we choose the interactions ABCE and BCDF. These are called the **design generators**. The **generalized interaction** of ABCE and BCDF is defined as their product, subject to squared terms being replaced by 1:

$$(ABCE)(BCDF) = ADEF$$

The complete defining relationship is said to be

$$I = ABCE = BCDF = ADEF$$

The aliases of any effect are found by multiplying into each **word** of the defining relation, again subject to squared terms being replaced by 1. For example, the main effect C has aliases

$$C = ABE = BDF = ACDEF$$

Table 10.8 Alias structure of 2^{6-2}; 1/4 replicate of 6 factors in 16 runs

$$I = ABCE = BCDF = ADEF$$

Aliases

$A = BCE = DEF$	$AB = CE$
$B = ACE = CDF$	$AC = BE$
$C = ABE = BDF$	$AD = EF$
$D = BCF = AEF$	$AE = BC = DF$
$E = ABC = ADF$	$AF = DE$
$F = BCD = ADE$	$BD = CF$
$ABD = CDE = ACF = BEF$	$BF = CD$
$ACD = BDE = ABF = CEF$	

The complete alias structure is shown in Table 10.8.

Although the composition of the meal is kept as consistent as possible, by thorough mixing during preparation, some natural variation is inevitable. Two concomitant variables were therefore monitored: the 'burnability' (x_7) and the water content (x_8). Both were based on four hourly spaced samples of raw meal during each run. Burnability is a measure of the potential of the kiln feed of being liquefied at high temperatures. It is a linear combination of the lime saturation factor, silica ratio, and the sum of magnesium oxide and alkalis. The runs were carried out in a random order, but the data are given in the logical order for constructing the design in Table 10.9. Our first analysis (Fig. 10.9) included all the main effects and quadratic effects together with all the interactions it is possible to fit (arbitrarily constructed using c1 or c2 first). We then decided to drop any quadratic terms or interactions with low t-ratios (Fig. 10.10). The three interactions are aliased but we thought it likely that feed rate and rotation speed would interact negatively with the fuel rate. An interaction between the proportion of fuel to oxidant and fuel rate was suspected as the dominant contribution to the $x_2 x_6$ interaction, but there is no way of checking this from the results of the experiment. The next step would be to use the fitted regression equation to maximize the ratio of feed rate to fuel rate, subject to the constraint that the free lime remain within the target range. Since the burnability and water content cannot be controlled, we had better assume they are fixed near the 'pessimistic' end of their ranges. The greatest value of the feed/fuel ratio will not necessarily lie at a mathematical maximum, and it is likely that it will correspond to at least some of the control variables being at the end of their ranges. Search algorithms are widely available (Press *et al.*, 1992), or a probably sub-optimum solution can be found by a mixture of common sense and trial and error.

Table 10.9 Percentage free lime (percentage above 1% times 100) in a cement product (other units coded)

Feed rate (x_1)	Rotation speed (x_2)	Fuel/ox (x_3)	Fuel rate (x_4)	Fan 1 speed (x_5)	Fan 2 speed (x_6)	Burn-ability (x_7)	Water content (x_8)	Free lime (y)
-1	-1	-1	-1	-1	-1	4	7	37
1	-1	-1	-1	1	-1	5	4	7
-1	1	-1	-1	1	1	7	7	131
1	1	-1	-1	-1	1	3	6	14
-1	-1	1	-1	1	1	11	-6	116
1	-1	1	-1	-1	1	1	4	91
-1	1	1	-1	-1	-1	2	-2	74
1	1	1	-1	1	-1	2	-24	48
-1	-1	-1	1	-1	1	3	16	102
1	-1	-1	1	1	1	3	-4	-12
-1	1	-1	1	1	-1	1	-18	-2
1	1	-1	1	-1	-1	0	12	-87
-1	-1	1	1	1	-1	10	5	150
1	-1	1	1	-1	-1	-2	10	-53
-1	1	1	1	-1	1	-4	-15	-12
1	1	1	1	1	1	-3	-3	-71
-2	0	0	0	0	0	4	-13	137
2	0	0	0	0	0	-4	-2	-16
0	-2	0	0	0	0	9	12	78
0	2	0	0	0	0	5	-15	15
0	0	-2	0	0	0	3	-3	49
0	0	2	0	0	0	5	1	88
0	0	0	-2	0	0	5	-1	136
0	0	0	2	0	0	2	-3	-5
0	0	0	0	-2	0	2	-22	-57
0	0	0	0	2	0	2	0	51
0	0	0	0	0	-2	-3	0	-23
0	0	0	0	0	2	4	-2	39
0	0	0	0	0	0	-4	11	63

Example 10.3

Experimental design techniques can sometimes have unexpected uses. A research student wished to calibrate a hydrodynamic model for the hydrological cycle (Système Hydrologique Européen (SHE)) to the River Brue catchment in Dorset, UK. Eight of the model's parameters could realistically be assigned to certain ranges but not to point values. A 2^{8-4} plus star composite design was tried, with the response being a measure of the goodness of fit of the model to the field data. Replication was not appropriate! The errors are the differences

```
MTB > read 'kildat' c1–c8 c100
        29 ROWS READ
ROW     C1     C2     C3     C4     C5     C6    C7    C8    C100
  1     −1     −1     −1     −1     −1     −1     4     7      37
  2      1     −1     −1     −1      1     −1     5     4       7
  3     −1      1     −1     −1      1      1     7     7     131

MTB > let c12 = c1*c2
MTB > let c13 = c1*c3
MTB > let c14 = c1*c4
MTB > let c15 = c1*c5
MTB > let c16 = c1*c6
MTB > let c24 = c2*c4
MTB > let c26 = c2*c6
MTB > let c11 = c1*c1
MTB > let c22 = c2*c2
MTB > let c33 = c3*c3
MTB > let c44 = c4*c4
MTB > let c55 = c5*c5
MTB > let c66 = c6*c6
MTB > regress c100 21 c1–c8 c12 c13 c14 c15 c16 c24 c26 c11 c22 c33 c44 c55 c66
```

The regression equation is
$$c100 = 62.0 - 36.7\ c1 - 5.22\ c2 + 14.2\ c3 - 28.1\ c4 + 15.6\ c5 + 10.5\ c6$$
$$+ 3.48\ c7 + 1.75\ c8 - 1.34\ c12 + 5.87\ c13 - 21.3\ c14 - 3.08\ c15$$
$$+ 3.01\ c16 - 17.4\ c24 - 12.0\ c26 + 2.64\ c11 - 9.58\ c22 - 1.68\ c33$$
$$- 1.56\ c44 - 13.4\ c55 - 13.8\ c66$$

Predictor	Coef	Stdev	t-ratio	p
Constant	61.98	16.15	3.84	0.006
C1	−36.659	6.504	−5.64	0.000
C2	−5.221	7.939	−0.66	0.532
C3	14.207	4.972	2.86	0.024
C4	−28.137	6.053	−4.65	0.000
C5	15.567	5.618	2.77	0.028
C6	10.479	4.923	2.13	0.071
C7	3.483	2.849	1.22	0.261
C8	1.7463	0.6509	2.68	0.031
C12	−1.336	6.751	−0.20	0.849
C13	5.873	6.451	0.91	0.393
C14	−21.348	5.975	−3.57	0.009
C15	−3.076	7.172	−0.43	0.681
C16	3.012	5.707	0.53	0.614
C24	−17.442	6.438	−2.71	0.030
C26	−11.977	5.906	−2.03	0.082
C11	2.640	4.900	0.54	0.607
C22	−9.576	6.801	−1.41	0.202
C33	−1.681	5.530	−0.30	0.770
C44	−1.559	5.329	−0.29	0.778
C55	−13.449	4.984	−2.70	0.031
C66	−13.758	4.837	−2.84	0.025

$s = 22.75$ R-sq = 97.0%   R-sq(adj) = 88.1%

Analysis of Variance

SOURCE	DF	SS	MS	F	p
Regression	21	118334.6	5635.0	10.89	0.002
Error	7	3622.6	517.5		
Total	28	121957.2			

Fig. 10.9 First analysis of kiln data.

MTB > regress c100 14 c1–c8 c14 c24 c26 c55 c66 c22

The regression equation is
C100 = 62.6 − 38.2 C1 − 6.98 C2 + 13.9 C3 − 29.4 C4 + 16.6 C5 + 11.0 C6
    + 2.53 C7 + 1.68 C8 − 21.1 C14 − 18.3 C24 − 12.0 C26 − 13.4 C55
    − 13.9 C66 − 8.14 C22

Predictor	Coef	Stdev	t-ratio	p
Constant	62.637	7.113	8.81	0.000
C1	−38.197	4.493	−8.50	0.000
C2	−6.977	4.973	−1.40	0.182
C3	13.857	3.898	3.56	0.003
C4	−29.442	4.234	−6.95	0.000
C5	16.627	4.157	4.00	0.001
C6	11.016	3.836	2.87	0.012
C7	2.528	1.427	1.77	0.098
C8	1.6778	0.4466	3.76	0.002
C14	−21.104	4.753	−4.44	0.001
C24	−18.343	4.824	−3.80	0.002
C26	−11.973	4.754	−2.52	0.025
C55	−13.373	3.645	−3.67	0.003
C66	−13.870	3.575	−3.88	0.002
C22	−8.143	4.045	−2.01	0.064

s = 18.39   R-sq = 96.1%   R-sq(adj) = 92.2%

Analysis of Variance

SOURCE	DF	SS	MS	F	p
Regression	14	117223.8	8373.1	24.77	0.000
Error	14	4733.4	338.1		
Total	28	121957.2			

Unusual Observations

Obs.	C1	C100	Fit	Stdev.Fit	Residual	St.Resid
6	1.00	91.00	57.69	13.13	33.31	2.59R
13	−1.00	150.00	123.59	14.41	26.41	2.31R

R denotes an obs. with a large st. resid.

Fig. 10.10 Follow-up analysis of kiln data.

between the fitted quadratic hypersurface and the surface that would be obtained if the SHE model were run at an extensive grid of points. It is not practical to do this because of limited time and computing resources.

10.3 Comparing several means

10.3.1 Completely randomized design

The production manager of a company which manufactures filters for liquids, for use in the pharmaceutical and food industries, wishes to compare the burst strength of four types of membrane. The first (A) is the company's own standard

Table 10.10 Burst strength of filter membranes (kPa)

Type A	95.5	103.2	93.1	89.3	90.4	92.1	93.1	91.9	95.3	84.5
Type B	90.5	98.1	97.8	97.0	98.0	95.2	95.3	97.1	90.5	101.3
Type C	86.3	84.0	86.2	80.2	83.7	93.4	77.1	86.8	83.7	84.9
Type D	89.5	93.4	87.5	89.4	87.9	86.2	89.9	89.5	90.0	95.6

membrane material, the second (B) is a new material the company has developed, and C and D are membrane materials from other manufacturers. The manager has tested five filter cartridges from ten different batches of each material. The mean burst strengths for each set of five cartridges are given in Table 10.10. The data can be analysed by setting up a multiple regression model. We let Y be the average burst strength for each set of five cartridges and x_1, x_2, x_3 be indicator variables coded as:

	x_1	x_2	x_3
Type A	0	0	0
Type B	1	0	0
Type C	0	1	0
Type D	0	0	1

The coefficients then represent the differences between the company's standard membrane and the others. This sets up four of the six possible comparisons. If we fit the model

$$Y_i = \beta_0 + \beta_1 x_{1i} + \beta_2 x_{2i} + \beta_3 x_{3i} + E_i$$

we obtain the following results

$$y = 92.84 + 3.24x_1 - 8.21x_2 - 2.95x_3$$

with $s = 3.901$ and the table of coefficients below:

Predictor	Coef	Stdev	t-ratio	p
Constant	92.84	1.234	75.27	0.000
Type B	3.24	1.744	1.86	0.071
Type C	-8.21	1.744	-4.71	0.000
Type D	-2.95	1.744	-1.69	0.099

The means for each membrane type are therefore:

	mean
Type A	92.84
Type B	96.08
Type C	84.63
Type D	89.89

The standard deviation of each of these means is 1.23, assuming the corresponding populations have the same variance, and the standard deviation for the difference between any two is 1.74. Notice that both of these standard deviations can be read from the table or calculated from $3.901/\sqrt{10}$ and $3.901/\sqrt{(1/10 + 1/10)}$, respectively. There is clear evidence that type C has a lower average burst strength than type A. There is some weak evidence that type B is stronger, and that type D has a lower strength than type A (Fig. 10.11).

The standard deviations of the Y for membrane types A, B, C and D are 4.83, 3.39, 4.29 and 2.76, respectively. Such differences can reasonably be attributed to chance (Exercise 10.3). If you square these standard deviations and average the variances you will obtain s^2, the estimated variance of the errors in the regression. There are two main sources contributing to the errors: differences between batches of the same membrane type and differences between cartridges from the same batch. Let σ_w^2 represent the variance of burst strength for cartridges made from the same batch of material, and σ_b^2 represent the variance of batch means for one material type. Each Y was the average of burst

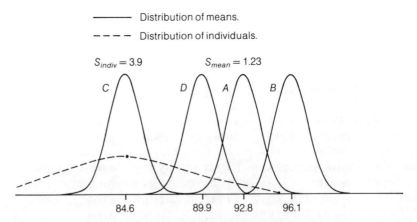

——— Distribution of means.

– – – – Distribution of individuals.

$S_{indiv} = 3.9$ $S_{mean} = 1.23$

C   D   A   B

84.6   89.9   92.8   96.1

Fig. 10.11 Means for cartridge burst strengths.

strengths for five cartridges made from the same batch of material, so the variance of the errors in the regression σ^2 is equivalent to

$$\sigma^2 = \sigma_b^2 + \sigma_w^2/5$$

Furthermore, the variance of each set of five measurements is an estimate of σ_w^2. I have not listed the original data, but the average of all 40 of these estimates was 11.62. The estimate of σ^2 is s^2 which equals 3.901^2. An estimate of σ_b^2 can therefore be found from the equation

$$(3.901)^2 = \hat{\sigma}_b^2 + 11.62/5$$

Summarizing:

$$s = 3.90 \quad \hat{\sigma}_b = 3.59 \quad \hat{\sigma}_w = 3.41$$

The standard deviation of a single cartridge about its mean is

$$\sqrt{(\sigma_b^2 + \sigma_w^2)}$$

which is estimated as 4.95.

Notice that it could be quite misleading to analyse individual burst strengths rather than the means for each set. It was suspected that cartridges made from the same batch of a material type would vary less than cartridges made from different batches, and this is supported by the data. We wish to compare material types, rather than particular batches of material of each type, and analysis of individual burst strengths would probably lead to an underestimate of the standard deviation of the errors, and spuriously small standard deviations for the coefficients.

10.3.2 Randomized block design

A mining company uses a process known as bacterial leaching to extract copper from ore. The company owns three mines and knows that the copper content of ores from them differ. A chemist has been asked to compare the yields obtained from four strains of thiobacillus A, B, C and D. If the experiment is restricted to ore from one mine we will not know how effective the strains are with ores from the other mines. If we randomly assign ores from all three mines to thiobacillus strain, and analyse the results in the manner of the previous section, we will probably swamp any differences between bacteria strains with the mine differences. Even if we allowed for the mines in the analysis we would not necessarily end up with an efficient design, and could find that some bacteria–mine combinations were not represented. A straightforward generalization of the paired comparison procedure can be used to ensure that each combination is represented and to reduce the variability in yield due to factors other than the bacteria strains. The company will take large batches of ore from one site in each of the three mines. Sites will be chosen where the local variability is relatively small. Each large batch is divided into four smaller

Table 10.11 Yields of copper from ores (kg per tonne)

| | Location of site | | |
Thiobaccilus strain	Mine 1	Mine 2	Mine 3
A	22	28	25
B	31	33	26
C	20	25	24
D	25	28	26

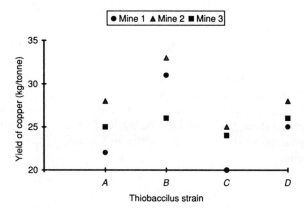

Fig. 10.12 Yields of copper from ore.

batches which are randomly allocated to the four strains of thiobacillus. This is an example of a randomized block design, where the 'blocks' are the three large batches of ore. Within each block there is a random allocation of ore to the four strains. Since randomization is only carried out within blocks, the blocks impose a restriction on randomization. The results of the experiment are given in Table 10.11 as kilogram yields per tonne, and are plotted in Fig. 10.12.

I have arbitrarily chosen to make comparisons relative to thiobacillus A and mine 1, so the model is

$$Y_i = \beta_0 + \beta_1 x_{1i} + \beta_2 x_{2i} + \beta_3 x_{3i} + \beta_4 x_{4i} + \beta_5 x_{5i} + E_i$$

with the following coding for the indicator variables.

	x_1	x_2	x_3
Thiobacillus A	0	0	0
Thiobacillus B	1	0	0
Thiobacillus C	0	1	0
Thiobacillus D	0	0	1

	x_4	x_5
Site in mine 1	0	0
Site in mine 2	1	0
Site in mine 3	0	1

The estimated coefficients, and their standard deviations, are:

Predictor	Coeff	Stdev	t-ratio	p
Constant	23.42	1.49	15.76	0.000
Type B	5.00	1.72	2.91	0.027
Type C	−2.00	1.72	−1.17	0.288
Type D	1.33	1.72	0.78	0.467
Block 2	4.00	1.49	2.69	0.036
Block 3	0.75	1.49	0.50	0.632

The value of s is 2.102.
  The means for each thiobacillus type are

	mean
Type A	25.00
Type B	30.00
Type C	23.00
Type D	26.33

and the standard deviation of any one of the six possible differences is 1.72. A 95% confidence interval for any difference will be:

$$\pm t_{6,0.025} \times 1.72 = \pm 4.21$$

Although type B differs from type A and type C by more than this we need to be a bit wary because these have been identified as the outstanding differences. If there really is no difference between thiobacillus types:

Pr (at least one of the six 95% confidence intervals for a difference excludes 0)
    $< 0.05 + 0.05 + 0.05 + 0.05 + 0.05 + 0.05 = 0.30$

The standard statistical approach to this problem is covered in more advanced books, but if we use the width of a 99% confidence interval as our criterion to judge differences, the corresponding probability will be less than 0.06. A 99% confidence interval is given by

$$\pm t_{6,0.005} \times 1.72 = \pm 6.38$$

and we would still conclude that type B is better than type C. If we had to recommend one thiobacillus type on the basis of this one experiment it would be type B, provided any cost differences were negligible, but it would be advisable to carry out a follow-up experiment including type A and type D.

The experiment could also have been set up as a completely randomized design with two factors. In this case the order of bacteria–mine combinations would have been randomized, and the practical difference would be that different batches of ore from the three mines would be used with each strain. The model would be the same algebraically but β_4 and β_5 would now represent mine effects rather than block effects – that is, a particular batch of ore from the one site in the mine. The standard deviation of the errors in the model would probably be larger and the comparison between bacteria types would be less precise. It would give us some empirical justification for assuming that differences apply to all ore from the three mines, but we might be prepared to assume this anyway as differences in bacteria are fairly consistent for ores from the different mines.

Yet another interpretation of β_4 and β_5 in the model is possible. Suppose the company owns a large number of mines and the blocks were random samples of ore from randomly selected mines. It would then be appropriate to treat the blocks as random effects (see Exercise 10.4).

10.3.3   Difference between completely randomized and randomized block designs

Anderson and McLean (1974) discuss a project to compare different designs of prosthetic cardiac valves. Four valve types were investigated in a mechanical apparatus which had been constructed to simulate the human circulatory system. Tests were carried out at six pulse rates, from 20 to 220 beats per minute. The response was maximum flow gradient (measured in millimetres of mercury).

A completely randomized design would need six valves of each type. There are 24 valve–pulse-rate combinations, and the allocation of valves of each type to the runs for each pulse rate, and the order of testing all combinations, should be randomized. A regression model for the experiment could be

$$Y_i = \beta_0 + \beta_1 x_{1i} + \beta_2 x_{2i} + \beta_3 x_{3i} + \beta_4 x_{4i} + \cdots + \beta_8 x_{8i} + E_i$$

where Y is the flow gradient, x_1 up to x_3 are indicator variables for the valve types, and x_4 up to x_8 are indicator variables for the pulse rate. Many variations are possible, interactions could be included, the pulse rate could be assumed linear or quadratic, allowing it to be modelled by one or two variables respectively, and so on. The errors include variation in valves of the same nominal type, as well as other deviations. The coefficients β_1, β_2 and β_3 represent the differences between valve type A and valve types B, C and D, respectively. The only snag with all this as an experimental programme is the need for six valves of each type. Prototypes of precision equipment are expensive to produce, and we might be tempted to treat valves as blocks and set up a randomized blocks design. We could then manage with one valve of each type. But we would have no information about the variability of valves of a given type. The regression model would look the same, but the β_1, β_2 and β_3 would now represent differences between the particular valves used as blocks. If we do find any significant differences we cannot tell from the experiment whether it is an exceptionally good valve of its type or whether its type is the best design.

A possible compromise, and the one adopted, was to design an experiment around two valves of each type. Each valve of each type was tested with all six pulse rates. This needed 48 runs, but the time taken for a run is negligible compared with that needed to make the prototype valves. The order of the runs was randomized, and the apparatus was reassembled after each run, even if the same valve was to be used. It would have saved time to try all pulse rates in sequence on the same valve but this would leave the possibility that an apparently poor performance of a valve was due to a poor assembly. It would be incorrect to use the previous model with an additional indicator variable for the second valve, because any difference between the first and the second valve is specific to their type and will be different for every type. Instead we use seven indicator variables for the valves. The coefficients represent the

Table 10.12 Coded flow gradients for prosthetic cardiac valves

Valve type	Pulse rate (beats per minute)					
	20	60	100	140	180	220
A(1)	12	8	4	1	8	14
A(2)	7	5	7	5	13	20
B(1)	20	15	10	8	14	25
B(2)	14	12	7	6	18	21
C(1)	21	13	8	5	15	27
C(2)	13	14	7	9	19	23
D(1)	15	10	8	6	10	21
D(2)	12	14	5	9	14	17

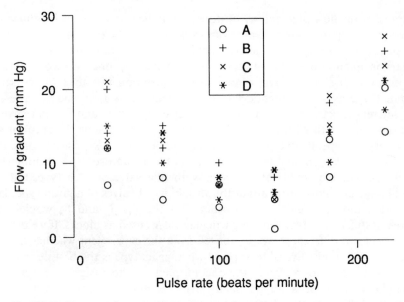

Fig. 10.13 Flow gradient against pulse rate for eight prosthetic cardiac valves.

differences between the first type A valve, A(1), and the others. The data are given in Table 10.12. The experiment is an example of a **split plot design**. A plot of the flow gradient against pulse rate (Fig. 10.13) suggests that a quadratic relationship with pulse rate is appropriate. The model is

$$Y_i = \beta_0 + \beta_1 x_{1i} + \cdots + \beta_7 x_{7i} + \beta_8 x_{8i} + \beta_9 x_{8i}^2 + E_i$$

where x_8 is pulse rate minus its mean of 120. The estimated coefficients and their standard deviations are as follows:

Predictor	Coef	Stdev	t-ratio	p
Constant	2.70	1.26	2.14	0.039
β_1 A(2)	1.67	1.65	1.01	0.319
β_2 B(1)	7.50	1.65	4.54	0.000
β_3 B(2)	5.17	1.65	3.13	0.003
β_4 C(1)	7.00	1.65	4.24	0.000
β_5 C(2)	6.33	1.65	3.83	0.000
β_6 D(1)	3.83	1.65	2.32	0.026
β_7 D(2)	4.00	1.65	2.42	0.020
β_8	0.029	0.006	4.77	0.000
β_9	0.0011	0.0001	10.62	0.000

Also $s = 2.86$ and $R^2 = 82\%$.

It is noticeable that there is not much variation between valves of the same type (Fig. 10.13). We can estimate this variance from the following argument.

Let $\mu_{k(j)}$ represent the flow gradient for a valve j of some particular type (k), standardized to a pulse rate of 120. For each valve type, k runs from A to D, we have two valves and estimates of the corresponding means $\hat{\mu}_{k(j)}$, with a standard deviation of 1.26. For instance

$$\mu_{A(1)} = 2.70$$
$$\mu_{A(2)} = 4.37$$
$$\vdots$$
$$\mu_{D(2)} = 6.70$$

Now

$$\text{var}(\hat{\mu}_{k(j)}) = \text{var}(\mu_{k(j)}) + \text{var}(\hat{\mu}_{k(j)}|\mu_{k(j)})$$

and the last term on the right-hand side is $(1.26)^2$. If we assume that the variance of valves of the same type is the same for all four types we can estimate the variance on the left-hand side:

$$\text{var}(\hat{\mu}_{k(j)}) = \left(\frac{2 \times 0.835^2}{2-1} + \frac{2 \times 1.165^2}{2-1} + \frac{2 \times 0.335^2}{2-1} + \frac{2 \times 0.085^2}{2-1} \right) \Big/ 4$$

$$= 1.04^2$$

An estimate of the variance of valves of the same type can be found by subtraction. In this case we obtain a physically impossible negative estimate and conclude that the variance is negligible. The negative estimate can reasonably be attributed to chance. If we combine our two estimates of the flow gradients for each valve type we obtain

Type A   3.535
Type B   9.035
Type C   9.365
Type D   6.615

We need to be quite careful when calculating the standard deviation of differences between valve types. The standard deviation of 1.65 is for the difference between A(1) and any one of the other valves. The design is balanced so that all valve effects are determined with the same accuracy, and the standard deviation for the difference between the averages of the two valves of each type is $1.65/\sqrt{2}$, which equals 1.17. But this is only appropriate for the differences between these particular pairs, and our concern is with all possible valves. To this standard deviation we must add an allowance for the uncertainty in the corresponding population means:

$$\sqrt{[1.17^2 + \text{var}(\mu_{k(j)})/2 + \text{var}(\mu_{k(j)})/2]}$$

Although the additional terms have been estimated as negligible in this example, it will not usually be so. Our conclusions are that type B and C are similar in respect of flow gradient and are both considerably higher than type A. Type D appears higher than type A and there is some rather weak evidence that it is lower than types B and C.

10.4 Experimental design for welded joints

Tony Greenfield has contributed this description of a research proposal to investigate the effects of geometry, weld, environment and corrosion protection techniques on the lifetimes of welded plate joints. It is based on his report to the company responsible for managing a multi-laboratory trial of welded joints used in the construction of offshore drilling platforms.

10.4.1 Introduction

The reason for using statistical methods in the design of experiments is to obtain as much information as possible with a reasonable, or affordable, amount of effort. An experiment should be designed so that the effects of contributing variables, on the specified response variable, can be estimated with adequate precision. The aim is to obtain values for these effects which are good enough to be used with some confidence in the design of working structures.

Lifetime, as measured by cycles to failure, is the response variable in which we are interested when studying the fatigue of welded joints. In the experiment different types of joint were subjected to harmonic oscillations in applied stress about some mean value. The amplitude and mean value, termed 'stress ratio' and 'applied stress', respectively, could be set at specified levels. The effect of shape, which is another explanatory variable, is measured by the difference in lifetimes of a T-joint and an X-joint, with everything else being constant at their average values. The effect of stress ratio is measured by the difference in lifetimes of joints submitted to different levels of stress ratio, with everything else being constant. These are examples of what are called **main effects**. However, the effect of stress ratio may be different in the two different shapes. Equally, the effect of shape may be different at two different levels of stress ratio. This difference in effect of one variable at different levels of another variable is called an **interaction**. Naive experimentation, in which only one variable is changed at each stage of the study, while all the other variables are held constant, is both inefficient and inadequate when we need to estimate main effects and interactions. **Factorial experiments**, in which all variables are changed together, are adequate. However, when we have many variables, or factors, each with several levels, the number of possible combinations is too great for a practical experiment.

For example, there are up to 12 contributing variables in this study of welded plate joints under constant loading. The numbers of specified levels of these variables range from two to five as follows:

Variable	Coded values	Number of levels
stress ratio	(four values)	4
environment	(air, water)	2
Pre-weld heat treatment (PWHT)	(absent, present)	2
thickness	(25, 38, 50, 75, 100 mm)	5
shape	(T or X)	2
cathodic protection	(0 to −1.5 V)	5
precorroded	(yes, no)	2
coating	(types 1, 2, 3, none)	4
temperature	(5°C, 20°C)	2
profile	(bad, standard, good)	3
weld improvement	(none, toe grind, hammer peened)	3
applied stress	(five values)	5

The number of possible combinations, including any impractical ones for the moment, is 576 000. This means that 576 000 welded joints would have to be made and held in test rigs until they all failed through fatigue, if we were to do a full factorial experiment. Our problem is to choose, from those 576 000 possible combinations, a relatively small number so that we can afford the cost of making and testing the welded joints. Can we, for example, choose about 200 combinations whose contrasting lifetimes will give us the information we need? Another question is how many we should choose. Or what would be the smallest sufficient number. In fact, there are many sets of 200 combinations that we could choose that would allow us to estimate the main effects and interactions, which is the information we need. We therefore ask another question: whether one of these many sets of 200 combinations (assuming that is the right number) is better than all the rest. Our criterion for deciding which of the combinations is the best is the **precision** of the estimates of the main effects and interactions.

As well as the main effects and interactions, there is another type of effect that must be studied with respect to some of the contributing variables. This is the **curvature effect**. For example, we suspect that the best cathodic protection is obtained with a potential of about −0.85 V, in which case the lifetime will be greater with that potential than with a greater or lesser potential. This is shown in Fig. 10.14. In this case, the curvature effect may be described approximately as a **quadratic effect**. As another example, consider the effect of applied stress. When the lifetime, expressed as the logarithm of the number of cycles to failure, is plotted against the logarithm of the applied stress, the graph is

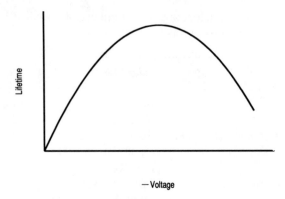

Fig. 10.14 Postulated quadratic effect of voltage on lifetime of joints.

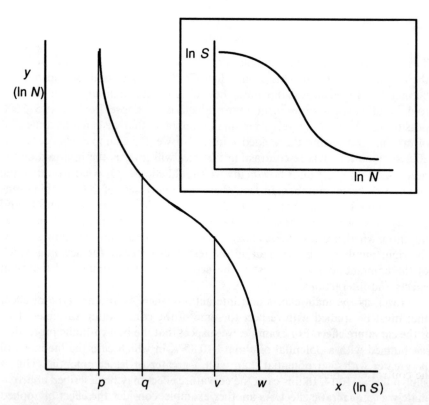

Fig. 10.15 Cubic effect of stress on lifetime and linear approximation over stress range from q to v.

approximately straight over a wide range. But it starts to curve towards the ends, as shown in Fig. 10.15. The overall relationship may be described approximately as a **cubic effect**. One of the aims of this research is to estimate this cubic effect, whatever the conditions of other variables, so that we can state a range over which the graph may be regarded as sufficiently near a straight line for the purpose of engineering design.

Another difficulty facing us, in the choice of the best small set of combinations, is that some of them are impractical. This is described in terms of **constraints**. In the case of welded plate joints under constant loading, the constraints are:

C1 Cathodic protection is used only with joints that are immersed in sea water.
C2 Cathodic protection is not used when a coating is used.
C3 Joints will be precorroded only if they are to be immersed in sea water.
C4 Coatings are used only with joints that are to be immersed in sea water.
C5 Joints that may be coated will not be precorroded.
C6 Temperatures will be controlled only during those tests in which the joints are immersed in sea water.

For the purpose of experimental design, it is convenient to deal with these constraints by splitting the full factorial experimental into three parts:

E1 Tests which are done in air.
E2 Tests which are done in sea water, where the joints may be coated, although for the purpose of estimating the effect of coating some of the joints are not coated.
E3 Tests which are done in sea water, where the joints may be precorroded and/or cathodic-protected, although for the purpose of estimating the effects of precorrosion and cathodic protection some of the joints are neither precorroded nor cathodic-protected.

Two other parts of the total experiment have been designed using statistical methods. These are:

E4 Tests on welded plate joints submitted to variable-amplitude loads applied in air.
E5 Tests on welded plate joints submitted to variable-amplitude loads applied in sea water.

All five parts to the research were described in the original report. Here we concentrate on the relatively small E5.

10.4.2 Design method

The control variables, and their codings, for the experimental subset E5 were as follows:

Pre-weld heat treatment (x_1) no (-1), yes $(+1)$
Weld improvement (x_2) none (-1), toe ground $(+1)$

Applied stress (x_3) coded -1.2, -1.0, 0.0, 1.0, 1.2
Cathodic protection (x_4) 0.0 V (-1), -0.85 V $(+1)$
Temperature (x_5) $5°C$ (-1), $20°C$ $(+1)$
Variation in amplitude of   narrow S1 (-1), wide S4 $(+1)$
    oscillatory load (x_6)

We thought that the applied stress might interact with any of the other variables, and we also wished to investigate any quadratic or cubic effects of applied stress. We therefore needed to fit a regression model which included:

$$x_7 = x_1 \times x_3$$
$$x_8 = x_2 \times x_3$$
$$x_9 = x_4 \times x_3$$
$$x_{10} = x_5 \times x_3$$
$$x_{11} = x_6 \times x_3$$
$$x_{12} = x_3^2$$
$$x_{13} = x_3^3$$

That is,

$$Y_i = \beta_0 + \beta_1 x_{1i} + \cdots + \beta_{13} x_{13i} + E_i$$

In matrix terms (Appendix A5)

$$Y = XB + E$$

where X is called the design matrix. The strategy is to choose a number of runs and then, conditional on this choice, find the corresponding values of $x_1, \ldots, x_6$ so that $\det(X^T X)$ is a maximum. This is known as a D-optimum design. The covariance matrix of $\hat{B}$ is

$$\sigma^2 (X^T X)^{-1}$$

and the criterion corresponds to minimizing the hypervolume of the hyper-ellipsoidal simultaneous confidence interval for the parameters. This is rather less daunting in the special case when the x are orthogonal. Then $X^T X$ is diagonal and the determinant of $X^T X$ is the reciprocal of the product of the elements along the diagonal of $(X^T X)^{-1}$, which are proportional to the variances of the estimators.

It had been decided that 20 joints would be tested for this experiment. There are $2 \times 2 \times 5 \times 2 \times 2 \times 2 = 160$ different values for the array of control variables $x_1, \ldots, x_6$, and the values of x_7 up to x_{13} will be determined by this. It remains to select the 20, from these 160 possibilities, that maximize the determinant. There are 1.4×10^{25} choices, so checking them all is not going to be possible! Some sort of algorithm, even if it cannot guarantee finding the overall maximum, is essential. We will refer to a particular array

$$x_i = \begin{bmatrix} 1 & x_{1i} & x_{2i} & \cdots & x_{13i} \end{bmatrix}$$

as a design point.

Greenfield suggested the following relatively simple algorithm.

1. Start with any 14 design points for which the rows of the 14×14 design matrix X are linearly independent. This can be achieved by using the Gram–Schmidt orthogonalization procedure which is described in most texts on linear algebra (for example, Lipschutz 1968).
2. Try adding each of the 146 remaining design points in turn, and choose the one for which the determinant criterion is a maximum. This is greatly facilitated by the following well-known and extremely useful result. Define

$$\Delta_m = \det(X_m^T X_m)$$

where m is the number of rows in the $m \times 14$ design matrix X. Now if

$$X_{m+1} = \begin{pmatrix} X_m \\ x \end{pmatrix}$$

is an $(m + 1) \times 14$ matrix, we have

$$\Delta_{m+1} = \Delta_m \det(1 + x(X_m^T X_m)^{-1} x^T)$$

Table 10.13 Experimental design for welded plate joints in sea water with cathodic protection and variable loading

	PWHT	Weld improvement	Applied stress	Cathodic protection (V)	Temp. (°C)	Variable load
1	N	N	−1.0	0.0	5	S1
2	Y	Y	1.2	0.85	5	S4
3	Y	N	−1.0	0.85	20	S1
4	Y	N	1.0	0.0	5	S4
5	N	Y	1.2	0.85	20	S1
6	N	Y	−1.2	0.85	20	S1
7	Y	N	1.2	0.85	5	S1
8	Y	Y	1.0	0.0	20	S1
9	N	Y	−1.2	0.0	5	S4
10	N	N	1.0	0.85	20	S4
11	N	Y	−1.0	0.85	5	S1
12	N	Y	1.0	0.85	5	S1
13	N	N	−1.2	0.85	20	S4
14	N	N	1.2	0.0	5	S1
15	Y	N	1.2	0.0	20	S4
16	Y	N	−1.0	0.0	5	S4
17	N	Y	1.0	0.0	20	S4
18	N	N	0.0	0.85	20	S4
19	Y	N	0.0	0.0	20	S4
20	Y	Y	0.0	0.85	5	S4

3. Now try leaving out each of the 15 design points, and drop the one which gives the least decrease in the criterion.
4. Iterate steps 2 and 3 until one of the design points which has been added in step 2 is dropped in step 3.
5. Add design points, one at a time so that the criterion is maximized, until the required 20 points have been selected.

This procedure will not generally lead to the absolute maximum of the criterion, but it should give a reasonable design. The design obtained for E5 is given in Table 10.13. There is no reason why some particular design points should not be fixed, and the algorithm implemented subject to these points remaining in the design. Atkinson and Donev (1992) cover the subject of optimum design in detail. Their procedure starts from a single design point, and avoids the problem of a singular matrix by adding $\varepsilon^2 I$, where ε is a small number and I is the identity matrix. I do not know how this compares with using the Gram–Schmidt orthogonalization procedure (step 1).

10.5 Summary

Factorial designs

A 2^n factorial design allows you to investigate the linear effects and interactions of n control variables on some response variable. Although the control variables are usually continuous they are restricted to two levels, 'high' and 'low'.

Fractional factorial design

If n is large the number of runs for a full factorial design soon becomes prohibitive. It is possible to generate fractions, involving 2^{n-m} runs, that can provide information on the linear effects and the low-order interactions which could be of practical importance.

Star design and composite design

A star design can be used as a follow-up to a factorial or fractional factorial design. It allows quadratic effects to be investigated. The overall design – factorial plus star – is known as a composite design. It is only practical for continuous control variables.

D-optimum design

A general technique for finding a design that can estimate chosen effects given a set of possible values for the control variables. There is no restriction on the number of levels.

Comparison of means

If we have several independent samples from different populations the regression procedure is a generalization of the two sample comparison of two means assuming the population variances are equal (Chapter 6 and Exercise 9.8).

The randomized block design is a generalization of the paired comparison procedure (Chapter 6).

Sample size

The principles that were explained in Chapter 6 apply.

Exercises

10.1 The yield (y) of a process can be altered by making small changes in temperature (x_1) and pressure (x_2) about the nominal operating point. The plant manager wishes to find the temperature and pressure which will give the highest yield.

  (i) Explain why a factorial design is preferable to changing one variable at a time.
  (ii) A regression plane

$$y = 50 + 3x_1 + x_2$$

  has been fitted to the results from one such experiment. Calculate the ratio of the change in y to the change in (x_1, x_2) if (x_1, x_2) moves from $(0, 0)$ to $(1, 1)$. If x_1 is changed by 1, what change in x_2 will correspond to moving in the direction of steepest ascent, and what is the ratio of the change in y to the change in (x_1, x_2)?

10.2 A researcher worker has identified four factors, A, B, C and D, which she thinks may influence the efficiency of an effluent treatment works. She only has time for eight experimental runs. Suggest a possible experimental procedure and explain its limitations.

10.3 Refer to the comparison of membrane types discussed in section 10.3.1. Construct 99% confidence intervals for the six differences in standard deviations of Y from the four types. Then use the argument that:

  Pr(at least one of the six 99% confidence intervals excludes 0|no differences in standard deviations of types)
  $\leqslant \sum \mathrm{Pr}(99\%$ confidence interval excludes 0|no difference in standard deviations)
  $= (1 - 0.99) \times 6 = 0.06$

to reach an overall conclusion.

10.4   Refer to the bacterial leaching example in section 10.3.2. Assume blocks are random effects and estimate the variation of yields between mines in the following three steps.

(i) Estimate the mean value of yield at each of the three mines.
(ii) Estimate the standard deviation of the estimates of the means in (i).
(iii) Hence estimate the standard deviation of yield between mines.

The following exercises need access to a multiple regression routine.

10.5   Eighteen limestone cores from the same quarry were prepared for a test of compressive strength. Six of the cores were selected at random and tested in their natural state. Another six were randomly selected and saturated in water at room temperature for two months before testing. The remaining six were dried in an electric oven before testing. All 18 cores were tested on the same machine in a random order. The measured compressive strengths, in newtons per square millimetre, were

Natural	60.6	65.1	96.9	57.9	85.6	82.7
Saturated	72.8	82.7	59.8	61.8	57.1	91.8
Dried	75.2	64.8	82.9	64.3	70.2	54.8

Use a regression analysis to summarize the results of the experiment. How does this differ from the multiple t-tests? What are the limitations of both procedures? (After Bajpai et al., 1968.)

10.6   Three replicate water samples were taken at each of four locations in a river to determine whether the quantity of dissolved oxygen varied from one location to another.

Location	Dissolved oxygen content		
A	4.0	5.1	5.9
B	5.8	6.2	6.9
C	6.5	7.3	8.1
D	8.2	9.4	10.3

(i) Is there any evidence of a difference between locations?
(ii) You are now told location A was adjacent to the wastewater discharge point for a certain industrial plant, and locations B, C and D were at points 10, 20 and 30 miles downstream from this discharge point. Bearing this in mind, suggest an alternative analysis of the data and discuss whether it would be preferable.

10.7   A metallurgist thinks that the rate of stirring an alloy may affect the grain size of the product. The plant has three furnaces, all of which have their own characteristics, so he plans a randomized block experiment using furnaces as blocks. The results of the experiment, from Montgomery (1991), where the grain size is given in coded units, are given below.

Stirring rate (rpm)	Furnace		
	A	B	C
5	49	58	45
10	74	72	67
15	83	85	86
20	78	88	70
25	60	72	55

Is there any evidence that stirring has an effect, or of a difference between furnaces?

10.8   The particle size of a powder used in the paint industry is an important characteristic. The powder is produced in a rotating kiln and is then drawn through ducting by powerful fans. It has been suggested that the rotation speed affects the particle size. An experiment has been designed to compare three rotation speeds: fast, medium and slow. There will be five runs at each speed. Each run will last for a day and the order will be randomized. However, ambient temperature may have an effect because the ducting is exposed, and it was measured as a concomitant variable (x, in degrees Celsius). Particle size was measured with a machine linked to a micro-computer which produced the distribution of sizes for the sample. The response was taken as the upper 1% quantile (y, coded units) of this distribution. Analyse the data:

(i) ignoring ambient temperature;
(ii) with ambient temperature.

What speed would you recommend if a small particle size is desirable?

Rotation Speed					
Fast		Medium		Slow	
x	y	x	y	x	y
12	133	17	85	18	84
14	109	14	102	13	119
15	104	12	131	12	128
13	122	12	122	14	106
14	112	13	109	13	119

Table 10.14 Designed experiment to investigate effects of additives on the tear strength of rubber

A in rubber (sachets)	B in resin (kg)	C modifier (sachets)	Tear strength (N)
1	13	1	427
3	13	1	341
1	25	1	410
3	25	1	352
1	13	3	424
3	13	3	373
1	25	3	428
3	25	3	365
0	19	2	417
4	19	2	286
2	10	2	395
2	28	2	380
2	19	1	410
2	19	3	431
2	19	2	428
2	19	2	419

10.9   A chemical engineer thinks the tear resistance of a rubber material used for car upholstery is likely to depend on the amount of compound A in the rubber, the amount of compound B in the resin, and the amount of modifier C. Compound A and modifier C are added in sachets. The data from a composite design, slightly modified because of the constraint of having to use whole sachets, are given in Table 10.14. Analyse the results of the experiment and write a brief report.

10.10 The cleanliness of carburettor assemblies is measured by the number of dirt particles in a sample of ten carburettors. This number is Y1, the response variable. There are six control variables, each with two levels:

		Low	High
A	Preclean wells in body	No	Yes
B	Prewash components	Yes	No
C	Flush time	Normal	Long
D	Double assembly tubes/jets	No	Yes
E	Environment	Clean	Extraclean
F	Speed of screwing	Normal	Slower

Table 10.15 Number of dirt particles in samples of ten carburettors, and values of the control variables*

x_1	x_2	x_3	x_4	x_5	x_6	Particles
-1	-1	-1	-1	-1	-1	145
1	-1	-1	1	-1	1	83
-1	1	-1	-1	1	1	63
1	1	-1	1	1	-1	131
-1	-1	1	1	1	1	111
1	-1	1	-1	1	-1	113
-1	1	1	1	-1	-1	145
1	1	1	-1	-1	1	73

*Fictitious data from DEX users' guide, example courtesy of the Ford Motor Company.

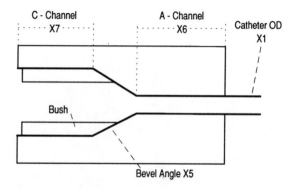

Fig. 10.16 Catheter valve assembly (courtesy of British Viggo).

The results from a 2^{6-3} fractional factorial experiment are given in Table 10.15. Analyse the results and write a short report. (Taken from DEX users' guide, with permission, and courtesy of Ford Motor Company.)

If the number of dirt particles approximately follows the Poisson distribution the assumption of equal variances of the errors is more nearly satisfied by analysing the square root of the response. Explain why this is so by using a Taylor expansion. Does it make any difference to the conclusions from this experiment?

10.11 Figure 10.16 shows how a catheter is fixed to a valve body. It enters through the A-channel and expands into the C-channel where it is held by a bush which is pressed into the end of the catheter. The purpose of the experiment was to discover the dimensions such that the catheter would be gripped with maximum security. The assembly is put into a tensile tester and force is gradually increased until the catheter is pulled out. The response variable (Y1) is the disassembly force measured in newtons.

There are some constraints in the dimensions. The C-channel ID must be greater than the bush OD. The C-channel ID must be greater than the A-channel ID. These constraints are avoided by using differences in variables.

The results of an experiment are given in Appendix D, Table D.11. Analyse these data and write a short report. (Taken, with permission, from the DEX users' guide, and courtesy of British Viggo.)

11

Modelling variability in time and space

Time and spatial variation are the common themes of the three case studies presented in this chapter. The first uses a Bayesian approach to data analysis, and some preliminary explanation may be helpful. So far, we have distinguished variables from unknown constant features of the population (parameters), such as the mean μ and variance σ^2. But the distinction became slightly blurred with our non-rigorous interpretation of confidence intervals. The Bayesian method takes this further and describes our uncertainty about parameters by assigning probability distributions to them. When we obtain further information we can update the probability distribution by using Bayes' theorem. The distribution before updating is called the **prior distribution** and the updated distribution is the **posterior distribution**. It is a conceptually attractive approach to statistical inference, but it does often lead to rather more complicated algebra which reduces its appeal to the harassed practitioner. It is also much less well supported with statistical software, although the situation is changing. However, Kamarulzaman bin Ibrahim and Metcalfe (1993) thought it was the most appropriate, and the easiest, way to assess the success of mini-roundabouts as a road safety measure. The following section is an edited version of our article (reproduced with permission).

11.1 Evaluation of mini-roundabouts as a road safety measure

11.1.1 Introduction

Road safety measures are implemented by experienced engineers who expect them to reduce accidents. Randomized trials are often difficult to justify, because highway authorities wish to treat all junctions with a bad accident record immediately, so evaluation usually relies on before and after studies. Although there is a legal requirement to record all accidents involving injury, detailed analyses of the data relating to particular road safety measures are limited. However, the County Surveyors' Society Standing Advisory Group on Accident Reduction (SAGAR) (1978) does aim to collect information on studies of selected safety measures. So there will often be several studies of the effectiveness of a particular measure, made at different times or in different areas. Furthermore, it is usually appropriate to allow for various other changes during the study period, and while measures of these changes may be based on many data, the precise allowances to be made depend on the analyst's judgement. Nevertheless, it is usually possible to provide limits within which most people can agree such allowances lie. It follows that assessments of road safety measures have to be made from data that have not been obtained under strictly controlled conditions, augmented with expert opinion backed up by national accident records. The Bayesian framework is ideal for such evaluations.

In this section we demonstrate the value of Bayesian overview techniques by combining information from studies of the effectiveness of mini-roundabouts, for reducing accidents at priority controlled junctions (without traffic lights), made during the 1970s and early 1980s. Two Bayesian methods for combining the results of the studies are compared. The first assumes a constant treatment effect for all the groups of sites. The studies are combined in chronological order with the posterior distribution of the treatment effect from one study becoming the prior distribution for the next. The second method is based on a hierarchical model (Lindley and Smith, 1972), which allows for treatment effects to vary from one group of sites to the next. DuMouchel (1989) gives a clear description of a 'standard' algorithm and demonstrates its use with a medical example. Although Bayesian methods are not routinely used in engineering, their versatility has been recognized by transport engineers for some time (for example, Abbess *et al.*, 1983; Hauer, 1983) and more recently by the water industry (O'Hagan *et al.*, 1992).

11.1.2 Studies on mini-roundabouts

Our objective is to assess the consequences of installing mini-roundabouts at priority controlled junctions in the hope of reducing accidents. The Department of Transport defines a mini-roundabout as 'a roundabout having flush or slightly raised circular marking less than 4 metres in diameter'. Small roundabouts are defined as having central islands, between 4 m and 30 m in diameter, kerbed or with street furniture to prevent over-run by vehicles.

After a reasonably thorough search we found five studies which have included data from mini-roundabouts, including one which can be described as 'desk-draw' because it was not published in the open literature. The data in the unpublished study were anomalous, and reservations about them were expressed in the internal report. While the distinction between small and mini-roundabouts is clear enough, it was not possible to isolate the results for mini-roundabouts from small and mini in one study, and in one other the investigators had decided to include roundabouts with diameters slightly greater than 4 m. This presents us with a typical dilemma: do we blur our definitions or eliminate some studies, thereby losing information and restricting the practical relevance of the overview? We have chosen to be slightly flexible, which seems to be in the spirit of Bayesian analysis, and have included these studies. It would be possible to repeat calculations without them, as part of a sensitivity analysis.

The Transport and Road Research Laboratory leaflet LF393 (TRRL 1975) provides a preliminary survey on the annual rate of accidents involving injury to road users before and after installation of roundabouts at various types of junction in Britain. The total number of sites surveyed was 110, consisting of 78 roundabouts which were formerly priority controlled junctions. Although accurate rates for the before and after period at each site were given, the number of accidents and the length of before and after periods at each site are not available. As the first mini-roundabout on a public road was installed in 1968 (Laurence, 1980), an average period of 3 years before and after is a reasonable guess. The lengths of the before and after periods are only needed to estimate allowances for trend, seasonal effects and regression-to-mean effect as well as variances of observed rates. It follows that reasonable guesses, while not ideal, will do. Since this analysis does not separate mini- and small roundabouts, we use the mixed category in our analysis.

Lalani (1975) investigated whether the results of the TRRL study could be extended to urban conditions with a before and after study, over an average of 19 months for each period, on the effect of installing roundabouts of various diameters at major and minor priority junctions in the Greater London Council area. The injury accident data in the study are limited to the period from January 1970 to May 1975. Lalani does not provide information on accidents at each site, but gives the total number of accidents during before and after periods for different categories of roundabout site. In our analysis, we use the total number of all road user accidents at the 20 single mini-roundabouts which were previously priority controlled junctions.

A nationwide study to investigate accident frequencies before and after installation of small and mini-roundabouts was reported by Green (1977). The times of installation, and therefore the lengths of before and after periods, of the 150 sites surveyed varied, but on average before and after periods were 3.4 and 2.5 years, respectively. Our analysis uses the total number of all road user accidents before and after at 54 mini-roundabouts. Four mini-roundabouts sites in Newcastle upon Tyne, which are also included in the study by Hanson and Wilson, were dropped from Green's study, in order to have independence between studies (not essential for DuMouchel's algorithm, but convenient).

Hanson and Wilson (1979) studied the safety records at 41 sites with recently installed small roundabouts in Newcastle. All the study sites had roundabouts installed between 1971 and 1975. Neither the detailed accident reports nor the roundabouts' dates of introduction are documented in the report. However, they do give estimates of lengths of before and after periods for each site, from which the average before and after periods were 3.7 years and 3.3 years, respectively. Although the authors did not distinguish mini- and small round-abouts in their analysis, we managed to separate these two categories by using information available from the Traffic Accident Data Unit (TADU) at Gateshead. Consequently, we will consider the accident data at 11 mini-roundabouts, which include three sites with island diameters slightly bigger than 4 m (up to 4.5 m).

Avon County Council (1984) conducted a study of the safety of mini-roundabouts following the petition of September 1984 objecting to the proposal of replacing the traffic signal controlled junction at Cheltenham Road and Zetland Road with a roundabout. The petitioners believed that this counter-measure could increase the danger to pedestrians and cyclists. The report by Avon County Council consists of two parts: one is the original before and after study at 12 junctions in the county, carried out in September 1977; and the other is the study at 23 mini-roundabouts conducted following the petition. In the first part the council only considered junctions which had been priority controlled and in the second it looked at junctions which had been controlled by traffic signals as a separate category. The report on the original study gives details of the number of accidents during before and after periods and their lengths. The average length of the before and the after period for this study is 20 months each. The second study gives the numbers of accidents during equal-length before and after periods at each site. An average length of 2.5 years is a reasonable estimate for both before and after periods. From the list of sites in the two studies, we could identify four which were common to both. Unfortunately, we cannot isolate the data for these specific sites and have estimated the correlation between the results from the two studies when using DuMouchel's algorithm.

The relevant data from all these studies are summarized in Table 11.1. They will be used to assess the effect of mini-roundabouts for each group of sites as well as for all the groups as a whole.

Table 11.1 The total number of sites in the group, average lengths of before and after periods (years) and the total numbers of accidents reported in six studies

		Before period		After period	
Study	Sites	Length	Accidents	Length	Accidents
TRRL Leaflet 393	78	3	171	3	117
Lalani	20	1.6	99	1.6	69
Green	54	3.4	405	2.5	192
Hanson and Wilson	11	3.7	49	3.3	63
Avon County Council					
(i) First study	12	1.7	41	1.7	32
(ii) Second study	23	2.5	62	2.5	41

11.1.3 Modelling the effect of mini-roundabouts

Nuisance factors
The observed rates of accidents during before and after periods are needed to evaluate the effectiveness of a road safety measure. However, these observations are influenced by many other factors which must also be considered before attributing any change to a treatment effect. Some of these factors are identifiable and examples include: traffic law; traffic volume; weather; hours of darkness; and more emphasis on safety in car design. Others, such as changes in driver behaviour, are less well defined. Scott (1986), for example, argues that it is virtually impossible to incorporate all components which contribute to seasonality in the time-series modelling of monthly accident figures in Britain. In our analysis, we use the monthly accident figures in Britain to estimate limits for a seasonality factor.

The studies by TRRL, Lalani and Avon Council did not use any control area and just compared before and after rates. In contrast to this, both Green and Hanson and Wilson used data from a control area to help infer whether the changes in the number of accidents at the treated sites should be attributed to a trend or the effect of mini-roundabouts. However, it is not very clear how one should identify the best control areas for comparison with the treated areas. Teasdale *et al.* (1991), for example, have investigated several definitions of control areas for assessing area-wide road safety schemes. The definitions include that of Hauer (1989a; 1989b), that there should be at least 200 accidents per period in the control area, and those of the IHT (Institution of Highways and Transportation, 1991) suggesting that control areas with accident frequencies similar to those in the treatment area should be used rather than control areas with substantially different accident frequencies. Teasdale *et al.* (1991) found little evidence to support either definition. In our analysis, we use the number of accidents in Britain to estimate limits for any trend factor.

It is also recognized that if sites were selected from a population of physically identical sites, because they happened to have the worst accident history, a decrease would be expected without applying any treatment. This is known as **bias by selection** or **regression to mean**, and has been discussed by Hauer (1980) and Wright *et al.* (1988), among others. Hauer (1980) derives a formula for bias by selection, assuming that accidents at sites have Poisson distributions. To demonstrate the possible size of this effect, suppose sites have a common underlying mean of 10 accidents per time period, and that one-tenth of sites are chosen because of their bad accident history over the last time period (so called **black-spot** sites). There would then be an expected reduction of 62%! However, even if sites were selected because of their records rather than their potential for accidents based on their engineering design, they certainly do not all have the same physical characteristics and the bias-by-selection effect rapidly decreases as the hypothetical Poisson site means become more diverse. Wright *et al.* (1988) estimate an expected reduction of 8% for a single black-spot site chosen from Hertfordshire (UK) junctions because of a bad record over the five-year period 1975–79. Here we will assume a bias-by-selection effect of at most 8% for a group of sites. Some of the studies are based on shorter periods than five years but this is offset, to some extent at least, by the fact that we have a group of sites rather than one on its own. We also assume that the bias-by-selection effect is uniformly distributed between 0% and its highest value of 8%. We consider this justified as we have no evidence that sites were selected from accident records alone. No allowance was made for the bias-by-selection effect in the published reports we have seen, and some of the sites had no accidents during the before period.

The number of accidents during complete years is not always particularly well modelled as a constant-mean Poisson process (Nicholson, 1985). This is not surprising as weather conditions, for example, can vary considerably between years. We only assume that the total number of accidents in the after period is a Poisson variable, and rely on the national trend to incorporate any effects of different weather patterns. Accidents, rather than numbers of vehicles in accidents or injuries, can reasonably be considered independent, so there is no strong argument against the Poisson distribution on this ground. It might be argued that a warning effect, after drivers hear of an accident, could lead to a slight reduction in variability, but we do not consider any such refinements.

General model for each study
We make several assumptions when applying Bayesian methods to a group of sites.

1. Accidents occur as a Poisson process.
2. The numbers of accidents at different mini-roundabouts are independent random variables, following Poisson distributions.
3. The total numbers of accidents at the group of mini-roundabouts follows the Poisson distribution with mean η per year, which is the sum of the mean rates for the different sites.
4. The prior density for the mean number of accidents (η) at the group of mini-roundabouts is a gamma distribution with pdf

$$f(\eta) = \frac{c^a \eta^{a-1} e^{-c\eta}}{\Gamma(a)} \quad \text{for } \eta > 0 \quad \text{with } a > 0, \quad c > 0 \qquad (11.1)$$

The gamma distribution is another continuous distribution which can take a range of shapes from near-symmetric to highly positively skewed (Fig. 11.1). Its definitions involves the gamma function (Exercise 2.15), and you may have already derived the pdf for the special case of the sum of k independent exponential variables in Exercise 4.19.

It is straightforward, remembering the definition of the gamma function, to show that the mean and variance of η are a/c and a/c^2, respectively, so we

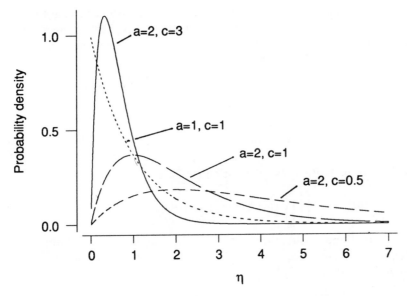

Fig. 11.1 Gamma distributions (after Devore 1982).

consider our prior information as equivalent to observing a accidents over c years.

The reason for choosing a gamma distribution in this context is that, if the observed number of accidents has a Poisson distribution, the posterior distribution for η is a gamma distribution with a modified mean and variance (similar to the results in Exercise 8.6). This neat algebraic result depends on the assumed gamma/Poisson pairing, and the Weibull distribution, for example, does not behave similarly. The gamma distribution is said to be a **conjugate prior** for the Poisson distribution. MINITAB provides the cdf and the inverse cdf of the gamma distribution, and can also generate random samples from it.

Suppose that in the period, of length t, following the introduction of mini-roundabouts at sites, y accidents are observed. For the after period, we assume the rate η will be changed by a multiplicative treatment effect (θ) and by the three 'nuisance factors'. These are seasonality (ψ), trend (v) and bias-by-selection effect (ξ). That is, the Poisson mean becomes $\eta t \theta v \xi \psi$, where θ is the treatment factor. We assume that the introduction of mini-roundabouts will result in something between reducing the underlying rate to zero and doubling it. That is, our prior distribution for θ is that it has a uniform distribution on $[0, 2]$. It is not very realistic, as we rather expect some improvement, but it is sufficiently diffuse to have little influence over the data. It was recently suggested to us that some lognormal distribution for θ would have been a better choice, and we will adopt this for further work. Let $f(\theta, \eta, t, v, \xi, \psi)$ be the joint prior distribution of parameters θ, η, v, ξ and ψ which are assumed to have mutually independent and, apart from η, uniform distributions. Let y denote the total number of accidents observed in the period after introduction of the mini-roundabouts. From Bayes' theorem, we can say

$$f(\theta|y,\eta,t,v,\xi,\psi) \propto f(y|\theta,\eta,t,v,\xi,\psi)f(\theta|\eta,t,v,\xi,\psi) \qquad (11.2)$$

The posterior density function of θ given y and t is

$$f(\theta|y,t) \propto \iiiint f(\theta|y,\eta,t,v,\xi,\psi)f(\eta,v,\xi,\psi)\,d\eta\,dv\,d\xi\,d\psi$$

and this can be written, using equation (11.2), as

$$f(\theta|y,t) \propto \iiiint f(y|\theta,\eta,t,v,\xi,\psi)f(\theta,\eta,t,v,\xi,\psi)\,d\eta\,dv\,d\xi\,d\psi \qquad (11.3)$$

The distributional assumptions are that the conditional distribution of y is Poisson, η is gamma (equation 11.1), and all the other variables are uniform. Therefore equation (11.3) becomes

$$f(\theta|y,t) \propto \iiiint \exp(-\eta t\theta v\xi\psi)(\eta t\theta v\xi\psi)^{v}\eta^{a-1}\exp(-c\eta)\,d\eta\,dv\,d\xi\,d\psi \qquad (11.4)$$

To simplify the integration, let $Q = t\theta v\xi\psi$, so that the integration with respect to η in equation (11.4), which can be recast as

$$\int_0^\infty \exp(-\eta((t\theta v\xi\psi)+c))(t\theta v\xi\psi)^v\eta^{a+y-1}d\eta$$

becomes

$$f(\theta|y,t) \propto \int_0^\infty \eta^{a+y-1}Q^y\exp(-\eta(Q+c))d\eta \qquad (11.5)$$

Now the integrand in equation (11.5) is proportional to a gamma pdf. By comparison with equation (11.1) we deduce

$$\int_0^\infty \frac{\eta^{a+y-1}(Q+c)^{a+y}\exp(-\eta(Q+c))}{\Gamma(a+y)}d\eta = 1$$

so the integral (11.5) actually equals

$$\frac{\Gamma(a+y)Q^y}{(Q+c)^{a+y}}$$

The factorial, $\Gamma(a+y)$, just becomes part of the implicit proportionality constant, but Q includes the other variables. Therefore, the posterior density function can now be written as

$$f(\theta \mid y, t) \propto \iiiint \frac{(\theta v \xi \psi)^{y}}{\left(\theta v \xi \psi + \dfrac{c}{t}\right)^{a+y}} \, dv \, d\xi \, d\psi \qquad (11.6)$$

The 'disappearing' factor of t^{-a} has been incorporated into the constant of proportionality.

11.1.4 Analysis of individual studies

We start by analysing the data from the individual studies for several reasons. We can assess whether or not the policy of replacing priority controlled junctions by mini-roundabouts for each particular group of sites has been successful, in a consistent manner which allows for nuisance factors. The results of these analyses can be used to decide whether or not an assumption of a constant treatment effect for all groups is plausible. Furthermore, the overview that does assume a constant effect proceeds in almost the same manner as the sequence of individual analyses. The only difference is that the posterior distribution of θ from one study becomes the prior distribution for the next. Individual results are also needed for the hierarchical overview.

Analysis of TRRL's study
We assume an average period of 3 years each for both the before and after periods, and use the total numbers of casualties for all road users throughout the years 1970 to 1975 as an estimate of the trend in the comparison areas. The total numbers of all road user casualties and their indices are given in Table 11.2 (based on data from *Road Accidents in Great Britain*). From these figures, we can say that the national figures indicate a reduction in casualties of 2% per year. However, we are unsure whether to relate changes in accident rates at the group of sites to national figures (2% reduction per year in the after period), or to the before period (no reduction per year in the after period) or to something in between. We therefore assume, as the difference between the mid-point of the before and after periods is 3 years, that

$v \sim U(0.94, 1)$

The choice of a uniform distribution is partly for convenience, but we also think it is a fair representation of our uncertainty in v, since both extremes seem as plausible as anything in the middle.

We also allow for the possibility that seasonal variation is affecting accident rates. If the before and after periods at each roundabout in the group are a whole number of years no allowance would have to be made. In this study our best guess of the length of both periods is 3 years, but variations at particular sites could lead to different months occurring more often in the before and after periods. For example, if the lengths of the before and after periods at a particular site were 2.5 years there would be 12 distinct possibilities, depending on the

Table 11.2 Yearly casualty indices of all road users (average yearly index is 100) from 1970 to 1979*

Year	Total number of all road user casualties	Index
1970	363 366	105
1971	352 017	102
1972	359 697	104
1973	353 778	103
1974	324 602	94
1975	324 863	94
1976	339 658	99
1977	348 061	101
1978	349 795	101
1979	334 513	97

*Details of yearly and monthly injury accidents for all road users are not given in *Road Accidents Great Britain*, but it seems reasonable to suppose a similar trend.

Table 11.3 Monthly casualty indices of all road users (average monthly index is 100) from 1970 to 1979

Month	1970	1971	1972	1973	1974	1975	1976	1977	1978	1979	Average index
Jan	89	98	97	94	90	90	86	88	98	85	91
Feb	88	87	92	97	80	84	83	78	74	74	86
Mar	92	92	95	90	87	95	85	93	92	94	92
Apr	89	90	92	98	87	88	87	89	89	90	90
May	99	101	101	100	101	100	103	96	98	103	100
June	96	98	101	103	102	99	100	96	102	97	99
July	108	106	103	104	102	105	107	108	106	103	105
Aug	105	108	96	102	104	108	101	106	106	108	104
Sep	102	98	94	108	114	105	101	101	105	107	104
Oct	110	109	102	102	110	103	115	111	105	110	108
Nov	111	108	114	102	110	103	115	111	105	110	108
Dec	111	105	113	100	100	116	116	116	108	116	110

month in which the mini-roundabout is installed. Straightforward arithmetic on the seasonal indices in Table 11.3 gives a maximum seasonal effect, averaged over the $2\frac{1}{2}$ years, between 0.94 and 1.06. This effect would be less for any other length of period between 2 and 3 years and similar results apply for periods longer than 3 years. We therefore feel justified in assuming that the seasonality factor for this study can be given by

$$\psi \sim U(0.94, 1.06)$$

This is a very rough assessment which probably overstates this source of variability. We model the bias-by-selection effect by

$$\xi \sim U(0.92, 1)$$

although this is an even rougher assessment! Bias by selection is an important issue which could be investigated by prospective research work. We do not have sufficient detail in our overview to make any more precise statement, and we use the same distribution for all the studies. There were 171 accidents in the 3-year before period, so the values of a and c in equation (11.1) are 171 and 3, respectively. There were 117 accidents during the 3-year after period. So $t = 3$ and equation (11.6) becomes

$$f(\theta|y = 117, t = 3) \propto \int_{0.94}^{1.06} \int_{0.92}^{1} \int_{0.94}^{1} \frac{(\theta v \xi \psi)^{117}}{(\theta v \xi \psi + 1)^{288}} \, dv \, d\xi \, d\psi \qquad (11.7)$$

The posterior density function of θ was obtained by numerical integration, and is shown in Fig. 11.2. Our point estimate of θ for this study is taken as the mode of this distribution, which equals 0.74.

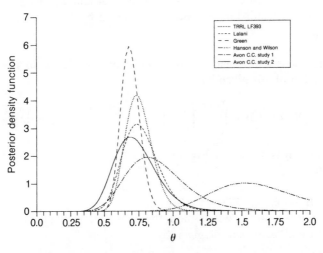

Fig. 11.2 Bayesian analysis of individual studies.

Analysis of Lalani's study

Since accident data in this study are limited to the period from January 1970 to May 1975, annual casualty totals for all road user accidents between 1970 and 1975 are a reasonable estimate for the trend in the comparison area. The trend factor is modelled by

$$v \sim U(0.97, 1)$$

The averages of the before and after periods are both 19 months, and the seasonality factor for this study is

$$\psi \sim U(0.9, 1.1)$$

There were 69 accidents during the 19-month after period and the posterior density function of θ becomes

$$f(\theta | y = 69, t = 1.6) \propto \int_{0.9}^{1.1} \int_{0.92}^{1} \int_{0.97}^{1} \frac{(\theta v \xi \psi)^{69}}{(\theta v \xi \psi + 1)^{168}} \, dv \, d\xi \, d\psi \qquad (11.8)$$

The estimated modal value of θ is 0.74 and its distribution is also shown in Fig. 11.2.

Analysis of Green's study

When deciding on the national trend for the data by Green, we consider the total number of years covered by both before and after periods for all sites and compute the trend factors accordingly and average them out. The trend factor is assumed to be uniformly distributed.

$$v \sim U(0.97, 1)$$

The lengths of the before and after periods are 3.4 and 2.5 years, respectively, so the seasonality factor for this study is modelled by

$$\psi \sim U(0.94, 1.06)$$

During the 2.5-year after period there were 192 accidents, so the posterior density function of θ is given by

$$f(\theta | y = 192, t = 2.5) \propto \int_{0.94}^{1.06} \int_{0.92}^{1} \int_{0.97}^{1} \frac{(\theta v \xi \psi)^{192}}{(\theta v \xi \psi + 1.36)^{597}} \, dv \, d\xi \, d\psi \qquad (11.9)$$

This distribution is shown in Fig. 11.2, and the modal estimate of θ for this study is 0.68.

Analysis of Hanson and Wilson's study

Since the study was carried out at about the same time as that of Green, we again assume the trend factor can be given by

$$v \sim U(0.97, 1)$$

The seasonality factor is assumed to be uniform between 0.96 and 1.04. These limits are based on average before and after periods of 3.7 years and 3.3 years, respectively. If these limits and the lengths of the periods are substituted into equation (11.6), the posterior density function of θ can be obtained from

$$f(\theta | y = 63, t = 3.3) \propto \int_{0.96}^{1.04} \int_{0.92}^{1} \int_{0.97}^{1} \frac{(\theta v \xi \psi)^{63}}{(\theta v \xi \psi + 1.12)^{112}} \, dv \, d\xi \, d\psi \qquad (11.10)$$

The modal estimate of θ is 1.52, and the distribution is shown in Figure 11.2.

Analysis of the first part of Avon County Council's study
The national trend was about 0.1% per year, so we assume

$$v \sim U(0.998, 1)$$

The before and after periods for this study were both about 20 months, and we model the seasonality factor by

$$\psi \sim U(0.9, 1.1)$$

The posterior density function of θ, given $y = 32$ and $t = 1.7$, is

$$f(\theta | y = 32, t = 1.7) \propto \int_{0.9}^{1.1} \int_{0.92}^{1} \int_{0.998}^{1} \frac{(\theta v \xi \psi)^{32}}{(\theta v \xi \psi + 1)^{73}} \, dv \, d\xi \, d\psi \qquad (11.11)$$

and the modal estimate of θ is 0.82. The distribution is again shown in Fig. 11.2.

Analysis of second part of Avon County Council's study
Since the details of accident data for each site are not available, the limits used in the first part of this study will be used as an indication of trend and seasonality for the second study. Thus the posterior density function of θ, given the 62 and 41 accidents during the before and after periods, is

$$f(\theta | y = 41, t = 2.5) \propto \int_{0.9}^{1.1} \int_{0.92}^{1} \int_{0.998}^{1} \frac{(\theta v \xi \psi)^{41}}{(\theta v \xi \psi + 1)^{103}} \, dv \, d\xi \, d\psi \qquad (11.12)$$

The modal estimate of θ is 0.68, and the distribution is shown in Fig. 11.2.

Collating our results, the studies by TRRL (1975), Lalani (1975), Green (1977), and Avon County Council (1984) estimate a reduction in accidents by 26%, 26%, 32%, 18% and 32%, respectively. In sharp contrast to these reports, Hanson and Wilson (1979) estimated an increase in accidents of 52%.

11.1.5 Bayesian overview

All but one of the studies analysed in section 11.1.4 give a consistent picture of a decrease in accidents due to the installation of mini-roundabouts. The main objective of this overview is to bring them together and thereby obtain a more precise estimate of the reduction on a nationwide basis. The anomalous study is not available in the open literature and the authors attribute its findings to 'acknowledged deficiencies' in engineering design. Nevertheless, any new measure does introduce risks of errors of judgement, and it may be appropriate to allow for this when calculating expected benefits. We therefore look at two techniques for combining the results from the different studies. The first assumes any effect will be the same for all groups of sites but the second makes no such assumption, and we apply it with and without the anomalous study. In the discussion we attempt to reconcile these different approaches and arrive at an overall conclusion.

Overview assuming the same effect for all groups of sites

It is unrealistic to suppose that any treatment effect will be the same, in proportional terms, at individual sites. But we are reviewing results from groups of sites, and are more prepared to accept it as a working hypothesis in this context. If we forget about the Hanson and Wilson study, it is not inconsistent with the results of section 11.4.1 displayed in Fig. 11.2. Tanner (1958) describes a formal test for the hypothesis. The overview proceeds by using the posterior density of θ from the TRRL study as the prior density of θ for analysing Lalani's data and so on. Since the distribution of θ is assumed independent of the other variables, the calculations follow the same lines as those described already, and the result is shown in Fig. 11.3. The mode of the final posterior distribution is 0.72, and the standard deviation is 0.06. A hypothesis of a common treatment effect is much less plausible if we include the anomalous study of Hanson and Wilson, but if we were to do so the estimate of the mode of the distribution would increase to 0.76 and the standard deviation would also increase slightly.

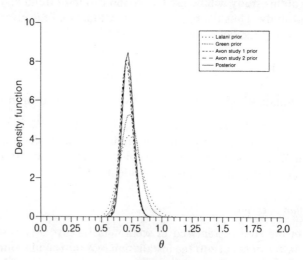

Fig. 11.3 Bayesian overview (excluding Hanson and Wilson study) (assuming constant treatment effect).

Hierarchical Bayesian model
The Bayesian hierarchical approach involves the construction of two different levels of parametric model. The first level of the model describes the distribution of the estimates of the treatment effect for each of the sites, while the second level allows for these effects to vary from site to site. These two levels of the model are combined by Bayes' theorem.

We start by letting y represent the 6×1 matrix of estimated treatment effects, that is, the modes of the posterior distributions from the individual analyses. We assume that y is normally distributed about the vector of actual effects (θ) with a covariance matrix C, a 6×6 matrix of estimated variances and covariances. The scalar τ^2, which multiplies C, models our uncertainty about these variances and covariances and is assumed to be proportional to an **inverse chi-square distribution** (see Lee (1989) for example) with df_τ degrees of freedom. The constant of proportionality is equal to df_τ so that τ^2 has an expected value of about 1. Thus the first stage of the model can be written as

$$(y \,|\, \theta, \tau) \sim N(\theta, \tau^2 C) \tag{11.13a}$$

$$\tau^{-2} \sim \chi^2(df_\tau)/(df_\tau) \tag{11.13b}$$

We now assume that the actual effects at the six sites are normally distributed about an unknown mean. This unknown mean is of the form

$$X\mu + d$$

where X and d are 6×1 matrices and μ is a scalar. If we set all the entries in the matrices X and d equal to 1, μ will represent the average change, defined as the average multiplicative effect of introducing mini-roundabouts minus the multiplicative effect corresponding to no change (1). Therefore a μ of 0 would correspond to no effect and negative values, between 0 and -1, correspond to improvements. The mean of the prior distribution of μ is zero, which corresponds to no effect. The algorithm is not sensitive to this assumption as μ has a diffuse prior distribution, that is, its variance tends to infinity. The variances of θ about μ are modelled by the 6×6 matrix V which contains our prior estimate of the variability of treatment effects between groups of sites. We are rather unsure about these and model our uncertainty by multiplying V by a scalar σ^2, which is again assumed to be proportional to an inverse chi-square distribution. The degrees of freedom are denoted by df_σ. Symbolically

$$(\theta \,|\, \mu, \sigma) \sim N(X\mu + d, \sigma^2 V) \tag{11.13c}$$

$$\sigma^{-2} \sim \chi^2(df_\sigma)/(df_\sigma) \tag{11.13d}$$

$$(\mu \,|\, \sigma) \sim N(0, D \to \infty) \tag{11.13e}$$

The hierarchical model is composed of the five statements (11.13). The algorithm used to calculate the posterior distributions of θ and μ (DuMouchel, 1989) is given in Appendix A6. The sizes of the matrices and vectors in the model are reduced by one when applying the hierarchical model to five studies (without Hanson and Wilson).

11.1.6 Fitting the hierarchical Bayesian model

The elements of y are the modes of the posterior distributions for the multiplicative treatment effects. For the six studies in chronological order:

$$y = (0.74\ \ 0.74\ \ 0.68\ \ 1.52\ \ 0.82\ \ 0.68)^{\mathrm{T}}$$

The covariance matrix C includes the relationship between the Avon studies as well as the uncertainty of the estimated value of θ in each study. The variances of the estimated values of θ are found from the posterior distributions, and a correlation of 0.25 is assumed between the two Avon County Council studies.

$$C = \begin{bmatrix} 70 & 0 & 0 & 0 & 0 & 0 \\ 0 & 128 & 0 & 0 & 0 & 0 \\ 0 & 0 & 33 & 0 & 0 & 0 \\ 0 & 0 & 0 & 686 & 0 & 0 \\ 0 & 0 & 0 & 0 & 319 & 58 \\ 0 & 0 & 0 & 0 & 58 & 168 \end{bmatrix} \times 10^{-4}$$

We base our choice of 50 degrees of freedom for τ on the following argument. Our observed frequencies ranged from 32 to 405 accidents in a before or after period at a given group of sites. Our estimates of the effects are based on functions of these observed frequencies and fairly narrow uniform variables. If we estimate a Poisson variance by the observed frequency, this estimator has a variance equal to itself. The estimator of the variance of a normal distribution, s^2, has an approximate variance of $2\sigma^4/n$ (Exercise 6.13). If the Poisson mean is 25, so is its variance which is equivalent to σ^2 in the approximating normal distribution, and then the degrees of freedom for τ would equal 50. This seems a fairly conservative choice. The elements of matrix V are decided by supposing that it is unlikely that mini-roundabouts would contribute more than a 30% change in accidents. If we take this 30% as equivalent to 2 standard deviations, the standard deviation is 15% and thus the variance is 0.0225. Therefore, the matrix V is taken to be

$$V = \begin{bmatrix} 225 & 0 & 0 & 0 & 0 & 0 \\ 0 & 225 & 0 & 0 & 0 & 0 \\ 0 & 0 & 225 & 0 & 0 & 0 \\ 0 & 0 & 0 & 225 & 0 & 0 \\ 0 & 0 & 0 & 0 & 225 & 0 \\ 0 & 0 & 0 & 0 & 0 & 225 \end{bmatrix} \times 10^{-4}$$

We will choose degrees of freedom for σ equal to 1 as we are rather unsure about the estimated variance and would prefer to have based it on a few past studies. Since the scalar μ represents the average difference between θ and 1 for the hypothetical population of all such groups of sites we set the matrices X and d equal to $(1, \quad 1, \quad 1, \quad 1, \quad 1, \quad 1)^T$.

Results

The expected value and the standard deviation of the posterior distribution of θ for each group of sites, with and without the anomalous study, are given in Table 11.4. The estimates of the hyperparameters of the model, that is, μ and its standard deviation together with the expected values of τ^2 and σ^2, are given in Table 11.5.

The Hanson and Wilson study is rather an embarrassment. One of the assumptions of the hierarchical model is that the θ for individual studies are normally distributed about μ. This pulls the posterior distribution of θ for the Hanson and Wilson group towards the mean for all groups, and the resultant has a mean of 0.93 with a large standard deviation of 0.16. If we assume a Poisson distribution for the number of accidents at this group the standard deviation of the observed increase, 63 compared with 44 (the 49 observed discounted by 3.3/3.7 to allow for the different-length time periods), is roughly 10 ignoring all nuisance factors. If we really believe the underlying rate has been reduced by a factor of 0.93 due to the mini-roundabouts, we have been

Table 11.4 Expected values of θ and the associated standard deviations

	$E(\theta\|y)$ and (standard deviation)	
Study	With H & W study	Without H & W study
TRRL	0.75 (0.08)	0.73 (0.07)
Lalani	0.75 (0.09)	0.73 (0.09)
Green	0.70 (0.06)	0.69 (0.05)
Hanson and Wilson	0.93 (0.16)	not applicable
Avon CC study (i)	0.79 (0.12)	0.75 (0.11)
Avon CC study (ii)	0.72 (0.10)	0.70 (0.09)

Table 11.5 Estimates of hyperparameters

Hyperparameter	With H and W study	Without H and W study
$E(\mu\|y)$	-0.23	-0.28
(Standard deviation)	(0.08)	(0.07)
$E(\tau^2\|y)$	1.10	0.98
$E(\sigma^2\|y)$	0.92	0.68

very unlucky to observe 63 accidents rather than the expected 40. But we have been told about the admitted deficiences in the engineering design and this crucial prior information has not been incorporated into the model. We do not, therefore, take the estimated 'improvement' of 0.93 seriously and would certainly not have advised the council to take no action! It is nevertheless worth noting that if investigation had not found design deficiencies at the sites, the observed results, although rather unlikely, would not be totally incompatible with the model. This could then be an example of the bias-by-selection problem. Another explanation would be that the θ are not normally distributed between groups of sites. Nevertheless, while the hierarchical model does tend to pull all the posterior distributions of θ together this is much less drastic than assuming an equal value for all groups of sites. A more interesting issue is whether the Hanson and Wilson study should be left in for estimating the hyperparameters. We think it should, because errors of judgement were made and others are likely to be made in the future. If we do include it, the expected value of the distribution of μ is -0.23 and the standard deviation is 0.08. We are quite confident of a reduction overall. The expected values of τ^2 and σ^2 should be approximately 1 if our prior estimates of variances and covariances are reasonable. If we exclude the Hanson and Wilson study, we naturally obtain a more impressive estimate of the reduction. The posterior expected value of σ^2 is rather low, suggesting that the distribution of θ-values is less dispersed than we originally thought.

11.1.7 Discussion and conclusion

All the investigators, except Hanson and Wilson, reported a reduction in the rate of accidents after installing mini-roundabouts at a group of sites. There is good reason to suppose that this is a consequence of the installations, and the policy appears to have been successful for these groups of sites. Hanson and Wilson were able to account for their anomalous findings, but the policy was not successful for this particular group.

The overviews implicitly assume that the groups investigated are representative of all sites that would be considered suitable for mini-roundabouts, and in the absence of random sampling we must make a subjective assessment of its reasonableness. Random sampling does not guarantee the sample will be representative, but the chance that it is not becomes quantifiable and can be reduced by appropriate stratification. Hanson and Wilson said in their report that some sites which experienced much higher number of accidents in the after period were unsuitable due to 'acknowledged deficiency' in engineering design. This criticism was only made at the end of study, and similar errors of judgement might arise if a general policy of replacing priority controlled junctions by mini-roundabouts is adopted. Whether or not to include it depends how likely similar deficiencies are to occur in the future, and this is best decided by transport engineers. If it is ignored both overviews give a predicted average reduction of

28%. The standard deviation of the posterior distribution is slightly less if we assume a common treatment effect for all groups, 6% compared with 7%. If we include the anomalous 'desk-draw' study the estimated average benefit is diminished to a reduction of 23% and the standard deviation increases to 8%.

Our overall conclusion is that a policy of replacing selected priority controlled junctions by mini-roundabouts is likely (90% chance) to lead to a reduction in accidents of at least 13% (1.3 standard deviations below the pessimistic mean), and that our best estimate of the benefit is a reduction of between 23% and 28%. We would also advise reopening the investigation of the Newcastle sites with the objective of identifying the precise problems with the engineering design, and modifying the general policy in the light of this experience.

11.2 Predicting short-term flood risk

11.2.1 Introduction

Seasonal variation in both the number and magnitude of peak flows has been observed by many research workers, for example Archer's (1981) investigations of flooding in North-East England. If this feature of hydrological processes is ignored the probabilities of high flows during the summer months, when flood risk is usually relatively low, will on average be overestimated. Furthermore, if the distributions of annual maxima or peaks over some threshold are modelled without allowing for seasonal changes, winter flood risk estimates will be subject to two sources of error: there will be a tendency to underestimate the mean and overestimate the variance. While these effects do at least work in opposite directions, it is rather optimistic to rely on their cancelling out!

Several methods for modelling seasonal variation have been proposed. Todorovic (1978) modifies peak-over-threshold methods by allowing the rate of exceedance and the distribution of exceedances to change with time of year. Ghani and Metcalfe (1985) reduce the threshold to include all peaks and use data from the required season only. This has a possible advantage of releasing more data, but the drawback is that they cannot be considered independent. We circumvented this problem by using **spectral analysis** (section 11.3) on the time series of midday observations. Tawn (1988) takes a more direct approach and presents an extreme-value theory model for dependent observations. Models which incorporate seasonable variation will give more accurate estimates of the flood risk which are valuable for long-term decisions concerning the levels of flood protection, or for the design of storm sewers.

Civil engineers are also expected to make decisions when the risk of flooding in the near future is of prime interest. For example, contractors working on a dam face, from barges or with floating cranes, will benefit from accurate estimates of flood risk. Water engineers responsible for reservoir operation, who need to balance the requirements of flood control, provision of domestic and industrial water supply, public amenity, and effluent dilution, will also benefit from up-to-date estimates of the risk of occurrence of high flows. In such cases the risk of flooding will be influenced by prevailing catchment conditions and weather forecasts, as well as the average seasonable variation. Insurance companies might also have an interest in estimating short-term flood risks. Ettrick *et al.* (1987) proposed a method for estimating short-term flood risk based on conditional distributions. Futter *et al.* (1991) compared this with an approach based on the Cox (1972) regression model. A modification W.-W. Tsang and I proposed was the replacement of two of the distributions in the Ettrick *et al.* formulation by a more realistic time-series model. The following uses (with permission) material from Tsang (1991).

11.2.2 Modelling short-term flood risk

Conditional distribution model

The model proposed by Ettrick *et al.* (1987) starts with a distribution of peakflows (q) conditional on some measure of catchment wetness (w) and depth of rainfall (y), given that a chosen rainfall threshold (y_0) has been exceeded (a rainfall event). The reason for choosing a rainfall threshold, rather than a flow threshold, to define independent peakflows was to extract information on relatively high-rainfall events which did not produce any notable peakflow because they fell on a dry catchment. The rainfall threshold is set low enough to include all rainfall events which cause substantial peakflows.

This conditional distribution was then elaborated to provide predictions 30 days ahead, for example, based on observed persistence of catchment wetness in several river basins. Baseflow – that is, the flow in a river before a rainfall event has had time to augment it – was used as a measure of catchment wetness in their applications. It has the advantage of being more readily available than, for example, soil moisture deficit measurements and, for the river basins with both measurements which they investigated, appeared to have a higher association with peakflows. A conditional distribution (k) of baseflows prior to rainfall events, for the 30 days after baseflow at start of period (w_s), was introduced. The conditional distribution for peakflows is then

$$F(q|w_s, y > y_0) = \int_0^q \int_{y_0}^\infty \int_0^\infty g(x|y, w, y > y_0)h(y|y > y_0)k(w|w_s, y > y_0)\,dw\,dy\,dx$$

$$(11.14)$$

if it is assumed that rainfall is independent of catchment wetness. A Poisson model was assumed for exceedances of the rainfall threshold, and flood risks were calculated once the parameters had been estimated for the required season. This approach ignores the fact that during the early days of the prediction period the baseflow is likely to be closer to the baseflow at the time of prediction, and towards the end of the prediction period the baseflow is likely to be closer to the seasonal average. It relies on an 'average distribution'. In general, a function of an average of a variable does not equal the average of that function of the variable. A time-series model for baseflow would be preferable, as it avoids the notion of 'averaging' in this way and is considerably easier to explain.

Time-series model
The following regression model was used to predict baseflow on day $t + 1 (w_{t+1})$ given previous baseflows and daily rainfall input sequences (r_t):

$$w_{t+1} = w_t - \Delta w + a_{-1}r_{t+1} + a_0 r_t + a_1 r_{t-1} + \cdots + a_q r_{t-q} + E_t \quad (11.15)$$

where E_t is a sequence of independent 'errors'. It should be noted that daily rainfall input is often zero, and, even when non-zero, is not necessarily the same as the rainfall associated with an event (y). For the River Dearne, y is the total rainfall over the two days preceding the event. The constant decrement was used to model the persistence found in baseflows series during dry periods. This cannot be achieved with low-order autoregressive models. **Fractionally differenced models** (Hosking, 1981; 1984) are an alternative, but preliminary results have not been any better than those obtained with equation (11.15). The rainfall sequence was modelled in two stages. A two-state **Markov chain model** was found adequate to describe transitions from dry to wet days – that is, the probability that tomorrow is wet depends only on whether today is dry or wet. The model is completely specified by the probabilities: tomorrow is wet given today is wet; and tomorrow is wet given today is dry. The probabilities that tomorrow is dry follow from the fact that the sum of the probabilities for dry and wet equals 1. A Weibull distribution was used for the amount of rainfall on wet days. Weather forecasts which include rainfall predictions could be used instead of the rainfall model if they are available. The regression part of the Ettrick *et al.* model is retained. That is

$$G(q|w, y, y > y_0) = \int_0^q g(x|w, y, y > y_0)\,dx \quad (11.16)$$

Seasonal variation is allowed for by estimating the parameters of the model for different seasons of the year and, if it seems appropriate, applying smoothing techniques.

The first step in estimating flood risk is to generate a sequence of wet and dry days for the required period (T), using the estimated transition probabilities. Next, a random sample of daily rainfalls is generated for the wet days. Equation (11.15) can then be used to generate a baseflow sequence. At this stage the number of rainfall events (m) is known. Now suppose that the probability of exceeding some critical flow (q_c) is required. For each rainfall event, the probability of not exceeding q_c can be calculated from equation (11.16). The risk of flooding during the period T is then estimated by:

Pr(at least one exceedance of q_c in T days)

$= 1 - \text{Pr}(q_c \text{ not exceeded in first event})$

$\times \cdots \times \text{Pr}(q_c \text{ not exceeded in event } m) \quad (11.17)$

This simulation process is repeated many times and the average value is taken as the final estimate of flood risk.

11.2.3 Fitting the time-series model

Description of catchment

The River Dearne flows through South Yorkshire, and its catchment contains agricultural, residential, commercial and industrial areas. The area of catchment above Barnsley is 119 km², and varies in height from 150 m OD to over 860 m OD on the edges. The geology of the area is dominated by lower and middle coal measures of the Upper Carboniferous. There are three rain gauges, which provide daily rainfall totals, at Cannon Hall, Emley Moor and Worsborough Dale. The daily maximum and minimum flows in the river are measured at Barnsley. The available records are coincident for 21 years between 1965 and 1986.

The catchment daily rainfall values were calculated using the Thiessen polygon method with weightings of 0.65, 0.26 and 0.09 given to the three stations, respectively. Occasionally data were missing, and the weightings were adjusted in the appropriate manner. Figure 11.4 outlines the catchment shape as well as giving the relative position of the flow gauging station and the rain gauges.

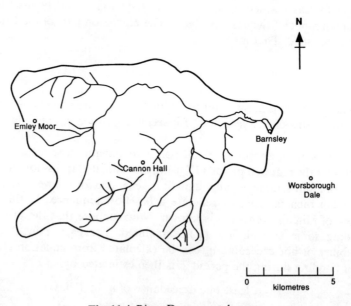

Fig. 11.4 River Dearne catchment.

Fitting the baseflow model

The baseflow series $\{w_t\}$ was taken as the daily minimum flows. Futter (1990) had analysed the same catchment and concluded that a time of concentration of 2 days and a rainfall threshold of 14 mm were best suited for the River Dearne. He found little evidence of any substantial variation in the parameters of interest during the period from March until October, which will be treated as a 'summer' season. The flood risk during the winter is affected by snow melt, and this would need to be incorporated into the modelling procedure. The treatment here is restricted to the assessment of short-term flood risk in the summer season, which is the most relevant for contractors working in rivers. The model of equation (11.15) was fitted using the regression procedure in MINITAB. The fitted model (where w is measured in cubic metres per second and r in millimetres) was:

$$w_{t+1} = w_t - 0.0202 + 0.055r_t - 0.0419r_{t-1} - 0.0059r_{t-2} \qquad (11.18)$$

and the standard deviation of the errors was estimated as 0.529. All the estimated coefficients were significantly different from zero at, or beyond, the 1% level. The negative coefficients for r_{t-1} and r_{t-2} can be accounted for by the fact that, when they are non-zero, they also influence w_t, which is itself included in the model. Addition of further terms on the right-hand side did not significantly reduce the estimated standard deviation of the errors.

Fitting the Markov chain

A wet day was defined as a day with rainfall exceeding 0.2 mm (anything less is likely to evaporate). Over the 21 years there were 1316 transitions from a wet day to a wet day (WW), 691 transitions from a wet day to a dry day (WD), 686 transitions from a dry day to a wet day (DW) and 1749 transitions from a dry day to a dry day (DD). The estimated transition matrix (M) was therefore

$$M = \begin{array}{cc} & \begin{array}{cc} W & D \end{array} \\ \begin{array}{c} W \\ D \end{array} & \begin{bmatrix} 0.656 & 0.344 \\ 0.282 & 0.718 \end{bmatrix} \end{array}$$

The transition matrix is a convenient summary of the probabilities. Also if $p^{(t)}$ contains the probabilities that day t is wet, $p_W^{(t)}$, and dry, $p_D^{(t)}$,

$$p^{(t)} = (p_W^{(t)} \quad p_D^{(t)})$$

then the probabilities for day $t+1$ are given by the equation

$$p^{(t+1)} = p^{(t)}M$$

Markov chains can have any number of states and Moran (1959) gives an interesting application to reservoir storage (see also the entry on 'Dam Theory' in Kotz and Johnson, 1982). This simple model was tested by comparing frequencies for different event durations, and the results shown in Fig. 11.5 suggest it is adequate.

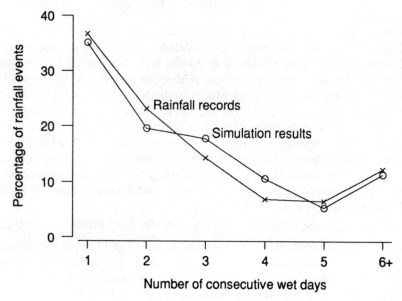

Fig. 11.5 Frequency of consecutive wet days.

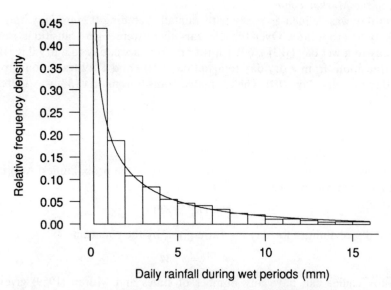

Fig. 11.6 Daily rainfalls during wet periods (mm).

Table 11.6 Comparison of mean daily rainfalls for different duration rainfall events

Duration (days)	Number of data	Mean (mm)	Standard deviation
1	254	2.92	4.69
2	160	4.12	4.55
3	100	4.78	4.36
4	49	5.13	4.63
5	44	4.70	3.06
6 or more	84	4.60	2.17

Fitting the rainfall distribution
A Weibull distribution with density

$$f(x) = \beta\alpha^\beta x^{\beta-1} \exp(-(\alpha x)^\beta) \qquad \text{for } x \geq 0 \qquad (11.19)$$

was found to give a good empirical fit to the amounts of rainfall (x) on wet days. Method-of-moments estimates for α and β were 0.2955 and 0.698, respectively. The fitted probability density function is compared with the histogram of daily rainfalls on wet days in Fig. 11.6.

The Weibull distribution provides a good fit to the marginal distribution of rainfall, but there was some evidence that the average amount of rainfall on the single wet days was less than the average daily rainfall during periods of more than one consecutive wet day (see Table 11.6). There was also some evidence that daily rainfall totals during periods of consecutive wet days were not independent. It can be seen in Table 11.7 that the standard deviations of rainfall totals, for periods of consecutive days from two to eight, appear to increase more rapidly than the square root of length of period. A 95% confidence interval for the power (b) in the relationship

$$\text{standard deviation} = \alpha(\text{number of consecutive days})^b \qquad (11.20)$$

is [0.54, 1.57].

Fitting the conditional distribution of peakflows
Futter (1990) fitted the Weibull distribution of equation (11.16) with α as a function of y (the total rainfall over the two days preceding the day of the event) and w (the baseflow on the day before the event). Several functional forms were tried and the final relationships, fitted by maximum likelihood, were

$$\hat{\alpha} = (0.38(y - y_0) + 4.78w)^{-1} \qquad (11.21a)$$

$$\hat{\beta} = 1.72 \qquad (11.21b)$$

where y_0 is the rainfall threshold (14 mm).

Table 11.7 The standard deviations of rainfall totals for events of different durations

Duration (days)	Number of data	Standard deviation of rainfall totals (mm)
1	254	4.7
2	160	9.1
3	100	13.1
4	49	18.5
5	44	15.3
6	21	35.2
7	17	40.1
8	19	32.2
9	9	20.2
10	3	23.7
11	5	39.3
12	4	82.3
13	1	–
14	1	–
15	3	17.6

11.2.4 Estimating the flood risk

A program was written to calculate the flood risks for 7 and 30 days ahead. It starts with a random number generation sub-routine (Press *et al.*, 1992), to generate a sequence of dry and wet days using the estimated parameters of the Markov chain. Rainfall magnitudes are next generated for all the wet days according to the fitted Weibull distribution. A series of baseflows can now be generated using the regression model, for any value of start of period baseflow. Figure 11.7(a) shows one realization of the rainfall inputs for 30 days ahead. The corresponding baseflows, for a start-of-month baseflow of $1\,m^3\,s^{-1}$, are shown in Fig. 11.7(b).

Events with 2-day rainfall total exceeding the rainfall threshold of 14 mm are then identified, and this is followed by the numerical integration of the peakflow distribution for these events. In the integrations, the values of baseflow and rainfall corresponding to that particular event are substituted into the conditional distribution of peakflow given in equations (11.19) and (11.21). The integrals are calculated using a subroutine for the Romberg method suggested by Gerald and Wheatley (1989). The flood risk for the simulation can now be estimated using equation (11.17). The process was repeated 1000 times for start-of-period baseflows from 1 to $5\,m^3\,s^{-1}$ in steps of $1\,m^3\,s^{-1}$, and the average risks are shown in Figs 11.8 and 11.9. With 1000 simulations the error attributable to variability between simulations was of the order of 2% for the lowest start-of-

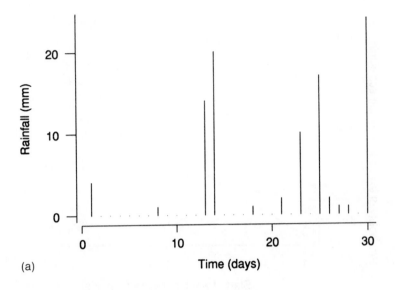

(a)

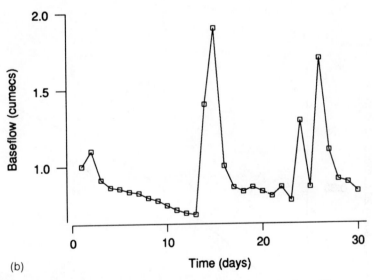

(b)

Fig. 11.7 Rainfall and baseflow: (a) realization of rainfall over 30 days; (b) corresponding baseflow series.

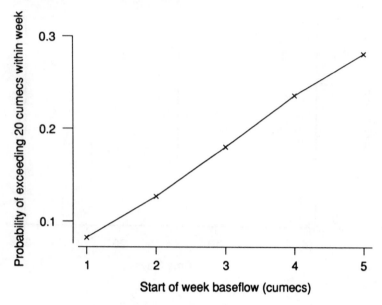

Fig. 11.8 Start of week baseflow ($m^3\,s^{-1}$).

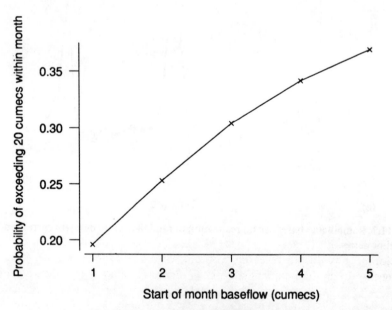

Fig. 11.9 Start of month baseflow ($m^3\,s^{-1}$).

period baseflow and 5% for the highest, and this is negligible when compared with uncertainty in parameter estimates.

11.2.5 Discussion

The results in Figs 11.8 and 11.9 show a relationship between the flood risk and start-of-period baseflow, which is more pronounced for the predictions seven days ahead. The model can be explained in physical terms and is reasonably straightforward to fit. It would be possible to refine it, for example by generating correlated random variables for rainfall sequences extending over more than one day, but this will only be worthwhile if it makes a substantial difference to risk estimates. The need for simulations to compute a flood risk is not a serious disadvantage given modern computing facilities.

11.3 Spectral analysis for design of offshore structures

11.3.1 Introduction

If you can hear the difference between Chopin and Jimi Hendrix you are responding to the different frequency composition of changes in air pressure. Radio waves, visible light and X-rays are all electromagnetic radiation, but they are at different frequencies. All life forms are sensitive to this difference, and it is not surprising that even inanimate structures are affected by both the magnitude and the frequency of disturbances. The collapse of the Tacoma Narrows Bridge emphasizes the need to understand the mechanism.

Oil rigs built in shallow water are relatively 'stiff'. That is, the frequencies at which they are most responsive (their natural frequencies) are all well above any predominant frequency ranges in the wave forces. However, in deep water the structures will be on longer legs and inevitably less stiff. The designer must ensure that the materials and geometry will result in natural freqencies which are high enough to avoid any resonances, or, if this is not practical, that the corresponding modes be heavily damped. The aim of this case study is to explain how we can describe the frequency composition of a random environment, such as the sea, and how this information can be used to estimate the response of a structure.

The statistical techniques for describing random environments, and predicting responses of structures in them, are known as 'spectral analysis'. They are an extension of the ideas of Fourier series, and much of the credit for their development is due to research workers who would describe themselves as engineers rather than statisticians. The standard theory is for linear systems, and it is essential to remain aware of this limitation. The crucial properties of a linear system are that the response to a disturbance at a given frequency is at the same frequency and proportional to the amplitude of the disturbance, and that the response to several disturbances is equal to the sum of the responses to each on its own.

11.3.2 Response to a disturbance at a single frequency

If a finite-element model of a design of oil rig is set up, it will be possible to determine the natural frequencies for the model, and the shape of the envelopes

Fig. 11.10 Offshore drilling platform.

of the induced vibrations at those frequencies (loosely speaking, the mode shapes). Some of these mode shapes will be torsional, and the response will in general depend on the wave direction. However, it is at least plausible that the lowest natural frequency of the rig sketched in Fig. 11.10 has the mode of vibration indicated in that sketch when forces are abeam. Given these assumptions, we can model the displacement of a point on the platform, due to disturbances up to this first natural frequency, by a linear second-order differential equation with the same frequency. The variable would be the angular deflection, and an approximate moment of inertia of the rig about the sea bed could be calculated from the geometry and the density of steel. Damping is difficult to estimate with any reliability and it is common to assume it is very low, which is tending towards the 'worst case'. The stiffness is then determined by the requirement that the natural frequencies be the same. The final design would be checked by constructing a full dynamic mathematical model, but it would save resources if obviously unsuitable designs could be screened out by the approximate method at an early stage. Shaw (1982) uses the same form of equation as a simple model for several wave energy devices. The clever bit is relating the stiffness and damping to the physical characteristics of the invention. The Bristol cylinder (Fig. 11.11) is a neutrally buoyant submerged cylinder which is constrained by a system of springs and dampers so that it absorbs the energy of incident waves. I shall describe a simplification which is constrained so that it can only move in a

Fig. 11.11 Tank model of the Bristol cylinder device (courtesy D. V. Evans, University of Bristol).

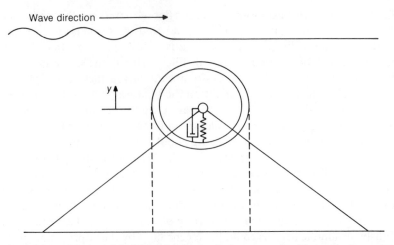

Fig. 11.12 Simplified model of Bristol cylinder device constrained to vertical movements.

vertical direction (Fig. 11.12). The vertical displacement from the equilibrium position is y, the mass per unit length M, the damping D, the stiffness K, and the harmonic hydrodynamic forces x. Then

$$M\ddot{y} + D\dot{y} + Ky = x \qquad (11.22)$$

Let x have unit amplitude and frequency ω. Then

$$x = \cos(\omega t)$$

and the steady-state response is

$$y = A\cos(\omega t + \phi)$$

where

$$A = \frac{1}{[(K - M\omega^2)^2 + D^2\omega^2]}$$

We will not need the phase shift (ϕ) for our calculations. Shaw (1982) discusses the energy extraction in some detail; it is a major contributor to the damping. We will just note that the power (Π) is proportional to the average of the square of the velocity ($\dot{y}$). That is, differentiate y and substitute for A to obtain

$$\Pi = \frac{\omega^2}{[(K - M\omega^2)^2 + D^2\omega^2]} \qquad (11.23)$$

where the units are proportional to watts. Sea waves are not harmonic, but we can treat them as if they are the sum of a large number of harmonic waves with randomly varying amplitudes.

11.3.3 *Frequency composition of waves*

The 397 data in Table D.9 were sampled by a probe situated at the centre of a wave tank which was equipped with sufficient wave makers to synthesize realistic sea states. The data are distances from the still water level measured in millimeters at 0.1 second intervals. The time ordered sequence, x_t, is known as a **time series**. If we fit the regression

$$x_t = \alpha_0 + \alpha_1 \cos(2\pi t/396) + \beta_1 \sin(2\pi t/396)$$
$$+ \alpha_2 \cos(4\pi t/396) + \beta_2 \sin(4\pi t/396)$$
$$+ \alpha_3 \cos(6\pi t/396) + \beta_3 \sin(6\pi t/396)$$
$$\vdots$$
$$+ \alpha_{198} \cos(\pi t) + \beta_{198} \sin(\pi t)$$

we will have an exact fit with no degrees of freedom for error. An algorithm known as the fast Fourier transform (see, for example, Press *et al.*, 1992) can be used to calculate the coefficients simultaneously, or each harmonic can be fitted on its own with MINITAB. The harmonics are orthogonal (zero correlations) so that the estimated coefficients do not depend on which other harmonics are being fitted. If you remember that

$$\alpha \cos \theta + \beta \sin \theta = r \cos(\theta + \psi)$$

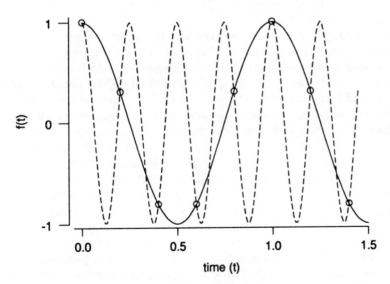

Fig. 11.13 Aliasing of harmonic signals. The sampling interval $\Delta = 0.2$ s. 1 Hz has alias frequencies $1 \pm 5k$ for any k. The case $k = -1$ gives frequency -4 equivalent to 4 Hz shown as a broken line. Nyquist frequency $1/2\Delta = 2.5$ Hz.

where

$$r = \sqrt{\alpha^2 + \beta^2}$$
$$\tan \psi = -\beta/\alpha$$

you can see that we have fitted 198 harmonic waves. The first corresponds to one wavelength over the whole record and is the lowest frequency we can fit. The last corresponds to a frequency of π radians per sampling interval and is the highest frequency we can fit. It is known as the **Nyquist frequency** and it is vital that it is higher than any frequencies that can reasonably be expected in the signal. This is because higher frequencies will be indistinguishable from some frequency in the range $(0, \pi]$ (see Fig. 11.13). In the figure a harmonic wave with a frequency of 2 Hz (cycles per second), or equivalently 4π rad s^{-1}, is sampled at 0.1 s intervals. This wave goes through $\frac{1}{5}$ of a cycle between sampling points and is therefore indistinguishable from waves that go through $(1.5 + k)$ cycles between sampling points, where k is any integer. The higher frequencies are known as **alias frequencies** because they are indistinguishable from the signal at 2 Hz. If we take $k = -1$ we have an alias frequency of -8 Hz, which is physically equivalent to 8 Hz. The Nyquist frequency is 5 Hz. The danger is that if the sampling rate is too slow, high frequencies which would affect the oil rig are mistaken for safer low frequencies. The variance of the signal, using a divisor n, can be decomposed as

$$\sum_{i=1}^{397} (x_i - \bar{x})^2/397 = \sum_{k=1}^{198} r_k^2/2$$

This is known as **Parseval's theorem** and the general regression theory in Appendix A5 is one proof of it. All the cosine and sine terms are uncorrelated (orthogonal) so an individual r_j^2 does not depend on which other frequencies have been fitted, and is the unique contribution to the variance from the harmonic of frequency $k2\pi t/396$ rad s^{-1}. A plot of r_k^2 against k is known as the **Fourier line spectrum**. However, this time series was just a sample of the waves produced when the wavemakers have been set to simulate a particular sea state and the tank has reached a steady state. Other 40-second time series would differ in detail. Furthermore, the fundamental frequency of $2\pi t/396$ rad s^{-1} is a direct consequence of the number of data. Therefore, we do not interpret our frequencies literally. Rather, we smooth out the line spectrum by taking a moving average of adjacent ordinates ($\bar{r}_k^2$). We plot the moving average, scaled by dividing by twice the increment between frequencies, against frequency. A smooth curve, $S(\omega)$, is drawn through the points. This curve is continuous over frequency, has a total area equal to the variance of the signal, and is known as the **spectrum**. In Fig. 11.14 I have averaged over 15 ordinates so

$$\bar{r}_k^2 = (r_{k-7}^2 + r_{k-6}^2 + \cdots + r_k^2 + \cdots + r_{k+6}^2 + r_{k+7}^2)/15 \qquad \text{for } k = 8, \ldots, 191$$

The frequency interval is $2\pi/396$ radians per sampling interval, and the band-

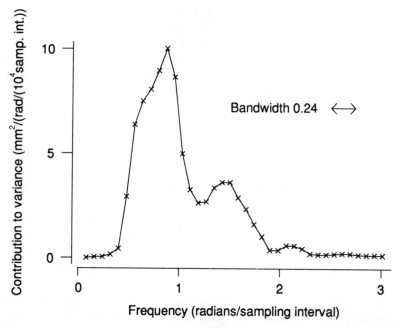

Fig. 11.14 Estimates of spectrum for wave data using Daniell and Parzen windows of bandwidth 0.24. The sampling interval $\Delta = 0.1$ s.

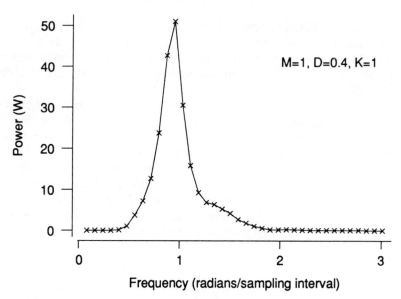

Fig. 11.15 Frequency composition of power from wave energy device.

width is defined as the product of the frequency interval with the number of ordinates averaged. In this case it is 0.24 radians per sampling interval, which is equivalent to $2.4\,\text{rad s}^{-1}$ or $0.38\,\text{Hz}$. The bandwidth is a measure of the resolution of the spectrum and we will not be able to separate frequencies closer than this. We inevitably lose resolution when we average, but individual ordinates are very unreliable so we would not have any confidence in apparent differences between one and the next.

11.3.4 Estimated power output of the device

If we combine the results of the previous sections we have an estimate of the distribution of power over frequency, $\Pi(\omega)$ as

$$\Pi(\omega) = \frac{\omega^2}{[(K - M\omega^2)^2 + D^2\omega^2]} S(\omega)$$

This is shown in Fig. 11.15. The next step is to increase the power output by tuning the device so that K, M and D have optimum values subject to design constraints.

In the case of oil rigs the objective would be to design it so that the response is kept well within safe limits. We have not actually solved the Tacoma Narrows Bridge problem, but spectral analysis is a part of the solution!

11.4 Endnote

These case studies have not been presented as methods for you to follow. You are unlikely to meet the same problems and you will no doubt have ideas of your own. The aim was to give you an indication of the versatility of probability theory and statistical techniques in engineering contexts. They are also only part of much larger projects.

Other issues arising in the evaluation of mini-roundabouts include: how they compare with traffic lights, how they affect traffic flows, and whether benefits are sustained. Recently, mini-roundabouts in the centre of Newcastle upon Tyne have been converted to junctions controlled by traffic lights. Reasons include possible improvements in traffic flow if there is a predominant direction, and technical improvements in traffic light systems. Flood prediction is an active research area and improving short-term rainfall forecasts, as part of warning and emergency procedures, is an important aspect of this. Some element of uncertainty is a feature of all engineering projects, and statistical methods enable us to account for this.

Appendix A: Mathematical explanations of key results

A1 Derivation of Poisson distribution

Assume that occurrences of events are random and independent, and that they occur at an average rate of λ per unit time. The Poisson distribution gives the probability of x occurrences in some length of time t, for integer values of x from 0 upwards.

The simplest derivation of the formula is to divide the time into a large number (n) of small increments of equal length (δt). The increments are so small that the probability of more than one event occurring within an interval is negligible. The exact result is obtained as the number of intervals tends to infinity, and their length tends to zero. Formally,

$$t = n\delta t$$

and

$$\Pr(1 \text{ event in time } \delta t) = \lambda\delta t$$

since the probability of more than one event is proportional to $(\delta t)^2$ and becomes negligible as δt tends to zero. It follows that

$$\Pr(0 \text{ events in time } \delta t) = 1 - \lambda\delta t$$

and X has an approximate $\text{Bin}(n, \lambda\delta t)$ distribution. That is,

$$P(x) = {}_nC_x(\lambda\delta t)^x(1 - \lambda\delta t)^{n-x}$$

$$= {}_nC_x\left(\frac{\lambda t}{n}\right)^x\left(1 - \frac{\lambda t}{n}\right)^{n-x}$$

$$= \frac{n(n-1)\cdots(n-x+1)(\lambda t)^x}{x!}\frac{1}{n^x}\left(1 - \frac{\lambda t}{n}\right)^n\left(1 - \frac{\lambda t}{n}\right)^{-x}$$

Now let n tend to infinity, remembering the result

$$\lim_{n\to\infty}\left(1 - \frac{\lambda t}{n}\right)^n = e^{-\lambda t}$$

which follows easily from Taylor expansions, to obtain the formula

$$P(x) = \frac{(\lambda t)^x e^{-\lambda t}}{x!} \qquad \text{for } x = 0, 1, \ldots$$

A2 Central limit theorem

A2.1 Moment generating function

Definition

A continuous probability distribution can be characterized by either its pdf or its cdf, and a discrete distribution can similarly be defined by its probability function or the cumulated probabilities. An alternative representation which is often useful for proving theoretical results is the **moment generating function** (mgf). The mgf of a variable X is defined by

$$M_X(\theta) = E[e^{\theta x}]$$

which holds for both discrete and continuous distributions. Provided it exists – which it does for the distributions we need – it determines the distribution uniquely. We will use it for two purposes. First, it can be a useful method for finding moments of a distribution.

$$E[e^{\theta X}] = \int_{-\infty}^{\infty} e^{\theta x} f(x)\, dx \quad \text{(or summation for a discrete distribution)}$$

$$= \int_{-\infty}^{\infty} \left(1 + \theta x + \frac{(\theta x)^2}{2!} + \cdots \right) f(x)\, dx$$

$$= 1 + \theta E[X] + \frac{\theta^2}{2!} E[X^2] + \cdots$$

It follows that

$$\left. \frac{d^k}{d\theta^k} M_X(\theta) \right|_{\theta=0} = E[X^k]$$

We can obtain the variance, for example, from the result

$$\begin{aligned} \sigma^2 &= E[(X - \mu)^2] \\ &= E[X^2 - 2\mu X + \mu^2] \\ &= E[X^2] - 2\mu E[X] + \mu^2 \\ &= E[X^2] - \mu^2 \end{aligned}$$

Similarly for other moments about the mean.

Example A2.1

For the binomial distribution

$$M_X(\theta) = \sum_{x=0}^{n} e^{\theta x} {}_nC_x p^x (1-p)^{n-x} = \sum_{x=0}^{n} {}_nC_x (e^\theta p)^x (1-p)^{n-x}$$

$$= [(1-p) + pe^\theta]^n$$

$$M_X^{(1)}(\theta) = n[(1-p) + pe^\theta]^{n-1} pe^\theta$$

$$\mu = M_X^{(1)}(0) = np$$

$$M_X^{(2)}(\theta) = n[(1-p) + pe^\theta]^{n-1} pe^\theta + n(n-1)[(1-p) + pe^\theta]^{n-2} p^2 e^{2\theta}$$

$$M_X^{(2)}(0) = np + n(n-1)p^2$$

$$\sigma^2 = np + n(n-1)p^2 - (np)^2 = np(1-p)$$

Our second reason for introducing the mgf is that it provides a relatively easy way of determining the distribution of a sum of independent random variables.

Preliminary theorems

Theorem 1
Let $X_1, \ldots, X_n$ be independent variables with mgf $M_{X_1}(\theta), \ldots, M_{X_n}(\theta)$ respectively. Define the sum T by

$$T = X_1 + \cdots + X_n$$

Then the mgf of T is the product of the mgfs of $X_1, \ldots, X_n$.

Proof

$$M_T(\theta) = E[e^{\theta T}] = E[e^{\theta(X_1 + \cdots + X_n)}]$$

$$= E[e^{\theta X_1} e^{\theta X_2} \cdots e^{\theta X_n}]$$

$$= \int \cdots \int e^{\theta x_1} e^{\theta x_2} \cdots e^{\theta x_n} f(x_1, \ldots, x_n) \, dx_1 \cdots dx_n$$

$$= \int e^{\theta x_1} f(x_1) dx_1 \cdots \int e^{\theta x_n} f(x_n) dx_n \qquad \text{(since they are independent)}$$

$$= M_{X_1}(\theta) \cdots M_{X_n}(\theta)$$

Example A2.2

Suppose independent variables X_1 and X_2 have binomial distributions with a common probability of success p, and n_1 and n_2 trials, respectively. Then, since

$$M_{X_1}(\theta) = [(1-p) + pe^\theta]^{n_1}$$

$$M_{X_2}(\theta) = [(1-p) + pe^\theta]^{n_2}$$

we have the result that

$$M_T = [(1 - p) + pe^\theta]^{n_1 + n_2}$$

This is as it should be, because it follows from the definition that T will be binomial with the same p but $n_1 + n_2$ trials.

Before we can proceed to prove the central limit theorem we need the following result.

Theorem 2
If a and b are constants, and b is positive, then

$$M_{(X+a)/b}(\theta) = e^{a\theta/b} M_X(\theta/b)$$

Proof
Define

$$W = (X + a)/b$$

By definition

$$\Pr(X < x) = F(x)$$

and it follows that

$$\Pr((X + a)/b < (x + a)/b) = F(x)$$

If we now use the substitution $w = (x + a)/b$

$$\Pr(W < w) = F(bw - a)$$

Now differentiate with respect to w to obtain

$$f_W(w) = b f_X(bw - a)$$

where the subscripts emphasize which pdf corresponds to which variable. Next, from the definition

$$M_W(\theta) = \int e^{w\theta} f_W(w)\, dw$$

$$= \int e^{w\theta} b f_X(bw - a)\, dw$$

now substitute $x = bw - a$ to obtain

$$\int e^{[(x+a)/b]\theta} f_X(x)\, dx$$

The result follows.

A2.2 Normal distribution

Something of the form

$$e^{-\frac{1}{2}z^2}$$

has the potential to be a useful pdf. It is symmetric and decays faster than the exponential distribution. However, it needs scaling so that it has an area of 1. To find the area, A, we argue as follows.

$$A = \int_{-\infty}^{\infty} e^{-\frac{1}{2}z^2} dz$$

$$A^2 = \int_{-\infty}^{\infty} e^{-\frac{1}{2}z^2} dz \int_{-\infty}^{\infty} e^{-\frac{1}{2}y^2} dy = \int_{-\infty}^{\infty} \int_{-\infty}^{\infty} e^{-\frac{1}{2}(z^2 + y^2)} dz \, dy$$

The double integral is represented by the volume under the surface,

$$f(z, y) = e^{-\frac{1}{2}(z^2 + y^2)}$$

and a typical element of volume is the product of this function with a typical element of area $\delta z \delta y$. In polar coordinates,

$$x = r \cos \theta \quad y = r \sin \theta$$
$$x^2 + y^2 = r^2$$

and an element of area (Figure A2.1a) is $r\delta\theta\delta r$

So

$$A^2 = \int_0^{2\pi} \int_0^{\infty} e^{-\frac{1}{2}r^2} r \, dr \, d\theta = \int_0^{2\pi} [-e^{-\frac{1}{2}r^2}]_0^{\infty} d\theta = 2\pi$$

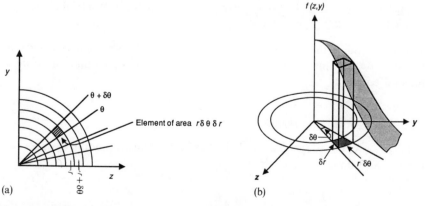

Fig. A2.1 Element of (a) area in domain of definition and (b) volume under surface.

and

$$A = \sqrt{2\pi}$$

Hence

$$\frac{1}{\sqrt{2\pi}} \int_{-\infty}^{\infty} e^{-\frac{1}{2}z^2} \, dz = 1$$

It is straightforward to show that

$$f(x) = \frac{1}{\sigma\sqrt{2\pi}} e^{-\frac{1}{2}((x-\mu)/\sigma)^2}$$

is the pdf of a similar distribution, with mean μ and standard deviation σ.

Theorem 3
The mgf of a normal distribution is

$$M_X(\theta) = e^{\mu\theta + \frac{1}{2}\sigma^2\theta^2}$$

Proof

$$M_X(\theta) = \frac{1}{\sigma\sqrt{2\pi}} \int_{-\infty}^{\infty} \exp(\theta x) \exp\left\{ -\tfrac{1}{2}[(x-\mu)/\sigma^2] \right\} dx$$

$$= \frac{1}{\sigma\sqrt{2\pi}} \int_{-\infty}^{\infty} \exp\left\{ -\tfrac{1}{2}[(x^2 - 2\mu x + \mu^2 - 2\theta\sigma^2 x)/\sigma^2] \right\} dx$$

$$= \frac{1}{\sigma\sqrt{2\pi}} \int_{-\infty}^{\infty} \exp\left\{ -\frac{1}{2}\left[\frac{(x-(\mu+\sigma^2\theta))^2 - 2\mu\sigma^2\theta - \sigma^4\theta^2}{\sigma^2} \right] \right\} dx$$

$$= e^{\mu\theta + \frac{1}{2}\sigma^2\theta^2} \int_{-\infty}^{\infty} \frac{1}{\sigma\sqrt{2\pi}} \exp\left\{ -\tfrac{1}{2}[(x-(\mu+\sigma^2\theta))^2/\sigma]^2 \right\} dx$$

The integrand in the last expression is a normal pdf., with mean $\mu + \sigma^2\theta$ and variance σ^2, the integration is over the interval $(-\infty, \infty)$ and therefore the integral equals 1.

A2.3 The central limit theorem

Let $X_1, \ldots, X_n$ be independent and identically distributed random variables with mean μ and finite variance σ^2. Define

$$U_n = \frac{\bar{X} - \mu}{\sigma/\sqrt{n}}$$

The distribution of U_n converges to the standard normal as $n \to \infty$.

Proof
Define

$$Y_i = \frac{X_i - \mu}{\sigma}$$

Then Y_i has zero mean and unit variance and its mgf can be written as

$$M_Y(\theta) = 1 + \frac{\theta^2}{2} + \frac{\theta^3}{3!} E\left[Y_i^3 \right] + \cdots$$

Also

$$U_n = \sum Y_i / \sqrt{n}$$

$$M_{U_n}(\theta) = \left(M_Y\left(\frac{\theta}{\sqrt{n}} \right) \right)^n = \left(1 + \frac{\theta^2}{2n} + \frac{\theta^3}{3! n^{3/2}} \xi + \cdots \right)^n$$

where

$$\xi = E[Y_i^3]$$

Now consider

$$\ln M_{U_n}(\theta) = n \ln\left(1 + \left(\frac{\theta^2}{2n} + \frac{\theta^3 \xi}{6n^{3/2}} + \cdots \right) \right)$$

and remember the Taylor series

$$\ln(1 + x) = x - x^2/2 + x^3/3 - \cdots$$

provided $|x| < 1$. Using this result

$$\ln M_{U_n}(\theta) = n \left(\left(\frac{\theta^2}{2n} + \frac{\theta^3 \xi}{6n^{3/2}} + \cdots \right) - \frac{1}{2}\left(\frac{\theta^2}{2n} + \frac{\theta^3 \xi}{6n^{3/2}} + \cdots \right)^2 + \cdots \right)$$

$$\lim_{n \to \infty} \ln M_{U_n}(\theta) = \frac{\theta^2}{2}$$

$$\lim_{n \to \infty} M_{U_n}(\theta) = e^{\theta^2/2}$$

which is the mgf for the standard normal random variable.

The theorem can be proved under considerably more general conditions, but the assumption that all the random variables have finite mean and finite variance is essential. In practical terms, the theorem justifies the following approximation. If we have a large random sample from any distribution with finite mean and variance the distribution of $\bar{X}$ will be near-normal. If we imagine errors as made up of a large number of small, additive, independent components then the errors will have a normal distribution. This provides some justification for assuming normality in many applications.

A3 Derivation of EVGI distribution

Let $X_1, \ldots, X_n$ be independent random variables with cdf $F(x)$. Define the maximum in the sample,

$$X_{(n)} = \max(X_1, \ldots, X_n)$$

and denote the cdf of this variable by $G(x)$. Then

$$
\begin{aligned}
G(x) &= \Pr(X_{(n)} < x) \\
&= \Pr(\max(X_1, \ldots, X_n) < x) \\
&= \Pr(X_1 < x, \ldots, X_n < x) \\
&= [F(x)]^n
\end{aligned}
$$

We now suppose that $G(x)$ will tend towards some standard form as n tends to infinity. If this is so, the distribution of the maximum of a random sample from G will have the same form of distribution, although the location and spread will change. That is,

$$[G(x)]^n = G(a_n x - b_n)$$

where the subscript on the a_n and b_n emphasizes that they depend on n.

The extreme value type I distribution (EVGI) is obtained by taking a_n as 1. It can be shown that the distribution of the maximum of samples of size n from distributions whose upper tail decays at least as fast as a negative exponential function does tend to a G such that

$$[G(x)]^n = G(x - b_n) \tag{A3.1}$$

Leadbetter *et al.* (1982) give a modern proof of this result.

It is relatively easy to show that if equation (A3.1) holds then $G(x)$ has the form given in section 4.2.5. The trick is to notice that the maximum of m maxima of n variables is the same as the maximum of all mn variables. Therefore

$$[G(x)]^{mn} = \{[G(x)^n]\}^m = [G(x - b_n)]^m = G(x - b_n - b_m)$$

But

$$[G(x)]^{mn} = G(x - b_{mn})$$

so

$$b_{mn} = b_m + b_n$$

and b_i is essentially a logarithm. That is,

$$b_n = \theta \ln(n) \tag{A3.2}$$

for some constant θ. The next step is to take logarithms of equation (A3.1) twice, remembering to multiply by -1 to obtain positive quantities between log operations:

$$\ln(n) + \ln(-\ln G(x)) = \ln(-\ln(G(x - b_b))) \tag{A3.3}$$

It is now convenient to define,

$$c(x) = \ln(-\ln G(x)) \qquad (A3.4)$$

From equation (A3.3) decreasing the argument of c by b_n leads to an addition of $\ln(n)$ which is the same as b_n/θ from equation (A3.2). Therefore decreasing the argument by x gives,

$$c(x) = c(0) - x/\theta$$

Finally, take exponential of equation (A3.4) twice to obtain

$$G(x) = \exp(-\exp(c(x)))$$
$$= \exp(-\exp(-(x-\xi)/\theta))$$

where ξ is the product of the constant θ and $c(0)$.

A4 Estimated variance of ratio estimator

Suppose we have a population of N pairs (x_j, y_j) and take a random sample of n of them. I will denote the random sample by

$$(X_i, Y_i) \qquad \text{for } i = 1, \ldots, n$$

Define θ as the ratio in the population,

$$\theta = \sum_{j=1}^{N} y_j \bigg/ \sum_{j=1}^{N} x_j$$

The sample ratio r is our estimator of θ and is defined by

$$r = \frac{\Sigma Y_i}{\Sigma X_i}$$

The error can be written in the form

$$r - \theta = \frac{\Sigma Y_i - \theta \Sigma X_i}{\Sigma X_i} = \frac{\Sigma(Y_i - \theta X_i)}{n\bar{X}}$$

We now make some simplifying assumptions. The $(Y_i - \theta X_i)$ are independent and the variance of $\bar{X}$ is negligible compared with their variance for large samples. So treating $\bar{X}$ as a constant,

$$\text{var}(r - \theta) = \frac{\Sigma \text{var}(Y_i - \theta X_i)}{n^2 \bar{X}^2}$$

Since θ is a constant the variance of $(r - \theta)$ is the same as the variance of r. The expected value of $(Y_i - \theta X_i)$ is 0, so

$$\text{var}(Y_i - \theta X_i) = E[(Y_i - \theta X_i)^2]$$

The expression for the variance can be turned into a useful estimator by replacing the unknown θ with the estimator r, and replacing $E[(Y_i - \theta X_i)^2]$ by $(Y_i - \theta X_i)^2$. So the approximate estimator is

$$\widehat{\text{var}}(r) = \frac{\Sigma(Y_i - rX_i)^2}{n^2 \bar{X}^2}$$

Barnett (1974) gives a more detailed proof (in a different notation) and ends up with $n(n-1)$ instead of n^2 in the denominator. A finite population correction factor should be included if the sample is a substantial proportion of the population. In the AMP unit cost exercise we had a hypothetical infinite population of the unit costs for all possible future schemes.

A5 Multiple regression model

A5.1 Matrix description of the model

The multiple regression model is

$$Y_i = \beta_0 + \beta_1 x_{1i} + \beta_2 x_{2i} + \cdots + \beta_k x_{ki} + E_i$$

where Y_i are the dependent variables, $x_{1i}, \ldots, x_{ki}$ are the explanatory variables and E_i are the errors. It is called a linear model because it is linear in the parameters β_i. The E_i will be assumed to have a zero mean and to be independent of the explanatory variables. They will also be assumed to have a constant variance, σ^2, and to be independent of each other, except in section A5.11. They will usually be assumed to have a normal distribution. A sample of size n can be modelled by

$$\begin{bmatrix} Y_1 \\ \vdots \\ Y_n \end{bmatrix} = \begin{bmatrix} 1 & x_{11} & \cdots & x_{k1} \\ \vdots & & & \vdots \\ 1 & x_{1n} & \cdots & x_{kn} \end{bmatrix} \begin{bmatrix} \beta_0 \\ \vdots \\ \beta_k \end{bmatrix} + \begin{bmatrix} E_1 \\ \vdots \\ E_n \end{bmatrix}$$

or, more concisely,

$$Y = XB + E$$

where Y, X, B and E are $n \times 1$, $n \times (k+1)$, $(k+1) \times 1$ and $n \times 1$ matrices, respectively. The sum of squared errors is

$$\psi = (Y - XB)^{\text{T}}(Y - XB)$$

where the superscript T denotes transposition.

A5.2 Calculus of several variables

The following general results are required before deriving the formula for the least-squares estimators. Let $\phi(v)$ be a scalar function of an $m \times 1$ array v, where

$$v^{\mathrm{T}} = [v_1, \ldots, v_m]$$

That is, ϕ is a function of m variables $v_1, \ldots, v_m$. Define the array of partial derivatives

$$\frac{\partial \phi}{\partial v} = \left[\frac{\partial \phi}{\partial v_1}, \ldots, \frac{\partial \phi}{\partial v_m} \right]^{\mathrm{T}}$$

The results given below are consequences of this definition and the usual rules of calculus. Let c be an $m \times 1$ array of constants and M be an $m \times m$ matrix of constants. Then:

R1 $\dfrac{\partial}{\partial v}(v^{\mathrm{T}}c) = c$

R2 $\dfrac{\partial}{\partial v}(v^{\mathrm{T}}Mv) = Mv + M^{\mathrm{T}}v$

R3   A necessary requirement for ϕ to have a maximum or minimum is that $\partial \phi / \partial v = 0$.

You are urged to verify these results for the case of $m = 2$.

A5.3 The normal equations

The problem is to minimize ψ:

$$
\begin{aligned}
\psi &= (Y - XB)^{\mathrm{T}}(Y - XB) \\
&= Y^{\mathrm{T}}(Y - XB) - B^{\mathrm{T}}X^{\mathrm{T}}(Y - XB) \\
&= Y^{\mathrm{T}}Y - Y^{\mathrm{T}}XB - B^{\mathrm{T}}X^{\mathrm{T}}Y + B^{\mathrm{T}}X^{\mathrm{T}}XB
\end{aligned}
$$

As all the terms on the right-hand side are 1×1 matrices, that is to say, scalars, they equal their transpose. In particular,

$$Y^{\mathrm{T}}XB = (Y^{\mathrm{T}}XB)^{\mathrm{T}} = B^{\mathrm{T}}X^{\mathrm{T}}Y$$

Therefore

$$\psi = Y^{\mathrm{T}}Y - 2B^{\mathrm{T}}X^{\mathrm{T}}Y + B^{\mathrm{T}}X^{\mathrm{T}}XB$$

Applying R1 and R2 and noting that $(X^{\mathrm{T}}X)^{\mathrm{T}} = X^{\mathrm{T}}X$, gives

$$\frac{\partial \psi}{\partial B} = -2X^{\mathrm{T}}Y + 2(X^{\mathrm{T}}X)B$$

Provided $X^{\mathrm{T}}X$ has an inverse, R3 leads to the **normal equations**

$$\hat{B} = (X^{\mathrm{T}}X)^{-1}X^{\mathrm{T}}Y$$

A5.4 Properties of the estimators

It was pointed out in section 9.11 that addition of more explanatory variables can only decrease the residual sum of squares. The practical question is whether this decrease is sufficient to give more accurate predictions. The properties of the estimators are needed to answer this. To begin with, the least-squares estimators are unbiased. This can be proved by taking expectations and using the assumption that the E_i are independent of the explanatory variables.

$$E[\hat{B}] = (X^{T}X)^{-1}X^{T}E[Y] = (X^{T}X)^{-1}X^{T}XB = B$$

The covariance matrix of the estimators, C, is defined by the equality

$$C = \begin{bmatrix} \text{var}(\hat{\beta}_0) & \text{cov}(\hat{\beta}_0, \hat{\beta}_1) & \text{cov}(\hat{\beta}_0, \hat{\beta}_2) & \cdots \\ \text{cov}(\hat{\beta}_0, \hat{\beta}_1) & \text{var}(\hat{\beta}_1) & \text{cov}(\hat{\beta}_1, \hat{\beta}_2) & \\ \vdots & & \text{var}(\hat{\beta}_2) & \\ & & & \ddots \\ & & & & \text{var}(\hat{\beta}_k) \end{bmatrix}$$

Remembering that

$$\text{cov}(\hat{\beta}_i, \hat{\beta}_j) = E[(\hat{\beta}_i - E[\hat{\beta}_i])(\hat{\beta}_j - E[\hat{\beta}_j])]$$

C can be written concisely as,

$$C = E[(\hat{B} - B)(\hat{B} - B)^{T}]$$

since

$$E[\hat{B}] = B$$

It is surprisingly easy to show that this equals

$$(X^{T}X)^{-1}\sigma^2$$

and the proof follows:

$$C = E[(\hat{B} - B)(\hat{B} - B)^{T}]$$
$$= E[(X^{T}X)^{-1}X^{T}(Y - E[Y])(Y - E[Y])^{T}X(X^{T}X)^{-1}]$$

Now note that $Y - E[Y]$ is a column of E_i. $E[E_i^2]$ is σ^2 and $E[E_iE_j]$ is 0 if i is different from j. So, continuing with the algebra, we have

$$C = (X^{T}X)^{-1}X^{T}\sigma^2 IX(X^{T}X)^{-1}$$

As σ^2 is a scalar it can take any position in the matrix product. Hence

$$C = (X^{T}X)^{-1}\sigma^2$$

In most applications σ^2 will not be known, so it will have to be estimated from the data. An unbiased estimator of σ^2 is

$$S^2 = RSS/(n - k - 1)$$

You may have anticipated this result. It is nearly the mean of the squared residuals; the denominator has been reduced by $k+1$ to account for the loss of degrees of freedom caused by estimating $k+1$ parameters. The least-squares procedure ensures that the RSS cannot exceed, and is almost certainly less than, ΣE_i^2. The proof that S^2 is an unbiased estimator of σ^2 uses the following matrix theory.

A5.5 Useful matrix results for statistical analysis

1. Let $A = (a_{ij})$ be a square $n \times n$ matrix. The **trace** of A is defined by

$$\text{tr } A = \sum_{i=1}^{n} a_{ii}$$

If B is another $n \times n$ matrix, it follows from the definition that

$$\text{tr}(A + B) = \text{tr } A + \text{tr } B$$

Also, if P and Q are $m \times n$ and $n \times m$ matrices, respectively then

$$\text{tr}(PQ) = \text{tr}(QP)$$

since

$$\sum_{i=1}^{m} \left\{ \sum_{\alpha=1}^{n} p_{i\alpha} q_{\alpha i} \right\} = \sum_{i=1}^{n} \left\{ \sum_{\alpha=1}^{m} q_{i\alpha} p_{\alpha i} \right\}$$

2. Let A be a square $n \times n$ matrix and c any $n \times 1$ matrix which is not identically equal to a column of zeros. The scalar expression $c^T A c$ is known as a **quadratic form** in the elements of c. The matrix A is **positive definite** if and only if

$$c^T A c > 0$$

and **positive semi-definite** if and only if

$$c^T A c \geqslant 0$$

For example, let Y_1 and Y_2 be random variables with variances σ_1^2 and σ_2^2 and covariance σ_{12}. The covariance matrix of Y_1 and Y_2 is

$$E[(Y - E(Y))(Y - E(Y))^T] = \begin{pmatrix} \sigma_1^2 & \gamma_{12} \\ \gamma_{12} & \sigma_2^2 \end{pmatrix}$$

where $Y^T = (Y_1 Y_2)$. The covariance matrix is positive semi-definite since

$$(c_1\, c_2) \begin{pmatrix} \sigma_1^2 & \gamma_{12} \\ \gamma_{12} & \sigma_2^2 \end{pmatrix} \begin{pmatrix} c_1 \\ c_2 \end{pmatrix} = \text{var}(c_1 Y_1 + c_2 Y_2) \geqslant 0$$

for any choice of c_1 and c_2. Also, if you write C for the covariance matrix,

$$\text{var}(c^T Y) = c^T C c$$

This is a generalization of the result for a single random variable. It is true for any number of random variables Y_i in an $n \times 1$ matrix Y. The essential steps in the proof are demonstrated by the 2×1 case considered above.

3. If an $n \times n$ matrix A is positive semi-definite and has an inverse, its inverse is also positive semi-definite. To prove this, set $c = A^{-1}v$ in the definition of positive semi-definiteness.

4. If Y is an $n \times 1$ matrix of random variables Y_i and A is a constant $n \times n$ matrix, then

$$E[Y^{\mathrm{T}}AY] = E[Y]^{\mathrm{T}}AE[Y] + \mathrm{tr}(AC)$$

where C is the covariance matrix of Y. This is a generalization of the result,

$$E[aY_i^2] = aE[Y_i]^2 + a\,\mathrm{var}[Y_i]$$

for a single random variable. You should verify the matrix result for $n = 2$; the general proof follows a similar argument.

A5.6 Expected value of the residual sum of squares

Recall that the residual sum of squares (RSS) is given by the equation

$$RSS = (Y - X\hat{B})^{\mathrm{T}}(Y - X\hat{B})$$

If you also remember, from the derivation of $\hat{B}$, that

$$(X^{\mathrm{T}}X)\hat{B} = X^{\mathrm{T}}Y$$

you can rewrite RSS as

$$RSS = Y^{\mathrm{T}}Y - \hat{B}^{\mathrm{T}}X^{\mathrm{T}}X\hat{B}$$

If you use result (4) of the previous section,

$$E[RSS] = E[Y^{\mathrm{T}}Y] - [B^{\mathrm{T}}X^{\mathrm{T}}XB + \mathrm{tr}((X^{\mathrm{T}}X)(X^{\mathrm{T}}X)^{-1}\sigma^2)]$$
$$= E[Y^{\mathrm{T}}Y] - B^{\mathrm{T}}X^{\mathrm{T}}XB - (k+1)\sigma^2$$

Finally, remembering the scalar version of the same result,

$$E[Y_i^2] = \mathrm{var}(Y_i^2) + E[Y_i]^2$$

the first term on the right-hand side

$$E[Y^{\mathrm{T}}Y] = n\sigma^2 + \sum_{i=1}^{n} (\beta_0 + \beta_1 x_{1i} + \cdots + \beta_k x_{ki})^2$$

$$= n\sigma^2 + B^{\mathrm{T}}X^{\mathrm{T}}XB$$

so

$$E[RSS] = (n - k - 1)\sigma^2$$

The estimator S^2 of σ^2 is $RSS/(n - k - 1)$, so it follows that

$$E[S^2] = \sigma^2$$

The expression $B^T(X^TX)B \geqslant 0$, because X^TX is the inverse of a covariance matrix multiplied by a positive constant. In later sections we will use the fact that S^2 is not correlated with $\hat{B}$. A formal proof will not be given, but there is no reason to suppose that $\hat{\beta}_k$ being above its mean β_k has any effect on the relationship of S^2 to its mean σ^2.

A5.7 Accuracy of calculations

The matrix approach allows the theory to be developed in an elegant fashion with a minimum of algebra, but there is no avoiding the solution of $k+1$ simultaneous linear equations to find the estimates of the coefficients for any given data set. Most people wishing to carry out a multiple regression analysis will have access to a micro-computer with statistical software. In principle it would not be difficult to write your own programme, but good software should be more efficient and less prone to rounding errors. Rounding errors can quickly become serious in multiple regression calculations and it may be advisable to mean correct the explanatory variables – that is, scale them to have a mean of zero. Three reasons for this advice are given below. Before reading them you should write out the normal equations in component form for the particular case of two explanatory variables.

1. The formulae for the $\hat{\beta}_k$ could involve subtractions of the form

$$\sum_{i=1}^{n} x_{ki}^2 - \left(\sum x_{ki}\right)^2 \bigg/ n$$

rather than the equivalent

$$\sum (x_{ki} - \bar{x}_{ki})^2$$

If the x_k give small coefficients of variation – that is, large means and relatively small standard deviations – the two terms in the first expression can be equal to several significant figures, and if the calculations are carried out on a machine which stores only 8 significant figures the final results will be quite unreliable. If the x_k are mean-corrected the second term will be zero. No respectable software should include such expressions, if they could be used with data which have small CV, but if you subtract $\bar{x}_k$ from all the x_{ki} at the beginning you do not rely on this. Conversely, adding a large number to all your explanatory variables would test the software.

2. If the x_k are mean-corrected the first normal equation reduces to

$$\hat{\beta}_0 = \bar{y}$$

and $\hat{\beta}_0$ does not appear in any of the other equations. This effectively reduces the number of simultaneous equations to k. It is also convenient for interpretation of the fitted regression equation.

3. While there is no requirement that the explanatory variables be independent,

if there is an exact linear relationship between any two of them the normal equations have no unique solution. Formally, the matrix $X^T X$ will be singular. For example, if

$$Y_i = \beta_0 + \beta_1 x_1 + \beta_2 x_2 + E_i$$

and

$$x_2 = a + b x_1$$

then

$$Y_i = (\beta_0 + a\beta_2) + (\beta_1 + b\beta_2) x_1 + E_i$$

and only two parameters $(\beta_0 + a\beta_2)$ and $(\beta_1 + b\beta_2)$ can be uniquely estimated. If two covariates are strongly correlated the matrix $X^T X$ will be nearly singular and inversion, whether carried out explicitly or not, will be unreliable. If x_k has a small coefficient of variation it will be highly correlated with x_k^2, while if x_k is mean-corrected it will not be highly correlated with x_k^2. In geometric terms, a small portion of a parabola a long way from the vertex is well approximated by a straight line, whereas the parabola near the vertex is not. This was shown in Fig. 9.4.

A5.8 ANOVA Tables

It was shown in section A5.6 that

$$E[\hat{B}^T X^T X B^T] = B^T X^T X B + (k+1)\sigma^2$$

The corrected sum of squares of the Y_i is

$$\sum_{i=1}^{n} (Y_i - \bar{Y})^2 = Y^T Y - n\bar{Y}^2$$

Bearing in mind that

$$E[\bar{Y}^2] = E[\bar{Y}]^2 + \sigma^2/n$$

these results can be summarised in an ANOVA table:

Source of variation	Corrected sum of squares	d.f.	Mean square	E[mean square]
Regression	$\hat{B}^T(X^T X)\hat{B} - n\bar{Y}^2$ $= CSS$	k	CSS/k	$\sigma^2 + (B^T X^T X B - nE[\bar{Y}]^2)/k$
Residual	$Y^T Y - \hat{B}^T(X^T X)\hat{B}$	$n-k-1$	s^2	σ^2
Total	$Y^T Y - n\bar{Y}^2$	$n-1$		

The expression $(B^T X^T X B - nE[\bar{Y}]^2)/k$ is a non-negative function of the β_j. It equals zero if all of $\beta_1, \ldots, \beta_k$ are zero. If the explanatory variables are mean-corrected β_0^2 can be used instead of $E[\bar{Y}]^2$ and the expression is easier to expand into component form. You should try this for $k = 2$.

A5.9 Assessing the fit

Two separate questions need to be addressed in considering the fit. The first is whether the assumptions underlying the model are plausible. The second concerns the number of explanatory variables in the model, and whether it could be improved by including more.

Examining the residuals

The residuals provide some answers to the first question. The necessary assumptions for the estimators to be unbiased are that the form of the model is correct, and that the errors have zero mean and are independent of the explanatory variables. If one only appears as a linear term, obvious curvature in the plot of residuals against it would indicate that the model should include its quadratic term. Since the sum of the residuals is given by

$$\sum r_i = \sum [Y_i - (\bar{Y} + \hat{\beta}_1(x_1 - \bar{x}) + \cdots + \hat{\beta}_k(x_k - \bar{x}))] = 0$$

they provide no information about the assumption that the errors have zero mean. Any difference between the mean and zero is incorporated into $\hat{\beta}_0$. Similarly, it is straightforward to deduce that

$$\sum r_i(x_{ji} - \bar{x}_j) = 0 \qquad \text{for } j = 1, \ldots, k$$

from the normal equationns. It follows that the residuals cannot indicate any correlation between the explanatory variables and the errors.

It is quite common for the variability of residuals to increase with increasing values of the dependent variable. A plot of the calculated residuals against the fitted values of the dependent variable, calculated from the fitted model, may suggest that assumptions of the form

$$\sigma_{E_i}^2 \propto E[Y_i]$$

or

$$\sigma_{E_i} \propto E[Y_i]$$

would be more appropriate than assuming a constant value. If the assumption of constant variance is inappropriate, the ordinary least-squares estimators will no longer be efficient and the estimators of their variances may be biased. The major practical problem is that limits of prediction for individual values about their mean could be misleading.

Sometimes consecutive residuals are correlated when they are considered in the same order as the data were obtained. A plot of the calculated residuals against time may indicate that this is occurring. Also the first autocorrelation of the residuals can be calculated – that is, the correlation between residuals at time t and those at time $t + 1$. This is usually defined as

$$\sum_{t=1}^{n-1} r_t r_{t+1} \Big/ \sum_{t=1}^{n} r_t^2$$

where r_t are the time-ordered residuals (remember that the mean of the residuals is constrained to equal zero). In fact, if all the assumptions of the model hold the residuals are slightly correlated, but this correlation decreases as the sample size increases and is unlikely to be interpreted as evidence of correlated errors. If the assumption that the errors are independent is inappropriate, ordinary least-squares estimators are no longer efficient and the estimators of their variances may be seriously biased. High correlations between residuals are quite common in economic data sets, and the Durbin–Watson statistic provides a formal test. If a particular covariance structure can be assumed for the errors, weighted least squares is an answer to both this problem and that of changing variance. Weighted least squares is described in section A5.11.

The assumption that the errors are normally distributed is required for the distribution of the $\hat{\beta}_j$ to be normal and for constructing limits of predictions for individual values about their mean. In practice the $\hat{\beta}_j$ will be approximately normally distributed for most distributions of the errors provided the sample is not too small. This is a consequence of the central limit theorem. By contrast, the limits of prediction would change considerably if the errors were to be modelled by an exponential distribution. A less important consideration is that the least-squares estimators will not necessarily be the minimum-variance linear estimators if the errors are not from a normal distribution. The residuals can be plotted against their normal scores to assess whether an assumption of normality is reasonable. When it is not a transformation of the dependent variable may improve matters.

Most statistical packages will print out the 'standardized residuals' as well as the residuals. These are the calculated residuals divided by their standard deviations. The latter vary, even though the standard deviation of the random variations is assumed constant, and are always slightly less than s. This is explained by the loss of degrees of freedom when fitting parameters. If a datum $(k + 1)$-tuple has a relatively strong influence on the parameter estimation, the standard deviation of its associated residual will be slightly less than the average because it tends to pull the hypersurface towards itself. If you return to the definition of the residuals, these comments can be explained in mathematical terms. The array of residuals is

$$\hat{Y} - Y = X(X^T X)^{-1} X^T Y - Y$$
$$= (X(X^T X)^{-1} X^T - I)Y$$

The second of the 'useful matrix results' of section A5.5 gives the covariance matrix of the residuals as

$$(I - X(X^T X)^{-1} X^T)\sigma^2$$

The variances lie along the diagonal and the off-diagonal terms give the covariances. The variance estimators are obtained by replacing σ^2 by S^2. You should consider the component form of the matrix, and verify that the terms along the leading diagonal are all between 0 and 1 and add up to $n - k - 1$.

From a practical point of view, it usually makes little difference whether the residuals or standardized residuals are used if subjective assessments are being made. If formal statistical tests are applied the standardized residuals should be used.

How many explanatory variables?
A regression model is just an empirical approximation to some underlying physical process, and there is no 'right' answer to this question. If the explanatory variables are significantly correlated among themselves several models, which may look quite different, can give similar results when making predictions within the domain of values used when fitting them. Extrapolation beyond this domain is inadvisable unless there are good physical reasons for doing so, and even then should not be relied on far beyond it. The following guidelines may be useful.

1. Keep variables which have an obvious physical relationship with the dependent variable in the model – for example, volume of rain in the regression of run-off for the Kentucky rivers.
2. Only add further explanatory variables if the estimated standard deviation of the errors, s, decreases sufficiently to compensate for the additional complexity. Note that unless the data set is large a small reduction in s can be accompanied by wider intervals of prediction, away from the centroid of the explanatory variables, because of the increased uncertainty in the estimates of the parameters.
3. It is appropriate to construct confidence intervals for the individual coefficients of variables of special interest such as use of the additive in the marine diesel example. A 95% confidence interval is given by:

$$\hat{\beta}_j \pm t_{n-k-1,0.025} sd(\hat{\beta}_j)$$

If the confidence interval does not include zero we have reasonable evidence that the variable has some effect.

A5.10 Making predictions

The most common reason for fitting multiple regression models is to make predictions for the dependent variables when the values of the explanatory variables are known. You should remember the distinction between confidence limits for the mean value of the dependent variable, and limits of prediction for a single value taken by it. For large samples $\pm 2s$ will give approximate, but rather too narrow, 95% limits of prediction. This 'rule of thumb' ignores the uncertainty in the estimation of parameters and therefore improves as the sample size increases.

Confidence interval for the mean value of Y given x
Let $x_p = (1 \quad x_{1p} \quad \dots \quad x_{kp})^T$. The predicted mean value of Y given that $x = x_p$ is

$$\hat{E}[Y|x = x_p] = x_p^T \hat{B}$$

The variance of this quantity is obtained from result 2 in section A5.5.

$$\text{var}(x_p^T \hat{B}) = x_p^T((X^T X)^{-1} \sigma^2) x_p$$

where $(X^T X)^{-1} \sigma^2$ is the covariance matrix of $\hat{B}$ found in section A5.4. Hence a $(1 - \alpha) \times 100\%$ confidence interval for $E[Y|x = x_p]$ is

$$x_p^T \hat{B} + t_{n-k-1, \alpha/2} S \sqrt{x_p^T (X^T X)^{-1} x_p}$$

These limits will be quite different if changing variances and different distributions are considered.

Limits of prediction for a single value of Y

$(1 - \alpha) \times 100\%$ limits of prediction for a single value of Y when $x = x_p$ are

$$x_p^T \hat{B} \pm t_{n-k-1, \alpha/2} S \sqrt{1 + x_p^T (X^T X)^{-1} x_p}$$

A5.11 Relaxing the assumptions

The assumptions that the E_i are independent of each other and have a constant variance can be replaced by any other assumed covariance matrix. Suppose that

$$Y = XB + E$$

where

$$E[E] = 0, \quad \text{cov}(E) = V \sigma^2$$

and the elements of E are independent of the covariates. This is a generalized linear regression model. Any covariance matrix V can be written in the form:

$$V = Q^2$$

Q can be found by a straightforward mathematical procedure. This is particularly easy when the covariances in V are all zero, in which case Q consists of the square roots of the diagonal of V. Now define

$$F = Q^{-1} E$$

then

$$\text{cov}(F) = E[FF^T] = E[Q^{-1} E E^T (Q^{-1})^T]$$

$$= Q^{-1} E[E E^T] Q^{-1} \qquad \text{since } Q \text{ is symmetric}$$

$$= Q^{-1} Q^2 Q^{-1} \sigma^2 = I \sigma^2$$

The original model premultiplied by Q^{-1}

$$Q^{-1} Y = Q^{-1} XB + Q^{-1} E$$

is of the form

$$W = UB + F$$

where W contains the dependent variables, U contains the explanatory variables, and the elements of F satisfy the usual assumptions. The practical problem is that V is unlikely to be known, and it is usual to make some assumptions about the form of V. As an example, the standard deviation of the random variations might be assumed to be in proportion to the expected value of the dependent variable. The expected values could be obtained from a preliminary unweighted analysis. In principle, the process could even be iterated.

In some cases there is a theoretical reason for assuming a variance structure. I was asked to analyse results from a computer model which simulated the transport of contaminants through the water table. The investigation was concerned with the effect of the position of the waste vault on, among other variables, the average times and standard deviation of times taken by particles to reach two waste pits. A simulation consisting of 1000 particles was repeated for different vault positions. The number of particles reaching the pits depended on the position of the vault. The variances of the mean travel times, and their standard deviations, could be assumed to be in inverse proportion to the numbers of particles.

A6 DuMouchel's algorithm

Let y be a k-vector of observations and μ a p-vector of parameters. The posterior mean vector is

$$E(\theta|y) = \int E(\theta|\phi, y) f(\phi|y) \, d\phi$$

where

$$E(\theta|\phi, y) = d + (I + \phi VC^{-1})^{-1}[Xm(y, \phi) + \phi VC^{-1}(y - d)]$$
$$m(y, \phi) = [X^T W(\phi)X]^{-1} X^T W(\phi)(y - d)$$
$$W(\phi) = (\phi V + C)^{-1}$$

and where $f(\phi|y)$, the posterior density of σ^2/τ^2, is

$$f(\phi|y) \propto \phi^{-(df_\sigma + 1)/2} |W|^{0.5} |X^T W(\phi)X|^{-0.5} \left[df_\tau + \frac{df_\sigma}{\phi} + S(y, \phi) \right]^{-[df_\sigma + df_\tau + k - p]/2}$$

where $|\cdot|$ denotes the determinant of a matrix, and where

$$S(y, \phi) = (y - d - Xm(y, \phi))^T W(\phi)(y - d - Xm(y, \phi))$$

The covariance matrix of θ is computed from the formulae

$$\text{cov}(\theta|y) = \int \{\text{cov}(\theta|\phi, y) + [E(\theta|\phi, y) - E(\theta|y)][E(\theta|\phi, y) - E(\theta|y)]^T\} f(\phi|y) \, d\phi$$

where

$$\text{cov}(\theta|\phi, y) = c(y, \phi)[\phi VW(\phi)C + CW(\phi)X(X^T W(\phi)X)^{-1} X^T W(\phi)C]$$
$$c(y, \phi) = \frac{[df_\tau + df_\sigma/\phi + S(y, \phi)]}{df_\sigma + df_\tau + k - p - 2}$$

The posterior expectation and covariance of μ are

$$E(\mu|y) = \int m(y, \phi) f(\phi|y) \, d\phi$$

$$\text{cov}(\mu|y) = \int \{ c(y, \phi) [X^T W(\phi) X]^{-1} + [m(y, \phi) - E(\mu|y)] \\ \times [m(y, \phi) - E(\mu|y)]^T \} f(\phi|y) \, d\phi$$

The posterior expectations of the hyperparameters τ^2 and σ^2 are

$$E(\tau^2|y) = \int c(y, \phi) f(\phi|y) \, d\phi$$

$$E(\sigma^2|y) = \int \phi c(y, \phi) f(\phi|y) \, d\phi$$

Appendix B: Reference guide

B1 Notation

B1.1 Introduction

The following is a list of the most commonly occurring symbols, described in their usual context. Some occur with different meanings (r in particular), but this should be clear from the context. There is no standard for statistical notation so I have tried to keep to the most common conventions. However, books rarely use identical notation and you do have to read the accompanying definitions.

B1.2 General notation

1. In most contexts a, b and c are constants.
2. Variables are usually

   $$W, X, Y, Z \text{ or } w, x, y, z$$

   It is sometimes useful to distinguish explicitly a random variable from the value it takes in a specific instance. Then the upper-case letter is the random variable and the lower-case is the specific value, as in

   $$\Pr(X = x)$$

   Adhering to this rule can become rather awkward, and I have chosen to rely on the context in many places. For example, $\hat{\beta}$ represents both the estimator of the slope of a regression line and the estimate for a particular data set.
3. The following are usually integers:

   $$i, j, k, l, m, n$$

   They always are if they appear as subscripts.
4. f is usually a pdf and ϕ tends to be used for other functions.
5. e is the base of natural logarithms ($2.718\ 28\cdots$). π is the area of the circle with radius one ($3.141\ 59\cdots$). Both notations for the exponential function are used:

   $$e^x, \quad \exp(x)$$

6. Multiplication is $\times$ or juxtaposition – usually the first for numbers and the second for letters. For example, 2×5 but ab rather than $a \times b$. Natural logarithms are used throughout: $\ln(y) = x$ if and only if $y = e^x$.
7. Matrices are in bold italic, superscript ^{T} is transpose, det stands for the determinant.
8. $\{x_i\}$ is a sample, and $x_{(i)}$ are the ordered data. That is,

$$x_{(1)} < x_{(2)} < \cdots < x_{(n)}$$

B1.3 Symbols

Pr( )	probability of
Pr$(B\|A)$	probability of B conditional on A
N, n	population, sample size
$n!$	n factorial
$\Gamma( )$	gamma function (in particular, $\Gamma(n+1) = n!$)
$_nP_r, _nC_r$	permutations (arrangements) and combinations (choices) of r from n
$\mu, \bar{x}$	population, sample mean
σ^2, σ	population variance, standard deviation
s^2, s	sample estimate of population variance, standard deviation
$CV, \widehat{CV}$	population, sample coefficient of variation
$\gamma, \hat{\gamma}$	population, sample skewness
$\kappa, \hat{\kappa}$	population, sample kurtosis
f_k	frequency of event referred to by k
$p, \hat{p}$	population, sample proportion
$P(x)$	probability function (for discrete variable)
$\sim$	distributed as
Bin(n, p)	binomial distribution: n trials, probability of success p
Poisson (λt)	Poisson distribution with rate λ, time interval t
$F(x), f(x)(f_X(x))$	cdf, pdf (subscript added if necessary)
$N(\mu, \sigma^2)$	normal distribution with mean μ and variance σ^2
$Z \sim N(0, 1)$	Z distributed standard normal
$\Phi(z), \phi(z)$	cdf, pdf of standard normal distribution
$U[a, b]$	uniform distribution
$M(\omega)$	exponential distribution
t_m	Student's t-distribution with m degrees of freedom
x_α	upper $\alpha \times 100\%$ point of distribution of X
$z_{\alpha/2}, t_{m,\alpha/2}$	percentage points of $N(0, 1)$ and t-distribution for $(1 - \alpha) \times 100\%$ confidence intervals
$E[\ ]$	expected value
var( )	variance
$P(x, y)$	bivariate probability function
$F(x, y), f(x, y)$	bivariate cdf, pdf
cov$(X, Y), \widehat{cov}$	population, sample covariance

ρ, r	population, sample correlation
$r, r_i, se(r),$	ratio estimator, individual ratio, estimated standard
$sd(r_i), z_k$	deviation of r, estimated standard deviation of r_i, and zone cost
$P(y\|x)$	conditional probability function
$f(y\|x)$	conditional pdf
$\alpha, \beta, \hat{\alpha}, \hat{\beta}$	population, sample intercept and slope of regression line
E_i, r_i	errors about line and their estimates (residuals)
σ^2, σ	variance, standard deviation of errors
s	the sample estimate of the standard deviation of the errors, known as the residual mean square
S_{xx}, S_{yy}, S_{xy}	corrected sums of squares, products
$\beta_0, \beta_1, \ldots, \beta_k$	population coefficients of multiple regression model
$\hat{\beta}_0, \hat{\beta}_1, \ldots, \hat{\beta}_k$	least squares estimators, and estimates, of $\beta_0, \ldots, \beta_k$
R^2	'R squared' (proportion of variance accounted for by regression)
R^2_{adj}	adjusted R^2

B2 Glossary

acceptance sampling
A random sampling scheme for goods delivered to a company. If the sample passes some agreed criterion, the whole consignment is accepted.

addition rule
The probability of one or the other or both of two events occurring is: the sum of their individual probabilities of occurrence less the probability that they both occur.

aliases
If there are a large number of factors to be investigated in an experimental programme, a full factorial design may not be feasible. If a fraction is used, observable effects can only be attributed to any one, or a linear combination of several, from sets of interactions and sometimes a main effect. The members of a set are said to be aliases. In practice, high-order interactions are often assumed to be negligible. A similar phenomenon arises in spectral analysis when the apparent contribution, at some frequency, to the variance of a signal could be due to members from a set of higher frequencies. The sampling interval must be short enough for these higher frequencies to be negligible on physical grounds. Spectral analysers also remove any frequencies – above about 25 KHz is standard – with analogue filters.

AOQL
The average proportion of defective material leaving an acceptance sampling procedure, average outgoing quality (AOQ), depends on the proportion of

defectives in the incoming material. The maximum value it could take is the AOQ limit (AOQL).

asset management plan (AMP)
A business plan for water companies. It includes estimates of the costs of maintaining, and where necessary improving, the water supply and sewerage systems over the following 20 years. UK companies have to produce an AMP every five years, and submit it to an independent government body.

Bayes' theorem
Bayes' theorem follows from the basic rules of probability. Its importance is that it enables us to update our knowledge, expressed in probabilistic terms, as we obtain new data.

bias
A systematic difference – which will persist when averaged over imaginary replicates of the sampling procedure – between the estimator and the parameter being estimated. Formally, the difference between the mean of the sampling distribution and the parameter being estimated. If the bias is small by comparison with the standard deviation of the sampling distribution, the estimator may still be useful. For example, s^2 is unbiased for σ^2 but s is slightly biased for σ.

binomial distribution
The distribution of the number of successes in a set number of trials.

class intervals
Before drawing a histogram the data are grouped into classes which correspond to convenient divisions of the variable range. Each division is defined by its lower and upper limits, and the difference between them is the length of the class interval.

coefficient of variation
The ratio of the standard deviation to the mean.

conditional probability
The probability of an event conditional on other events having occurred or an assumption they will occur. (All probabilities are conditional on the general context of the problem.)

confidence interval
A 95%, or whatever, (frequentist) confidence interval for some parameter is an interval constructed in such a way that on average, if you imagine millions of random samples of the same size, 95% of them will include the parameter. From a practical point of view, I imagine there is a 95% chance the interval contains

the parameter, and Bayesian confidence intervals are properly interpreted in this way.

correlation
A dimensionless measure of linear association between two variables that lies between -1 and 1. Zero represents no association and negative values correspond to one variable increasing as the other decreases.

covariance
A measure of linear association between two variables, that equals the average value of the mean-corrected products.

crown race
The part of the bearing between the front forks and frame on a bicycle which provides the curved surface for the ball bearings to run in.

cumulative distribution function
A function which gives the probability that a continuous variable is less than any particular value. It is the population analogue of the cumulative frequency polygon. Its derivative is the pdf.

cumulative frequency polygon
A plot of the proportion, often expressed as a percentage, of data less than or equal to any variable value against that value.

degrees of freedom
Starting from a sample of n data, suppose m statistics are considered fixed. Then the number of data values that could be arbitrarily assigned subject only to the m 'constraints' is the degrees of freedom $(n-m)$. In multiple regression, the number of constraints equals the number of coefficients to be estimated, which is one more than the number of explanatory variables.

error
A deviation from the deterministic part of a model.

expected value
A mean value in the population.

explanatory variable
In a multiple regression the dependent variable, usually denoted by Y, is expressed as a linear combination of the explanatory variables. In designed experiments I have sub-divided the explanatory variables into **control** variables, whose values are chosen by the experimenter, and **concomitant** variables, which can be monitored but not preset. I have also called the dependent variable the **response.**

exponential distribution
A continuous distribution of the times between occurrences of an event, when occurrences are random and independent.

factorial experiment
An experiment designed to examine the effects of two or more factors. Each factor is applied at two or more levels and all combinations of these factor levels are tried in a full factorial. This allows interactions to be investigated, and is also an efficient means of estimating the main effects. The usual number of levels is two, high and low, so a full factorial for n factors requires 2^n runs. If this is prohibitive, fractional factorial designs, which retain most of the benefits, can be used.

frequency
The number of times an event occurs. Also, when talking about vibration, the number of cycles per second (hertz) or radians per second.

gamma distribution
The gamma function is defined as an integral from 0 to infinity. If the upper limit is treated as the value of a variable the integral is proportional to a cdf. It can exhibit a wide variety of shapes from near-normal to highly positively skewed. It is algebraically convenient in Bayesian applications, and also arises as a sampling distribution.

gamma function
A generalization of the factorial function to values other than positive integers.

histogram
A chart consisting of rectangles drawn above class intervals with areas equal to the proportion of data in each interval. It follows that the heights of the rectangles equal the relative frequency density, and the total area equals 1. If all the class intervals are of the same length the heights are proportional to the frequencies.

imaginary infinite population
The population sampled from is often imaginary and arbitrarily large. A sample from a production line is thought of as a sample from the population of all items that will be produced if the process continues on its present settings. An estimate is thought of as a single value from the imaginary distribution of all possible estimates, so that we can give its precision.

independent
Two events are independent if the probability that one occurs does not depend on whether the other does.

indicator variable
A means of incorporating categorical variables into a regression. Suppose some members of a population have an attribute and the rest do not. The indicator variable is 1 for members who have the attribute and 0 for the others. The coefficient of this variable is the effect of that attribute. For example, a study of the effect the alcohol on drivers' reaction times might code women 0 and men 1. The estimated coefficient of this indicator variable would be the estimated difference between women's and men's reaction times if it is constant for any given alcohol consumption. The possibility of the difference depending on alcohol consumption can be allowed for by including an interaction between the indicator variable and the amount of alcohol. The device can be extended to deal with several categories. Suppose three oil companies, A, B, and C, supply a chemical plant. Two indicator variables would be needed to distinguish between them. Both 0 corresponds to A, the first 1 and the second 0 to B, and the first 0 with the second 1 to C. The coefficients represent differences relative to A which should be the company for which most data are available.

interaction
Two explanatory variables interact if the effect of one depends on the value of the other. Their product is then included as an explanatory variable in the regression. If their interaction effect depends on the value of some third variable a third-order interaction exists, and so on.

kurtosis
In most engineering contexts it can be thought of as a measure of weight in the tails. It is 2, 3 and over 20 for a uniform, normal and t-distribution with 3 degrees of freedom, respectively.

marginal distribution
The marginal distribution of a variable is the distribution of that variable. The 'marginal' refers to the fact that multivariate data are available, or being modelled, but information on the other variables has been ignored.

meal
A mixture of powders used as raw material for a chemical process.

mean
The sum of several quantities divided by their number. Also used as an alternative to 'average' in 'average value of …'.

mean-corrected
Data are mean-corrected if their mean is subtracted from them. Mean-corrected data therefore have an average value of 0.

median
The middle value if data are put into ascending order.

mode
The most commonly occurring value. Also the value of the variable at which the pdf has its maximum.

multiple regression
One variable, called the dependent variable or the response, is expressed as a linear combination of explanatory variables plus random error. The coefficients of the variables in this combination are the unknown parameters of the model and are estimated from the data. The explanatory variables can be nonlinear functions of each other, for example, y is a linear combination of x_1, x_2, x_1^2, x_2^2, and $x_1 x_2$ represents a quadratic surface.

multiplicative rule
The probability of two events both occurring is the product of the probability that one occurs with the probability that the other occurs conditional on the first occurring.

mutually exclusive
Two events are mutually exclusive if they cannot occur together. A set of events is mutually exclusive and exhaustive if exactly one must occur.

normal distribution
A bell-shaped pdf which is a plausible model for random variation if it can be thought of as the sum of a large number of smaller components. It is also important as a sampling distribution, especially of the sample mean.

or
In probability 'A or B' is conventionally taken to include both.

orthogonal
In a designed experiment the values of the control variables are usually chosen to be uncorrelated, when possible, or nearly so. If they are uncorrelated they are said to be orthogonal. (If the values are put in columns and thought of as 'vectors', the vectors are orthogonal.)

parameter
A constant which is a salient feature of a population. Its value is usually unknown.

paver
A paving block. Modern ones are made from concrete in a variety of shapes and colours.

percentage point
The upper alpha percentage point of a pdf is the value beyond which a proportion $\alpha\%$ of the area under the pdf lies. The lower point is defined similarly.

Poisson distribution
The number of occurrences in some length of continuum if occurrences are independent.

Poisson process
Occurrences in some continuum, often time, form a Poisson process if they are random, independent, and occur at some constant average rate (in the standard case at least).

precision
The precision of an estimator is a measure of how close replicate estimates are to each other. Formally, it is the reciprocal of the variance of the sampling distribution.

priority controlled junction
A road junction which is controlled by 'Give Way' signs and road markings, rather than by traffic lights.

probability
A measure of how likely some event is to occur on a scale ranging from 0, representing impossibility, to 1, representing certainty. It can be thought of as the long-run proportion of times the event would occur if the scenario were to be repeated.

probability density function
A curve such that the area under it between any two values represents the probability that a continuous variable will be between them. The population analogue of a histogram.

probability function
A formula that gives the probability that a discrete variable takes any of its possible values. The population analogue of a line diagram.

process capability index
Six times the ratio of the difference between the upper and lower specification (spec) to the process standard deviation (C_p). Provided the process is near-normal, almost all production will be within spec when the process mean is at the mid-point of the spec range. The process performance index (C_k) takes account of any known deviation of the process mean.

pseudo-random numbers
A sequence of numbers, generated by a deterministic algorithm, which appear to be random and are indistinguishable from genuine random numbers by empirical tests. Computer-generated random numbers are actually pseudo-random.

quantiles
The lower $x\%$ quantile is the value of the variable below which $x\%$ of the data lie.

quartiles
The upper (lower) quartile, UQ (LQ), is the datum above (below) which one-quarter of the data lie.

random digits
A sequence in which each one of the digits $0, 1, \ldots, 9$ is equally likely to occur next.

random numbers
A sequence of numbers from a specified distribution, such that the next is independent of the existing sequence.

random sample
A sample which has been selected so that every member of the population has a known, non-zero, probability of appearing.

range
Difference between the largest datum and the smallest datum when the data are put into numerical order.

relative frequency
The ratio of the frequency of occurrence of some event to the number of scenarios in which it could potentially have occurred. That is, the proportion of times on which it occurred.

relative frequency density
Relative frequency divided by the length of the class interval.

replication
The use of two or more experimental units for each experimental treatment. The execution of an entire experiment more than once so as to increase precision and obtain a more precise estimate of sampling error.

run-out
A measurement of buckle of a bicycle wheel. If it is rotated with a clock gauge held against the rim, the range of the reading is the 'run-out'.

sample space
A list of all possible outcomes of some operation which involves chance.

sampling distribution
An estimate is thought of as a single value from the imaginary distribution of all possible estimates, known as the sampling distribution. The idea of a sampling distribution is necessary to measure the precision of an estimator. The sampling distributions in this book are assumed to be approximately normal or *t*-distributions.

session window
MINITAB Release 9 is a windows version. The session window displays numeric and text output. It also allows you to type in session commands from a prompt, just like using the earlier releases, as an alternative to using the mouse.

simple random sample
A sample chosen so that every possible choice of *n* from *N* has the same chance of occurring. It implies all members of the population have the same probability of selection, but many other sampling schemes also have this property (see **stratified sampling**).

skewness
A measure of asymmetry of a distribution. Positive values correspond to a tail to the right.

standard deviation
The positive square root of the variance.

standard error
The standard deviation of some estimator. Commonly used, without qualification, for the standard deviation of the sample mean.

standard normal distribution
The normal distribution scaled to have a mean of 0 and a standard deviation of 1. Its cdf and percentage points are given in Tables E.2 and E.3.

statistic
A summary number calculated from the sample, usually to estimate the corresponding population parameter.

Student's t-distribution
The sampling distribution of many statistics is normal and can therefore be scaled to standard normal. If the mean of the sampling distribution is the parameter of interest, and the unknown standard deviation is replaced by its sample estimate, with v degrees of freedom, the normal distribution becomes a t-distribution with v degrees of freedom. If v exceeds about 30 there is little practical difference.

t-ratio
The ratio of an estimate to its standard deviation. If the absolute value of the t-ratio is less than 1 a 66% confidence interval for the parameter will include 0. If the t-ratio exceeds about 1.7 the 90% confidence interval will exclude 0.

unbiased estimator (estimate)
An estimator is unbiased, for some parameter, if the mean of its sampling distribution is equal to that parameter. That is, averaged over imagined replicates of the sampling procedure, we would obtain the parameter value. An unbiased estimate is a particular value of the estimator.

uniform distribution
A variable has a uniform distribution between two limits if the probability that it lies within some interval between those limits is proportional to the length of the interval. Limits of 0 and 1 are by far the most usual, and arise in the context of random number generation.

variable
A quantity that varies from one member of the population to the next. It can be measured on some continuous scale, be restricted to integer values (discrete), or be restricted to descriptive categories.

variance
Average of the squared deviation from the mean.

weighted mean
An average in which the data are multiplied by numbers called **weights**, summed, and then divided by the sum of the weights. If the weights are all the same this is the usual mean.

B3 Problem-solving guide

B3.1 Overview

I think it is useful to give a broad schematic and then look at the two strands in more detail.

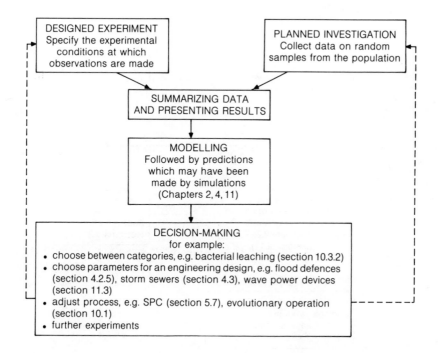

B3.2 *Planned investigation*

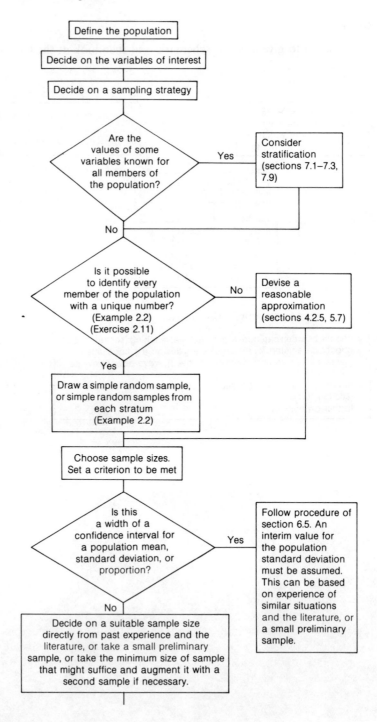

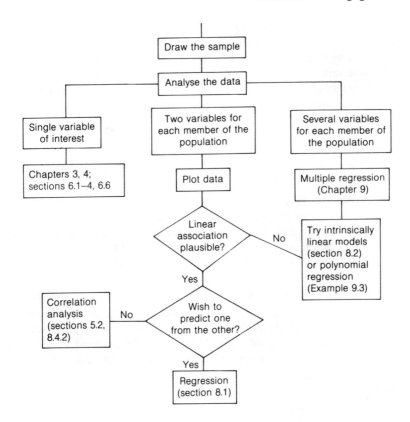

B3.3 Designed experiment

The following is a categorization of common designs, and the appropriate sections in the book are indicated.

(a) Descriptive
(i) A sample of several test pieces, all randomly selected from a standard material, is tested to determine the elementary characteristics of that material. You may, for example, wish to report the mean and the standard deviation of tensile strength.
(Chapter 3, sections 6.1, 6.2 as it is a planned investigation)

(b) Comparative
(ii) Comparison against a standard.
You may wish to compare the characteristics of a new material against a specified industry standard. You would test a sample of several pieces and ask if there was sufficient evidence to conclude that the measured characteristic of this material was different from the standard specification.
(Chapter 3, sections 6.1, 6.2 as it is a planned investigation)

(iii) Comparison of two materials with independent samples.
You may have two materials, perhaps of different compositions, or made by slightly different processes, or, even if they are claimed to be of the same composition and made by exactly the same process, they are made at different places. You wish to determine if they have the same or different properties so you test a sample of several pieces from each material. These samples are to be selected independently of each other.
(section 6.4.1)

(iv) Comparison of several materials with independent samples.
Completely randomized design.
(section 10.3.1)

(v) Comparison of two materials by paired samples.
As above, you may have two materials and you wish to determine if they have the same or different properties. However, you wish to ensure, in the presence of uncontrollable outside influences, a fair comparison. For example, you may wish to expose samples of rubber to the weather and measure their deterioration. One approach would be to expose test pieces in pairs, each pair comprising one piece of each material, thus ensuring that both members of the pair experience the same weather conditions. The data to be analysed would be the difference in deterioration measured between each pair.
(section 6.4.2)

(vi) Comparison of several materials in blocks.
Randomized block design
(section 10.3.2).

(vii) Comparison of several types of component under different conditions when there are insufficient components of each type for a completely randomized design.
(section 10.3.3).
(The possibility of significant variability between components of the same type precludes the use of a randomized block design with one component of each type being a block.)

(c) Response

(viii) Factorial experiments
When new materials or manufacturing processes are being developed there are usually several variables, or factors, that can influence a material property. Experiments to investigate the effects of several variables should be designed to allow all of those variables to be set at several levels. It is inefficient to change one variable at a time, with the others fixed, and it can be misleading if there are interactions. A factorial experiment with m factors all at two levels will need 2^m runs (section 10.2 up to 10.2.1). If m is large this may be impractical and fractional factorial designs can be used.

(ix) Response surface exploration with composite designs.
In the final stage of a development study, when you are seeking the conditions (such as the values of composition and process variables) that will yield the best value of a material property (such as the highest value of tensile strength), additional points must be added to factorial experiments so that curvature of the response can be estimated. These designs are known as **augmented** or **composite** designs (section 10.1., 10.2.1).

(x) *D*-optimum designs.
These allow the experimenter to choose which interactions, quadratic effects and cubic effects are to be estimated (section 10.4).

B4 Suggested short course

It may be that you are looking at this book because you are expected to design an experiment or survey as part of your research programme, or because you have been advised to use a 'multiple regression analysis' on your data. If this is the case you will be keen to reach the relevant material as soon as possible, and the following is suggested reading.

Chapter 1   A brief introduction.
Chapter 2   You need the general ideas of probability, but only the simplest examples are necessary for what follows. Read sections 2.1–4 as far as Example 2.10. The formula for choices in section 2.5 is used for the binomial distribution.
Chapter 3   Presenting and summarizing data is fundamental. However, the example on fatigue damage and the section on skewness and kurtosis can be omitted. Read section 3.1 and 3.2.
Chapter 4   The binomial, normal and uniform distribution are the most used, and the concept of expectation occurs throughout. Read sections 4.1, 4.2 up to the end of section 4.2.4. Also read section 4.2.8 on the normal score plot, and the paragraphs on the probability plot in section 4.2.5 to which it refers.
Chapter 5   Linear combinations of variables, and the idea of correlation are essential. Read sections 5.1, 5.2, 5.4, 5.5, 5.6 up to Example 5.8.
Chapter 6   Confidence intervals are a measure of precision of our estimates, and confidence intervals for means, and differences in means, are the priority. So read sections 6.1 and 6.4.
Chapter 7   The concept of stratification is essential for survey work, and is really just 'common sense'. If nothing else, read sections 7.1, 7.2, 7.3, 7.8.1, and the entry on stratification in the summary.
Chapter 8   Regression on a single explanatory variable is covered in section 8.1 and this is essential reading. Most of the summary refers to this.

Chapter 9   This chapter covers multiple regression.
Chapter 10 This chapter is an introduction to designed experiments. The case study on the multi-laboratory trial of welded joints is rather more advanced.

Having done this, the problem-solving guide (Appendix B3) may point you to other areas relevant to your needs.

Appendix C: Summary of MINITAB commands used in text

This is an excerpt from the Quick Reference section (Appendix A) of the MINITAB (Release 9) Reference Manual. It covers only the facilities mentioned in this book.

Worksheet and commands

Minitab consists of a worksheet for data and about 200 commands which allow you to enter data into the worksheet; manipulate and transform your data; produce graphical and numerical summaries; and perform a wide range of statistical analyses.

The worksheet contains columns of data denoted by column numbers C1, C2, C3..., or by column names such as 'Age' or 'Height'. Enclose column names in single quotation marks in session commands. The worksheet can also contain stored constants denoted by K1, K2, K3..., and matrices denoted by M1, M3, M3...

To execute a command, choose it from a menu or enter it into the Session window at the Minitab prompt, MTB⟩. When you execute a menu command, Minitab lists available options in dialog boxes. When you execute a session command, you need to know the correct command syntax.

Subcommands

Some Minitab session commands have subcommands (shown indented). To use a subcommand, put a semicolon at the end of the main command line. Then type the subcommands, starting each on a new line and ending each with a semicolon. When you are done, end the last subcommand with a period. The subcommand ABORT cancels the whole command.

Notation

K	denotes a constant such as 8.3 or K14
C	denotes a column, such as C12 or 'Height'
E	denotes either a constant or column
M	denotes a matrix, such as M5
[ ]	encloses an optional argument

General information

HELP	explains Minitab commands, can be a command or a subcommand
INFO	[C...C] gives the status of worksheet
STOP	ends the current session

Input and output of data

READ   data into C...C
   TAB   columns are separated with tabs
      NONAMES   columns are not named (use with TAB)
      ALPHA   columns are in K...K (use with TAB)
SET   data into C # abbreviations allow you to enter patterned data
                1:3/.5 expands to 1, 1.5, 2, 2.5, 3
                3(1:3) expands to 1, 2, 3, 1, 2, 3, 1, 2, 3
                (1:3)2 expands to 1, 1, 2, 2, 3, 3
                3(1:2)2 expands to 1, 1, 2, 2, 1, 1, 2, 2, 1, 1, 2, 2
INSERT   data [between rows K and K] of C...C
READ, SET, and INSERT have the subcommands:
   FILE   'filename'
   FORMAT   (format statement)
   NOBS   = K
   SKIP   K lines
END   of data
NAME   for E is 'name', for E is 'name'... for E is 'name'
PRINT   the data in E...E
WRITE   [to 'filename'] the data in C...C
   TAB   separate columns with tabs
      NONAMES   omit column names (use with TAB)
   REPLACE   replace existing file
   NOREPLACE   do not replace existing file
PRINT and WRITE have the subcommand:

FORMAT	(format statement)
SAVE	[in 'filename'] a copy of the worksheet
PORTABLE	
REPLACE	replace existing file
NOREPLACE	do not replace existing file
LOTUS	save in Lotus format
RETRIEVE	the Minitab saved worksheet [in 'filename']
PORTABLE	
LOTUS	retrieve Lotus file
OUTFILE	'filename' put all output in file
OW	=K output width of file
OH	=K output height of file
NOTERM	no output to terminal
NOOUTFILE	output to terminal only
NEWPAGE	start next output on a new page
JOURNAL	['filename'] record Minitab commands in this file
NOJOURNAL	cancel JOURNAL
TYPE	'[path]filename.extension' display contents of a file
CD	[path] change directory
DIR	[path] list files
SYSTEM	[systemcommand] provides access to operating system commands
RESTART	begin fresh Minitab session

Editing and manipulating data

LET	$C(K)=K$ # changes the number in row K of C
DELETE	rows K...K of C...C
ERASE	all data in E...E
INSERT	data [between rows K and K] of C...C
FORMAT	(format statement)
NOBS	=K
COPY	C...C into C...C
COPY	C into K...K
USE	rows K...K
USE	rows where $C=K...K$
OMIT	rows K...K
OMIT	rows where $C = K...K$
COPY	K...K into C
CODE	(K...K) to K...(K...K) to K for C...C, put in C...C
STACK	(E...E)...on (E...E), put in (C...C)
SUBSCRIPTS	into C
UNSTACK	(C...C) into (E...E)...(E...E)
SUBSCRIPTS	are in C

CONVERT	using table in C C, the data in C, and put in C
CONCATENATE	C...C put in C
SORT	C [carry along C...C] put into C [and C...C]
BY	C...C
DESCENDING	C...C
RANK	the values in C, put ranks into C

Arithmetic

LET = expression

Expressions may use
   Arithmetic operators $+$ $-$ $*$ / $**$(exponentiation)
   Comparison operators $=$ $\sim$ $=$ $<$ $>$ $<=$ $>=$
   Logical operators   & | $\sim$
and any of the following: ABSOLUTE, SQRT, LOGTEN, LOGE, EXPO,
ANTILOG, ROUND, SIN, COS, TAN, ASIN, ACOS, ATAN, SIGNS,
NSCORE, PARSUMS, PARPRODUCTS, COUNT, N, NMISS, SUM,
MEAN, STDEV, MEDIAN, MIN, MAX, SSQ, SORT, RANK, LAG, EQ,
NE, LT, GT, LE, GE, AND, OR, NOT.

EXAMPLES: LET C2 = SQRT(C1 − MIN(C1))
             LET C3(5) = 4.5

Simple arithmetic operations

ADD	E to E...	to E, put into E
SUBTRACT	E from	E, put into E
MULTIPLY	E By E...	by E, put into E
DIVIDE	E by	E, put into E
RAISE	E to the power	E, put into E

Columnwise functions

ABSOLUTE value	of E, put into E
SQRT	of E, put into E
LOGE	of E, put into E
LOGTEN	of E, put into E
EXPONENTIATE	E, put into E
ANTILOG	of E, put into E
ROUND to integer	E, put into E
SIN	of E, put into E
COS	of E, put into E

TAN	of E, put into E
ASIN	of E, put into E
ACOS	of E, put into E
ATAN	of E, put into E
SIGNS	of E, put into E
PARSUMS	of C, put into C
PARPRODUCTS	of C, put into C

Normal scores

| NSCORES | of C, put into C |

Columnwise statistics

COUNT the number of values in	C [put into K]
N (number of nonmissing values) in	C [put into K]
NMISS (number of missing values) in	C [put into K]
SUM of the values in	C [put into K]
MEAN of the values in	C [put into K]
STDEV of the values in	C [put into K]
MEDIAN of the values in	C [put into K]
MINIMUM of the values in	C [put into K]
MAXIMUM of the values in	C [put into K]
RANGE of the values in	C [put into K]
SSQ (uncorrected sum of sq.) for	C [put into K]

Rowwise statistics

RCOUNT	of E...E put into C
RN	of E...E put into C
RNMISS	of E...E put into C
RSUM	of E...E put into C
RMEAN	of E...E put into C
RSTDEV	of E...E put into C
RMEDIAN	of E...E put into C
RMINIMUM	of E...E put into C
RMAXIMUM	of E...E put into C
RRANGE	of E...E put into C
RSSQ	of E...E put into C

Indicator variables

| INDICATOR | variables for subscripts in C, put into C...C |

Basic statistics

DESCRIBE	C...C
BY	C
ZINTERVAL	[K% confidence] assuming sigma = K for C...C
ZTEST	[of mu = K] assuming sigma = K for C...C
ALTERNATIVE	= K
TINTERVAL	[K% confidence] for data in C...C
TTEST	[of mu = K] on data in C...C
ALTERNATIVE	= K
TWOSAMPLE	test and c.i. [K% confidence] samples in C C
ALTERNATIVE	= K
POOLED	procedure
TWOT	test and c.i. [K% confidence] data in C, groups in C
ALTERNATIVE	= K
POOLED	procedure
CORRELATION	between C...C [put into M]
COVARIANCE	between C...C [put into M]
CENTER	the data in C...C put into C...C
LOCATION	[subtracting K...K]
SCALE	[dividing by K...K]
MINMAX	[with K as min and K as max]

Regression

REGRESS	C on K predictors C...C
NOCONSTANT	in equation
WEIGHTS	are in C
SRESIDUALS	put into C
FITS	put into C
MSE	put into K
COEFFICIENTS	put into C
XPXINV	put into M
RMATRIX	put into M
HI	put into C (leverage)
RESIDUALS	put into C (observed – fit)
TRESIDUALS	put into C (deleted studentized)
COOKD	put into C (Cook's distance)
DFITS	put into C
PREDICT	for E...E
VIF	(variance inflation factors)
DW	(Durbin–Watson statistic)

PURE	(pure error lack-of-fit test)
XLOF	(experimental lack-of-fit test)
TOLERANCE	K [K]
STEPWISE	regression of C on the predictors C...C
FENTER	= K (default is four)
FREMOVE	= K (default is four)
FORCE	C...C
ENTER	C...C
REMOVE	C...C
BEST	K alternative predictors (default is zero)
STEPS	= K (default depends on output width)
BREG	C on predictors C...C
INCLUDE	predictors C...C
BEST	K models
NVARS	K [K]
NOCONSTANT	in equation
NOWARN	suppress warning for model with many predictors
NCONSTANT	in REGRESS, STEPWISE and BREG commands that follow
CONSTANT	fit a constant in REGRESS, STEPWISE and BREG
BRIEF	K

Distributions and random data

RANDOM	K observations into C...C
BERNOULLI	trials p = K
PDF	for values in E [put results in E]
CDF	for values in E [put results in E]
INVCDF	for values in E [put results in E]

RANDOM, PDF, CDF, INVCDF have the subcommands:

BINOMIAL	n = K, p = K
POISSON	mu = K
INTEGER	discrete uniform on integers K to K
DISCRETE	dist. with values in C and probabilities in C
NORMAL	[mu = K [sigma = K]]
UNIFORM	[continuous on the interval K to K]
T	degrees of freedom = K
F	df numerator = K, df denominator = K
CAUCHY	[a = K [b = K]]
LAPLACE	[a = K [b = K]]
LOGISTIC	[a = K [b = K]]

LOGNORMAL	[mu = K [sigma = K]]
CHISQUARE	v = K
EXPONENTIAL	[b = K]
GAMMA	a = K b = K
WEIBULL	a = K b = K
BETA	a = K b = K

SAMPLE	K rows from C...C put into C...C
REPLACE	(sample with replacement)
BASE	for random number generator = K

Matrices

READ	[from 'filename'] into a K by K matrix M
PRINT	M...M
TRANSPOSE	M into M
INVERT	M into M
DEFINE	K into K by K matrix M
DIAGONAL	is C, form into M
DIAGONAL	of M, put into C
COPY	C...C into M
COPY	M into C...C
COPY	M into M
USE	rows K...K
OMIT	rows K...K
ERASE	M...M
EIGEN	for M put values into C [vectors into M]

In the following comands E can be either C, K or M

ADD	E to E, put into E
SUBTRACT	E from E, put into E
MULTIPLY	E by E, put into E

Miscellaneous

OH	= K number of lines for height of output
OW	= K number of spaces for width of output
IW	= K number of spaces for width of input
BRIEF	= K controls amount of output from REGRESS, GLM, DISCRIMINANT, ARIMA, RLINE
TSHARE	operate interactively
BATCH	operate in batch mode

Plotting data

These are character graph commands, documented in the *MINITAB Graphics Manual*.

You must type the command GSTD to enable these commands. To re-enable Professional Graphics commands, type GPRO. The commands DOTPLOT, STEM-AND-LEAF, GRID, and CONTOUR are available in both Standard Graphics mode (GSTD) and Professional Graphics mode (GPRO).

GSTD	enable Standard Graphics
GPRO	enable Professional Graphics
HISTOGRAM	C...C
DOTPLOT	C...C

HISTOGRAM and DOTPLOT have the subcommands:

INCREMENT	=K
START	at K [end at K]
BY	C
SAME	scales for all columns

STEM-AND-LEAF	display of C...C
TRIM	outliers
INCREMENT	=K
BY	C
BOXPLOT	for C
INCREMENT	=K
START	at K [end at K]
BY	C
LINES	=K
NOTCH	[K% confidence] sign c.i.
LEVELS	K...K
PLOT	C vs C
SYMBOL	='symbol'
MPLOT	C vs C, and C vs C, and ...C vs C
LPLOT	C vs C using tags in C
TPLOT	C vs C vs C

PLOT, MPLOT, LPLOT, and TPLOT have the subcommands:

TITLE	='text'
FOOTNOTE	='text'
YLABEL	='text'
XLABEL	='text'
YINCREMENT	=K
YSTART	at K [end at K]
XINCREMENT	=K
XSTART	at K [end at K]

TSPLOT	[period K] of C
ORIGIN	=K
MTSPLOT	[period K] of C...C
ORIGIN	=K for C...C [...origin K for C...C]

TSPLOT and MISPLOT have the subcommands:

INCREMENT	=K
START	at K [end at K]
TSTART	at K [end at K]

GRID	C [K to K] C [K to K]
CONTOUR	C vs C and C
BLANK	bands between letters
YSTART	=K [up to K]
YINCREMENT	=K
WIDTH	of all plots that follow is K spaces
HEIGHT	of all plots that follow is K lines

Symbols

* Missing Value Symbol. An * can be used as data in READ, SET and INSERT, in data files and in the Date Editor. Enclose the * in single quotes in session commands and subcommands.

    Example: CODE (-99) to '*' in C1, put into C3
    EXAMPLE: Copy C6 INTO C7;
            OMIT C6 = '*7.

\# Comment Symbol. The symbol # anywhere on a line tells Minitab to ignore the rest of the line.
& Continuation Symbol. To continue a command onto another line, end the first line with the symbol &.

Other facilities

Other facilities in Release 9 include:

- ANALYSIS OF VARIANCE
- MULTIVARIATE ANALYSIS
- NONPARAMETRICS
- TABLES
- TIME SERIES
- STATISTICAL PROCESS CONTROL

and design of experiment facilities in the MINITAB QC Manual, which covers aspects of quality control.

Appendix D: Data sets

Table D.1 Compressive strengths of 180 concrete cubes (in newtons per square milli-metre). Data from the Department of Civil Engineering, University of Newcastle upon Tyne.

57.4	59.5	62.1	53.0	56.5	64.5	56.6	55.6	58.2
59.6	57.5	60.3	65.2	65.2	62.1	62.0	68.2	56.9
68.5	62.1	43.8	57.0	58.2	59.6	58.6	58.7	59.1
63.1	55.6	62.1	63.0	63.0	61.4	59.1	56.7	58.6
60.6	63.7	57.5	62.6	54.8	62.1	68.2	63.6	60.8
56.5	65.6	62.0	60.7	65.7	59.0	55.6	58.0	59.4
64.3	60.5	66.1	54.6	66.6	66.1	66.2	65.8	57.8
62.2	61.7	56.0	53.2	61.7	56.8	57.9	61.0	64.6
55.6	66.8	61.5	62.2	63.8	63.7	64.8	57.7	63.8
65.2	65.2	59.9	58.9	61.0	57.8	59.2	67.4	66.3
59.2	60.9	60.9	49.8	60.9	60.5	56.1	61.8	61.0
63.4	61.7	58.8	64.4	55.3	58.6	67.0	65.6	62.8
66.9	61.2	60.8	64.8	60.7	60.3	61.2	61.2	62.6
58.2	57.5	57.5	57.8	59.6	48.2	63.4	61.6	62.5
61.0	57.0	56.8	64.3	68.2	63.6	62.4	57.8	60.1
61.5	63.5	65.6	67.8	56.2	61.5	57.3	59.9	59.6
68.7	64.4	65.5	64.8	64.2	65.7	68.4	65.1	60.1
59.7	61.3	56.8	61.3	65.6	61.2	62.0	57.0	66.2
62.6	61.8	58.8	61.1	56.4	61.7	58.6	66.0	68.8
66.3	63.8	61.7	58.5	56.5	62.0	61.3	61.3	64.7

Table D.2 Annual maximum daily mean discharges ($m^3 s^{-1}$, naturalized) for the River Thames at Kingston (Teddington) (Data supplied by the National Rivers Authority)

Year[a]	Month	Rank	Flood ($m^3 s^{-1}$)
1884	2	84	231
1885	2	86	229
1885	12	79	244
1887	1	65	284
1888	3	95	207
1889	3	81	237
1890	1	96	204
1891	2	103	171
1891	10	42	339
1893	2	58	299
1894	2	101	173
1894	11	1	1065[b]
1896	3	98	201
1897	2	38	351
1998	1	102	171
1899	2	=70	262
1900	2	7	533
1901	4	99	200
1901	12	104	162
1903	6	24	386
1904	2	10	517
1905	3	87	229
1906	1	77	249
1907	1	94	220
1907	12	28	375
1909	3	97	204
1910	2	85	231
1910	12	17	428
1912	1	36	366
1913	1	75	255
1914	3	74	256
1915	1	4	585
1916	3	31	373
1916	12	46	327
1918	1	39	351
1919	3	43	334
1920	2	76	251
1921	1	80	240
1922	3	100	198
1923	4	83	231

Table D.2 (*continued*)

Year[a]	Month	Rank	Flood $(m^3 s^{-1})$
1924	1	61	298
1925	1	9	522
1926	1	32	370
1927	3	30	374
1928	1	8	526
1928	12	82	235
1929	12	6	551
1930	12	88	228
1932	5	66	274
1933	2	11	479
1934	3	108	94
1935	3	90	227
1936	1	12	478
1937	1	16	437
1937	12	78	247
1939	2	33	369
1940	2	19	410
1940	11	26	384
1941	1	60	298
1943	2	13	457
1944	1	107	115
1944	12	72	261
1945	12	73	256
1947	3	2	714
1948	1	89	227
1949	1	59	299
1950	2	48	324
1951	2	25	384
1951	11	27	376
1952	12	69	263
1954	6	92	222
1954	12	15	452
1956	2	52	315
1957	2	53	314
1958	2	51	316
1959	1	29	375
1960	1	56	308
1960	11	14	456
1962	1	41	344
1963	3	64	285
1964	3	34	369
1965	9	106	122
1966	2	=49	324

Table D.2 (*continued*)

Year[a]	Month	Rank	Flood ($m^3 s^{-1}$)
1967	3	55	313
1968	9	3	600
1968	12	35	369
1970	1	91	224
1971	1	37	362
1972	3	45	330
1972	12	68	266
1974	2	22	396
1974	11	5	559
1975	12	105	152
1977	1	44	334
1978	1	47	326
1979	4	=49	324
1979	12	23	393
1981	3	62	289
1981	12	54	314
1982	12	40	345
1984	3	63	286
1984	11	67	270
1985	12	20	408
1987	4	57	304
1988	1	21	402
1989	2	=70	262
1990	2	18	427
1991	1	93	220

[a]Water year (1st October until 30th September)
[b]Subsequent investigations show that the correct figure in probably about $789\,m^3 \pm s^{-1}$.

Table D.3 Volumes of shampoo (millilitres) in 100 plastic containers with nominal contents of 200 ml (fictitious but realistic)

201.11	200.92	201.22	200.98	201.09
200.91	200.86	201.04	200.87	200.89
200.97	201.08	201.15	201.14	200.88
201.04	201.01	201.14	201.02	201.17
200.93	201.08	200.97	201.02	201.52
201.06	200.96	200.83	201.12	200.98
201.02	200.97	201.08	201.06	200.88
201.01	201.11	201.11	200.84	200.68
200.97	201.03	200.93	200.85	200.79
201.02	200.91	201.01	200.91	201.11
201.17	201.00	200.98	200.95	201.46
200.99	201.01	200.93	201.01	200.64
201.12	201.04	201.02	201.02	201.02
201.22	201.00	200.87	201.12	201.30
201.03	201.05	200.79	201.12	201.00
200.99	201.09	201.04	201.08	200.55
200.92	200.98	201.15	201.09	200.84
200.96	201.03	201.00	200.95	200.86
200.97	200.98	200.91	200.96	200.87
201.00	200.97	201.02	200.91	201.05

Table D.4 Noise level exceeded 10% of exposure time (decibels) at 46 sites near the Tyne and Wear Metro. Data from Transport Operations Research Group, University of Newcastle upon Tyne

73.8	63.8	62.5
72.8	58.0	58.3
57.8	60.0	57.0
60.8	64.5	59.5
54.8	56.8	65.1
57.8	60.5	58.0
54.3	58.8	
58.0	59.3	
55.8	60.0	
52.3	60.5	
56.5	66.0	
75.3	61.3	
60.0	68.8	
72.8	57.3	
59.8	55.3	
58.0	55.3	
59.3	57.5	
61.0	57.0	
59.0	56.8	
59.0	59.5	

Table D.5 Simulated waiting times (minutes) of 90 passengers for tube trains which run every 10 minutes, if passengers have not referred to the timetable (MINITAB random sample from $U[0, 10]$)

9.46	5.26	6.59	6.00	7.79
9.68	2.84	5.65	8.17	6.82
1.44	7.15	7.69	9.85	1.96
7.60	7.88	8.07	0.16	5.29
5.07	5.71	8.02	3.84	3.68
7.67	1.16	5.97	7.10	6.72
2.06	6.28	3.51	0.92	8.67
0.48	3.74	8.81	5.90	8.85
4.65	0.34	5.95	9.46	4.21
3.43	6.84	3.64	9.37	3.62
3.52	3.74	6.68	4.24	
5.72	3.08	8.00	8.58	
4.45	4.54	4.72	9.55	
5.60	7.13	8.65	0.24	
9.38	3.45	4.10	5.06	
2.40	1.98	2.79	3.63	
0.11	7.11	1.93	6.93	
8.41	0.42	8.05	1.59	
9.11	2.84	4.46	4.15	
0.07	4.41	3.84	4.83	

Table D.6 Fuel used (cubic metres) by a ferry for 141 passages between the same ports. Data from a shipping company

52.37	49.43	50.05	48.24	48.74	50.94	48.35	50.14
47.59	55.91	47.41	48.97	45.24	50.81	48.76	
48.88	51.39	51.26	49.02	48.22	49.49	53.74	
51.16	66.13	49.53	41.94	55.03	51.79	53.17	
52.50	50.70	49.98	46.97	48.45	53.09	49.95	
47.87	47.66	47.04	46.27	50.36	52.43	49.55	
50.99	48.59	49.78	50.02	47.10	49.98	50.27	
51.51	50.56	49.58	47.95	48.87	47.41	49.19	
51.45	51.49	52.01	49.20	48.25	50.56	47.60	
48.69	50.23	55.31	48.86	49.38	50.03	47.49	
49.91	56.65	46.54	50.24	45.72	49.47	41.50	
50.96	52.06	49.17	47.50	48.27	48.21	48.71	
48.14	51.62	50.52	48.07	46.96	49.22	49.92	
50.77	55.61	47.87	44.72	47.69	48.91	50.79	
50.32	48.95	49.73	45.56	50.69	48.46	40.74	
49.28	51.11	51.75	48.22	49.74	48.30	52.94	
50.05	55.11	49.30	47.60	49.42	51.13	50.93	
44.72	52.90	49.17	46.94	49.65	47.98	49.18	
50.23	52.85	46.45	48.66	49.61	71.75	49.13	
51.43	48.43	43.57	47.05	49.66	58.95	49.33	

Table D.7 Costs of schemes to prevent flooding (monetary units adjusted to current prices). Data provided by Northumbrian Water Ltd.

Cost of scheme	Number of properties affected
56 770	10
180 769	8
2 133	7
15 267	6
70 644	5
20 708	5
252 351	5
4 300	4
61 234	4
14 609	4
179 567	4
135 796	3
12 060	3
16 501	3
12 101	3
4 291	3
30 275	3
3 031	2
29 746	2
5 420	2
677	1
6 327	1
3 308	1
16 278	1
30 862	1
15 911	1
13 276	1
8 807	1
7 861	1
4 008	1
2 547	1

Table D.8 Temperature (x_1, degrees
Celsius), percentage calcium in salt
phase (x_2), percentage calcium in
metal phase (y) for 28 runs of a
chemical process. Data from ICI,
Runcorn, UK

x_1	x_2	y
547	3.02	63.7
550	3.22	62.8
550	2.98	65.1
556	3.90	65.6
572	3.38	64.3
574	2.74	62.1
574	3.13	63.0
575	3.12	61.7
575	2.91	62.3
575	2.72	62.6
575	2.99	62.9
575	2.42	63.2
576	2.90	62.6
579	2.36	62.4
580	2.90	62.0
580	2.34	62.2
580	2.92	62.9
591	2.67	58.6
602	3.28	61.5
602	3.01	61.9
602	3.01	62.2
605	3.59	63.3
608	2.21	58.0
608	2.00	59.4
608	1.92	59.8
609	3.77	63.4
610	4.18	64.2
695	2.09	58.4

Table D.9 Trips per occupied dwelling unit day, average car ownership, average household size, socio-economic index, urbanization index, for 57 traffic analysis zones. Data from the Transportation Centre at Northwestern University, Chicago

Trips	Carown	Size	Socind	Urbind
3.18	0.59	3.26	28.32	60.10
3.89	0.57	3.13	20.89	65.71
3.98	0.61	3.02	25.99	63.19
4.16	0.61	3.14	28.52	66.24
3.60	0.63	3.75	27.18	58.36
4.10	0.66	3.24	27.95	59.58
4.36	0.71	2.77	39.91	64.64
4.87	0.77	2.74	48.36	67.88
5.85	0.84	3.02	42.15	56.86
4.97	0.74	2.84	38.14	62.44
3.54	0.67	2.93	51.30	68.67
4.31	0.64	3.87	43.90	59.49
4.54	0.73	3.16	30.27	57.76
4.82	0.86	3.42	32.18	63.06
4.04	0.66	3.54	34.45	47.73
4.60	0.64	3.49	43.32	59.36
3.40	0.50	2.76	75.32	75.81
4.65	0.58	2.91	62.20	75.26
3.02	0.53	1.83	82.53	83.66
9.14	1.11	3.00	67.31	38.21
4.30	0.70	2.94	64.01	55.51
4.24	0.80	3.19	51.16	52.44
5.00	0.77	2.61	59.15	59.38
5.93	0.96	3.24	48.51	46.51
5.11	0.86	2.95	47.44	51.17
5.84	0.92	2.95	57.34	58.60
4.70	0.80	3.00	62.60	62.40
4.54	0.79	2.71	73.00	67.23
5.51	0.91	3.46	33.96	41.29
5.10	0.75	3.38	43.67	56.64
4.70	0.83	3.11	52.74	54.02
5.17	0.76	3.20	52.29	58.35
5.41	0.87	3.24	43.42	47.78
6.46	1.16	3.60	45.94	51.21
6.03	0.90	3.02	61.53	54.92
4.79	0.53	3.09	49.37	58.63
4.83	0.75	2.46	87.38	65.67
6.30	0.78	3.36	55.85	59.00
4.94	0.69	2.94	50.15	61.09
6.01	0.96	3.27	67.01	48.39

Table D.9 (*continued*)

Trips	Carown	Size	Socind	Urbind
6.39	0.86	3.32	62.18	50.04
5.82	1.09	3.29	45.58	46.47
6.25	1.15	3.58	60.85	26.36
6.13	0.90	3.09	55.59	43.58
6.70	1.02	3.02	75.73	35.89
7.10	1.00	3.33	57.84	28.28
7.89	1.32	3.58	79.69	25.37
7.80	1.06	3.17	57.01	31.97
8.02	1.02	3.35	50.93	38.17
7.20	0.98	3.43	49.75	34.69
5.14	0.82	3.31	36.36	46.98
5.56	0.94	3.21	62.27	36.27
5.74	0.90	3.52	42.64	26.15
6.77	0.62	3.92	21.66	24.08
4.94	0.77	3.02	49.18	51.39
7.64	0.93	3.37	34.74	44.54
7.25	0.75	4.50	26.21	44.80

Table D.10 Number of snags, time
taken (person-days), and initial assessment of difficulty of setting up 40
computer systems

snags (y)	time (x_1)	initasd (x_2)
148	14	2
161	8	1
9	10	2
72	25	4
295	18	2
142	14	3
10	25	2
128	19	3
16	5	2
75	6	1
76	33	5
54	4	1
175	16	1
131	28	4
227	17	2
198	22	3
52	10	1
362	30	3
230	10	2
51	10	2
90	8	1
10	1	1
100	15	1
41	20	3
5	11	2
22	5	1
23	15	4
247	20	2
22	28	5
114	20	3
79	5	2
58	15	2
8	23	4
237	24	4
90	23	4
70	20	1
95	14	2
92	13	3
137	13	2
83	13	1

Table D.11 Disassembly of catheter. The control variables are catheter outside diameter (OD) (X_1), bush OD (X_2), difference between C-channel inside diameter (ID) and bush OD (X_3), difference between C-channel ID and A-channel ID (X_4), bevel angle (X_5), A-channel length (X_6), C-channel length (X_7). The response is the disassembly force in newtons. Fictitious data from DEX users' manual (the example is courtesy of British Viggo)

X_1	X_2	X_3	X_4	X_5	X_6	X_7	Force
1.7	1.6	0.25	0	45	3.5	1.5	23.3
1.8	1.6	0.25	0	45	3.5	1.5	24.3
1.7	1.6	0.35	0	45	3.5	2.5	30.5
1.8	1.6	0.35	0	45	3.5	2.5	27.8
1.7	1.6	0.25	0.1	45	4.5	2.5	29.9
1.8	1.6	0.25	0.1	45	4.5	2.5	29.9
1.7	1.6	0.35	0.1	45	4.5	1.5	22.9
1.8	1.6	0.35	0.1	45	4.5	1.5	21.0
1.7	1.6	0.25	0	75	4.5	2.5	26.3
1.8	1.6	0.25	0	75	4.5	2.5	27.8
1.7	1.6	0.35	0	75	4.5	1.5	25.3
1.8	1.6	0.35	0	75	4.5	1.5	24.6
1.7	1.6	0.25	0.1	75	3.5	1.5	22.8
1.8	1.6	0.25	0.1	75	3.5	1.5	22.1
1.7	1.6	0.35	0.1	75	3.5	2.5	28.0
1.8	1.6	0.35	0.1	75	3.5	2.5	27.2
1.7	1.8	0.25	0	45	4.5	2.5	28.9
1.8	1.8	0.25	0	45	4.5	2.5	28.6
1.7	1.8	0.35	0	45	4.5	1.5	24.0
1.8	1.8	0.35	0	45	4.5	1.5	25.6
1.7	1.8	0.25	0.1	45	3.5	1.5	21.5
1.8	1.8	0.25	0.1	45	3.5	1.5	22.3
1.7	1.8	0.35	0.1	45	3.5	2.5	30.5
1.8	1.8	0.35	0.1	45	3.5	2.5	29.2
1.7	1.8	0.25	0	75	3.5	1.5	24.9
1.8	1.8	0.25	0	75	3.5	1.5	23.5
1.7	1.8	0.35	0	75	3.5	2.5	30.6
1.8	1.8	0.35	0	75	3.5	2.5	29.3
1.7	1.8	0.25	0.1	75	4.5	2.5	27.9
1.8	1.8	0.25	0.1	75	4.5	2.5	29.3
1.7	1.8	0.35	0.1	75	4.5	1.5	24.7
1.8	1.8	0.35	0.1	75	4.5	1.5	24.0
1.7	1.7	0.30	0.05	60	4.0	2.0	28.4
1.8	1.7	0.30	0.05	60	4.0	2.0	25.0
1.75	1.6	0.30	0.05	60	4.0	2.0	27.9
1.75	1.8	0.30	0.05	60	4.0	2.0	25.8
1.75	1.7	0.20	0.05	60	4.0	2.0	26.4
1.75	1.7	0.40	0.05	60	4.0	2.0	29.4
1.75	1.7	0.30	0	60	4.0	2.0	25.7
1.75	1.7	0.30	0.1	60	4.0	2.0	29.1
1.75	1.7	0.30	0.05	30	4.0	2.0	24.1
1.75	1.7	0.30	0.05	90	4.0	2.0	25.0
1.75	1.7	0.30	0.05	60	3.0	2.0	27.4
1.75	1.7	0.30	0.05	60	5.0	2.0	28.3
1.75	1.7	0.30	0.05	60	4.0	0.5	13.4
1.75	1.7	0.30	0.05	60	4.0	3.5	32.2
1.75	1.7	0.30	0.05	60	4.0	2.0	28.2

Table D.12 Wave heights (in millimetres relative to still water level) from a wave tank. Sampled at 0.1 s intervals over 39.7 s. Read along the rows to obtain the time series. Data courtesy of G. E. Hearn, University of Newcastle.

367	407	−255	−515	−500	−342	−188	77	494	737	375
−221	−313	−301	−311	−109	−10	150	178	47	47	767
−74	−594	−541	−133	148	116	169	417	295	−317	−472
−266	−12	314	241	375	59	−550	−439	−121	367	478
4	−315	−3	194	−45	50	136	−42	−296	−394	−145
209	536	116	136	−167	−244	18	−33	−204	−20	120
−89	−83	176	160	166	94	−65	−311	−430	−398	199
659	488	22	−258	−400	−162	−196	29	721	118	−356
−340	1	166	138	93	93	−148	−326	−95	279	−10
−99	96	227	18	61	−125	−460	−337	59	211	73
95	119	−1	−123	−56	−26	92	125	−362	−324	116
438	235	−120	−209	36	−132	−201	192	310	−116	−282
−172	−41	204	34	−12	276	167	−328	−260	163	−62
−204	−107	172	469	50	−96	−62	−40	342	−205	−488
−363	138	323	189	120	64	326	−71	−306	−150	17
−112	−296	−238	190	578	268	−33	12	−148	−132	−163
−389	−81	328	19	−32	268	380	−26	−245	−309	−289
30	125	170	141	−7	−114	−29	134	104	−65	−365
−128	221	76	−29	113	64	32	296	−201	−513	−339
167	167	255	271	122	−500	−416	129	195	9	105
343	−136	−522	−196	480	159	−77	−54	−107	−102	−14
324	137	−312	−236	−28	239	176	−24	29	−41	−184
−331	76	276	−111	−171	−53	226	4	−114	171	354
−133	−345	−218	−307	−270	93	574	622	45	−162	−51
12	−312	−517	−265	−33	474	529	−97	−160	127	191
−281	−315	−20	24	−120	−34	260	178	164	322	−293
−522	−468	28	425	365	41	−114	6	−117	−134	−173
−19	180	239	−56	93	−94	−340	−362	189	756	600
−120	−622	−752	−526	115	875	607	−44	−349	−478	−321
30	253	511	30	−249	−262	−56	46	115	183	2
15	−85	−234	−245	72	441	282	−92	−237	−222	−37
90	−20	−78	−56	139	530	−68	−437	−351	94	351
124	0	46	48	−366	−429	−78	396	235	172	−161
−186	10	283	58	−152	−160	−131	−52	38	116	124
−48	−19	276	−29	−331	−382	−158	162	316	375	119
−174	−120	−173	−41	−50	−251	−262	154	391	266	231
177										

Appendix E: Statistical tables

Table E.1 Random digits

19211	73336	80586	08681	28012	48881	34321	40156	03776	45150
94520	44451	07032	36561	41311	28421	95908	91280	74627	86359
70986	03817	40251	61310	25940	92411	34796	85416	00993	99487
65249	79677	03155	09232	96784	17126	50350	86469	41300	62715
82102	03098	01785	00653	39438	43660	02406	08404	24540	80000
91600	94635	35392	81737	01505	04967	91097	02011	26642	38540
20559	85361	20093	46000	83304	96624	62541	41722	79676	98970
53305	79544	99937	87727	32210	19438	58250	77265	02998	02973
57108	86498	14158	60697	41673	18087	46088	11238	82135	79035
08270	11929	92040	37390	71190	58952	98702	41638	95725	22798
90119	23206	75634	60053	90724	29080	69423	66815	11896	18607
45124	69607	17078	61747	15891	69904	79589	68137	19006	19045
83084	02589	37660	63882	99025	34831	92048	23671	68895	73795
06485	31035	93828	16159	05015	54800	76534	22974	13589	01801
61349	04538	89318	27693	02674	34368	24720	40682	20940	37392
14082	65020	49956	01336	41685	01758	49242	52122	01030	60378
82615	53477	58014	62229	72640	32042	73521	14166	45850	02372
50942	78633	16588	19275	62258	20773	67601	93065	69002	03985
76381	77455	81218	02520	22900	80130	61554	98901	26939	78732
05845	35063	85932	22410	31357	54790	39707	94348	11969	89755
78591	83750	46137	74989	39931	33068	35155	49486	28156	04556
31945	87960	04852	41411	63105	44116	95250	04046	59211	07270
08648	89822	04170	38365	23842	61917	57453	03495	61430	20154
32511	07999	18920	77045	44299	85057	51395	17457	24207	02730
79348	56194	58145	88645	84867	41594	28148	84985	89949	26689
61973	03660	32988	70689	17794	61340	58311	32569	23949	85626
92032	60127	34066	28149	22352	12907	53788	86648	57649	07887
74609	71072	63958	58336	67814	40598	12626	30754	75895	42194
98668	76074	25634	56913	88254	41647	05398	69463	49778	31382
85248	72078	58634	88678	21764	67940	45666	84664	35714	43081

Table E.1 (*continued*)

82002	96916	94138	74739	99122	03904	46052	97277	60234	37424
79100	55938	23211	10111	17115	90577	94202	01063	85522	64378
30923	71710	70257	05596	42310	02449	31211	50025	99744	78084
90513	50966	78981	70391	45932	13535	21681	66589	94915	08855
94474	79358	16098	65806	79252	14190	88722	39887	15553	58386
65236	62984	19968	22071	49898	96140	80264	57580	56775	63138
80502	04192	84287	32589	50664	63846	71590	67220	71503	27942
01315	04632	50202	89148	41556	11584	35916	13979	25016	32511
81525	76670	88714	28681	56540	84963	85543	69715	86192	79373
19500	41720	79214	20079	42053	29844	02294	11306	78537	65098
25812	77090	45198	98162	13782	60596	99092	50188	65405	63227
80859	94220	92309	01998	45090	24815	13415	86989	01677	39092
41107	33561	04376	40072	78909	61042	04098	73304	21892	63112
00465	00858	22774	80730	07098	80515	09970	40476	10314	24792
58137	02454	15657	24957	48401	02940	92828	26372	31071	58192
32013	97147	69725	78867	73329	74935	69276	46001	04181	38838
17048	84788	12531	01773	43551	34586	61239	87927	03232	31312
33935	07944	98456	11922	96174	24100	00307	85697	06527	34381
47633	49394	38673	22281	68096	76599	38462	16662	81959	03358
82161	92521	10712	58839	18546	32920	89220	90493	73725	22327
99050	30876	80821	14955	11495	25666	37656	91874	93051	64664
08090	84688	36332	86858	73763	62534	93378	54809	97076	09077
67619	00352	32735	56954	97851	57350	33068	35393	75938	86086
63779	66008	02516	93878	67930	38445	44166	20168	55128	65337
03259	72119	04797	95593	02754	87120	68167	04455	75318	93127
92914	02066	97320	00328	51685	89729	27446	32599	82486	01718
80001	70542	01530	63033	64348	01306	75419	90348	34717	05147
38715	09824	86504	14817	74434	80450	95086	73824	40550	14266
15987	74578	12779	69608	76893	94840	36853	00568	35697	00783
06193	94893	24598	02714	69670	06153	97835	71087	58193	97912
40134	12803	33942	46660	05681	35209	65980	77899	38988	75580
88480	27598	48458	65369	81066	02000	68719	90488	50062	10428
49989	94369	80429	97152	67032	62342	96496	91274	71264	45271
62089	52111	92190	85413	95362	33400	03488	84666	99974	01459
01675	12741	94334	86069	71353	85568	16632	97577	18708	99550
04529	19798	47711	63262	06316	00287	86718	33705	31645	70615
63895	63087	91886	43467	55559	35912	39429	18933	75931	18924
17709	21642	56384	85699	24310	85043	00405	59820	54228	58645
11727	83872	22553	17012	02849	39794	50662	32647	67676	95488
02838	03160	92864	29385	63585	46055	41356	96398	70904	87103

Source: Chatfield (1983)

Table E.2 Areas under the normal distribution pdf

	0.00	0.01	0.02	0.03	0.04	0.05	0.06	0.07	0.08	0.09
0.0	0.5000	0.5040	0.5080	0.5120	0.5160	0.5199	0.5239	0.5279	0.5319	0.5359
0.1	0.5398	0.5438	0.5478	0.5517	0.5557	0.5596	0.5636	0.5675	0.5714	0.5753
0.2	0.5793	0.5832	0.5871	0.5910	0.5948	0.5987	0.6026	0.6064	0.6103	0.6141
0.3	0.6179	0.6217	0.6255	0.6293	0.6331	0.6368	0.6406	0.6443	0.6480	0.6517
0.4	0.6554	0.6591	0.6628	0.6664	0.6700	0.6736	0.6772	0.6808	0.6844	0.6879
0.5	0.6915	0.6950	0.6985	0.7019	0.7054	0.7088	0.7123	0.7157	0.7190	0.7224
0.6	0.7257	0.7291	0.7324	0.7357	0.7389	0.7422	0.7454	0.7486	0.7517	0.7549
0.7	0.7580	0.7611	0.7642	0.7673	0.7703	0.7734	0.7764	0.7793	0.7823	0.7852
0.8	0.7881	0.7910	0.7939	0.7967	0.7995	0.8023	0.8051	0.8078	0.8106	0.8133
0.9	0.8159	0.8186	0.8212	0.8238	0.8264	0.8289	0.8315	0.8340	0.8365	0.8389
1.0	0.8413	0.8438	0.8461	0.8485	0.8508	0.8531	0.8554	0.8577	0.8599	0.8621
1.1	0.8643	0.8665	0.8686	0.8708	0.8729	0.8749	0.8770	0.8790	0.8810	0.8830
1.2	0.8849	0.8869	0.8888	0.8906	0.8925	0.8943	0.8962	0.8980	0.8997	0.9015
1.3	0.9032	0.9049	0.9066	0.9082	0.9099	0.9115	0.9131	0.9147	0.9162	0.9171
1.4	0.9192	0.9207	0.9222	0.9236	0.9251	0.9265	0.9279	0.9292	0.9306	0.9319
1.5	0.9332	0.9345	0.9357	0.9370	0.9382	0.9394	0.9406	0.9418	0.9429	0.9441
1.6	0.9452	0.9463	0.9474	0.9484	0.9495	0.9505	0.9515	0.9525	0.9535	0.9545
1.7	0.9554	0.9564	0.9573	0.9582	0.9591	0.9599	0.9608	0.9616	0.9625	0.9633
1.8	0.9641	0.9648	0.9656	0.9664	0.9671	0.9678	0.9686	0.9693	0.9699	0.9706
1.9	0.9713	0.9719	0.9726	0.9732	0.9738	0.9744	0.9750	0.9756	0.9761	0.9767
2.0	0.9772	0.9778	0.9783	0.9788	0.9793	0.9798	0.9803	0.9808	0.9812	0.9817
2.1	0.9821	0.9826	0.9830	0.9834	0.9838	0.9842	0.9846	0.9850	0.9854	0.9857
2.2	0.9861	0.9864	0.9868	0.9871	0.9875	0.9878	0.9881	0.9884	0.9887	0.9890
2.3	0.9893	0.9896	0.9898	0.9901	0.9904	0.9906	0.9909	0.9911	0.9913	0.9916
2.4	0.9918	0.9920	0.9922	0.9924	0.9927	0.9930	0.9930	0.9932	0.9934	0.9936
2.5	0.9938	0.9940	0.9941	0.9943	0.9945	0.9946	0.9948	0.9949	0.9951	0.9952
2.6	0.9853	0.9955	0.9956	0.9957	0.9959	0.9960	0.9961	0.9962	0.9963	0.9964
2.7	0.9865	0.9966	0.9967	0.9968	0.9969	0.9970	0.9971	0.9972	0.9973	0.9974
2.8	0.9974	0.9975	0.9976	0.9977	0.9977	0.9978	0.9979	0.9979	0.9980	0.9981
2.9	0.9981	0.9982	0.9982	0.9983	0.9984	0.9984	0.9985	0.9985	0.9986	0.9986
3.0	0.9986	0.9987	0.9987	0.9988	0.9988	0.9989	0.9989	0.9989	0.9990	0.9990
3.1	0.9990	0.9991	0.9991	0.9991	0.9992	0.9992	0.9992	0.9992	0.9993	0.9993
3.2	0.9993	0.9993	0.9994	0.9994	0.9994	0.9994	0.9994	0.9995	0.9995	0.9995
3.3	0.9995	0.9995	0.9995	0.9996	0.9996	0.9996	0.9996	0.9996	0.9996	0.9996
3.4	0.9997	0.9997	0.9997	0.9997	0.9997	0.9997	0.9997	0.9997	0.9997	0.9998
3.5	0.9998	0.9998	0.9998	0.9998	0.9998	0.9998	0.9998	0.9998	0.9998	0.9998
3.6	0.9998	0.9998	0.9998	0.9999	0.9999	0.9999	0.9999	0.9999	0.9999	0.9999

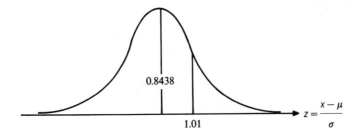

0.8438

1.01

$$z = \frac{x - \mu}{\sigma}$$

Source: Chatfield (1983)

Table E.3 Percentage points of the standard normal distribution

100α	z_α	100α	z_α	100α	z_α	100α	z_α	100α	z_α	100α	z_α
50	0.000	5.0	1.645	3.0	1.881	2.0	2.054	1.0	2.326	0.10	3.090
45	0.126	4.8	1.665	2.9	1.896	1.9	2.075	0.9	2.366	0.09	3.121
40	0.253	4.6	1.685	2.8	1.911	1.8	2.097	0.8	2.409	0.08	3.156
35	0.385	4.4	1.706	2.7	1.927	1.7	2.120	0.7	2.457	0.07	3.195
30	0.524	4.2	1.728	2.6	1.943	1.6	2.144	0.6	2.512	0.06	3.239
25	0.674	4.0	1.751	2.5	1.960	1.5	2.170	0.5	2.576	0.05	3.290
20	0.842	3.8	1.774	2.4	1.977	1.4	2.197	0.4	2.652	0.01	3.719
15	1.036	3.6	1.799	2.3	1.995	1.3	2.226	0.3	2.748	0.005	3.891
10	1.282	3.4	1.825	2.2	2.014	1.2	2.257	0.2	2.878	0.001	4.265
5	1.645	3.2	1.852	2.1	2.033	1.1	2.290	0.1	3.090	0.0005	4.417

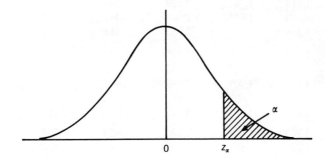

Table E.4 Percentage points of Student's t-distribution

α	0.10	0.05	0.025	0.01	0.005	0.001
ν						
1	3.078	6.314	12.706	31.821	63.657	318.310
2	1.886	2.920	4.303	6.965	9.925	22.327
3	1.638	2.353	3.182	4.541	5.841	10.215
4	1.533	2.132	2.776	3.747	4.604	7.173
5	1.476	2.015	2.571	3.365	4.032	5.893
6	1.440	1.943	2.447	3.143	3.707	5.208
7	1.415	1.895	2.365	2.998	3.499	4.785
8	1.397	1.860	2.306	2.896	3.355	4.501
9	1.383	1.833	2.262	2.821	3.250	4.297
10	1.372	1.812	2.228	2.764	3.169	4.144
11	1.363	1.796	2.201	2.718	3.106	4.025
12	1.356	1.782	2.179	2.681	3.055	3.930
13	1.350	1.771	2.160	2.650	3.012	3.852
14	1.345	1.761	2.145	2.624	2.977	3.787
15	1.341	1.753	2.131	2.602	2.947	3.733
16	1.337	1.746	2.120	2.583	2.921	3.686
17	1.333	1.740	2.110	2.567	2.898	3.646
18	1.330	1.734	2.101	2.552	2.878	3.610
19	1.328	1.729	2.093	2.539	2.861	3.579
20	1.325	1.725	2.086	2.528	2.845	3.552
21	1.323	1.721	2.080	2.518	2.831	3.527
22	1.321	1.717	2.074	2.508	2.819	3.505
23	1.319	1.714	2.069	2.500	2.807	3.485
24	1.318	1.711	2.064	2.492	2.797	3.467
25	1.316	1.708	2.060	2.485	2.787	3.450
26	1.315	1.706	2.056	2.479	2.779	3.435
27	1.314	1.703	2.052	2.473	2.771	3.421
28	1.313	1.701	2.048	2.467	2.763	3.408
29	1.311	1.699	2.045	2.462	2.756	3.396
30	1.310	1.697	2.042	2.457	2.750	3.385
40	1.303	1.684	2.021	2.423	2.704	3.307
60	1.296	1.671	2.000	2.390	2.660	3.232
120	1.289	1.658	1.980	2.258	2.617	3.160
∞	1.282	1.645	1.960	2.326	2.576	3.090

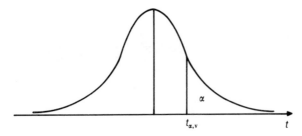

Source: Chatfield (1983)

Table E.5 Control chart factors for the sample range

Sample size	Lower percentage factors		Upper percentage factors	
	0.1%	2.5%	2.5%	0.1%
2	0.00	0.04	3.17	4.65
3	0.06	0.30	3.68	5.06
4	0.20	0.59	3.98	5.31
5	0.37	0.85	4.20	5.48
6	0.53	1.07	4.36	5.62
7	0.69	1.25	4.49	5.73
8	0.83	1.41	4.60	5.82
9	0.97	1.55	4.70	5.90
10	1.08	1.67	4.78	5.67

Multiply the estimate of the standard deviation by the tabled factors, which are based on a normal distribution.

Appendix F: Answers to selected exercises

2.1 (i) 4/25 (ii) 1/10 (iii) These three outcomes are not equally likely.

2.2 (i) 0.33 (ii) 0.24 (iii) 0.50 (iv) 0.43

2.3 (i) 12/17 (ii) 132/272 (iii) 20/272 (iv) 120/272 (v) 252/272

2.4 (i) 0.0199 (ii) 0.0297 (iii) $1 - 0.99^n$

2.5 (i) 0.80 (ii) 5/6 (iii) 5/7

2.6 $1 - c(a + b - ab)$

2.7 $c(1 - 2q^2 + q^4) + (1 - c)(1 - (2p - p^2)^2)$

2.8 The probabilities of successful flights for two and four engines are $1 - q^2$ and $1 - 4q^3 + 3q^4$. The former would only be greater if q exceeded $1/3$, which is rather unlikely!

2.9 $P(0) = 1 - 2q + aq$, $P(1) = 2q(1 - a)$, $P(2) = aq$.

2.10 7/17, 4/16 and 5/16 for A, B and C respectively.

2.13 (i) 5040 (ii) 2520 (iii) 720 (iv) 360

2.14 35

3.1 $\bar{x} = 83.3$, $\hat{\sigma}^2$ (divisor n) $= 12.51$, s^2(divisor $n - 1$) $= 13.17$, $\hat{\sigma} = 3.54$, $s = 3.63$, median $= 84$, range $= 14$.

3.2 (i) Remember that heights of bars should be relative frequency divided by the length of the class interval. The heighest is 0.0208.

   (ii) Remember to use right-hand end points of intervals – for example, 2.5% less than 80,..., 100% less than 240.

   (iii) Remember to use mid-points; $\bar{x} = 118.3$ and $s = 25.4$.

   (iv) median $= 115$, $LQ = 103$, $UQ = 129$, interquartile range $= 26$.

3.3 $49.1\ \text{km h}^{-1}$ (harmonic mean of 30, 60 and 90).

3.5 The geometric mean $((1.1)(1.5)(1.2))^{1/3} = 1.256$ irrespective of the order.

3.9 (ii) $\bar{x} = 0$, $s = 1$, $\hat{\gamma} = 1.2/(1)^3$, $\hat{k} = 4.2/1)^3$

3.10 (ii) $\bar{x} = 16.5$, $s = 10.6$, $\hat{\gamma} = 1.2$. (iii) skewness becomes -0.4.

4.1 0.092

4.2   (i) 0.071   (ii) 0.982   (iii) 0.302

4.5   0.1167, 0.1148 (the error is only 1.6%), 0.0041.

4.6   0.199

4.7   (i) 0.0107   (ii) 0.9599   (iii) 0.9544   (iv) 0.0026   (v) 1.282   (vi) 2.326
      (vii) 2.575   (viii) 0.9878   (ix) 0.102   (x) 0.9487

4.8   (i) 58%   (ii) 67% and 55%

4.9 $14.6 \, \text{N} \, \text{mm}^{-2}$

4.10 72.68 mm

4.12 (i) 0.096   (ii) 0.395   (iii) 0.634   (iv) 69   (v) $1 - (1 - 1/T)^n$

4.14 0.2212

4.16 C_p is 0.58 but the process is 'capable'.

4.17 (i) $\alpha = 6/A^3$ (iii) $219A^2$ (iv) $183A^2$

4.19 $f(t) = \lambda(\lambda t)^{k-1} e^{-\lambda t}/(k-1)!$
      If $k = 1$ this is the exponential distribution.

5.1 $r = -0.61$

5.2   0.996

5.3   0.090

5.4   0.057

5.5   0.672

5.6 $\alpha = 0.125$

5.7   Mean and standard deviation are 23.3 m and 9.4 m, respectively. A
      distance of 45 m will be exceeded by 1% of cars. If a lognormal distribution
      is assumed the natural logarithm of stopping distance has a mean of 3.074
      and a standard deviation of 0.3866. The distance becomes 53 m.

5.8   It is reasonable to assume E is independent of L and it follows that

$$\sigma_M^2 = \sigma_L^2 + \sigma_E^2$$

Hence

$$\sigma_L^2 = \sigma_M^2 - \sigma_E^2$$

In the numerical case, σ_L equals 17 and $\rho(M, E)$ equals $\frac{1}{3}$.

5.10 (i) $\frac{1}{8}$
      (ii) $f(x) = x + 0.5$, $0 \leqslant x \leqslant 1$, and similarly for y by the symmetry of the
          problem.
      (iii) $\frac{3}{8}$

5.14 Approximately 0.75% in both cases.

5.16 $f(x) = 4x^3$, $0 \leqslant x \leqslant 1$
      $f(y) = 4y(1 - y^2)$, $0 \leqslant y \leqslant 1$

5.17 (i) $3/(2e)$ (ii) $F(t) = 1 - (1 + t/2 + t^2/8) e^{-t/2}$ (iii) $(t^2/16) e^{-t/2}$

6.1   95% and 90% confidence intervals are 2.047 ± 0.047 and 2.047 ± 0.040, respectively.

6.2   A 95% confidence interval for the mean is [1.65, 2.55] assuming a random sample from a near-normal distribution. A rather approximate confidence interval for the standard deviation is [0.27, 0.81].

6.3   (i) 80%
      (ii) $n = 325$. It is only an approximation because we are estimating the population standard deviation by s.

6.4   [0.52, 0.63]

6.5   (i) 7.4 ± 15 (ii) $[-90, -46]$ (iii) $[-91, -45]$

6.6 $[-0.90, -0.02]$. Test 25 pavers.

6.7 0.67 ± 0.13. There is some evidence that the standard deviation exceeds specification.

6.8   [0.010, 0.158]

6.9   If the campaign has had no effect the number changing preference $X \sim \text{Bin}(70, 0.5)$. Then

$$\Pr(X \geqslant 42) = 0.06$$

There is only very weak evidence of any effect.

6.10 (a) -15 ± 14 (b) $(3.7 - 1.5) \pm 1.7\sqrt{(0.37 + 0.75)}$, i.e. 2.2 ± 1.8

6.11 (a) $\sigma^2(1/n + 5.43/(2n))$
      (b) (i) 2.88 ± 0.90 [8, 39]
          (ii) -1.56 ± 1.01 (using $t_{8, 0.05}$ which equals 1.86).

8.1   (a) (ii) $y = -8.24 + 10.511x$ (iv) [46, 114]
      (b) (ii) $\ln y = 1.6569 + 0.34674x$
          (iv) [26, 358] (predicted median 97)
      (c) $\hat{b} = \Sigma xy / \Sigma x^2$
          $y = 9.258x$
          prediction for 8400 properties is 78

8.2   (a) (ii) $r = 0.572$, 90% confidence interval for ρ is $[-0.09, 0.88]$
          (iii) 6.64 ± 7.55
      (b) (ii) $r = 0.510$, 90% confidence interval for ρ is $[-0.17, 0.86]$,
          (iii) 5.64 ± 7.55

8.3 $P(x|y=2) = \begin{cases} 0.2 & x = 0, 1, 2 \\ 0.4 & x = 3 \end{cases}$

8.4 $P(x) = (0.5)^{x+1}$, for $x = 0, 1, 2, \ldots$

8.5   (i) $P(x, y) = 1/x$, for $0 \leqslant y \leqslant x \leqslant 1$
      (ii) $\frac{1}{2}$
      (iii) $\ln 2 / \ln 4 = \frac{1}{2}$

8.7   Estimate the logarithms of costs from a linear regression. When you take exponential, remember to multiply by $\exp(s^2/2)$: this allows for the difference between the median and mean of the lognormal distribution (see section 4.2.6).

8.8   (i) $\hat{\theta}_1 = (2Y_1 + Y_2 + Y_3)/3$, $\hat{\theta}_2 = (Y_1 + 2Y_2 - Y_3)/3$
$\hat{\theta}_3 = (Y_1 - Y_2 + 2Y_3)/3$. Variance is $2\sigma^2/3$.

9.1   1.000 (to 3 decimal places), 0 (exactly).

9.2   (ii) $y = 9.70 - 0.123x$
(iii) $y = 12.1 - 1.73x + 0.160x^2$

(iv) $dy/dx = 0$ when $x = 5.4$

9.4   Regression on x_1: $s = 60.52$, $R^2 = 54.4\%$, $R^2_{adj} = 52.8\%$
Regression on x_1 and x_2: $s = 36.49$, $R^2 = 84.0\%$, $R^2_{adj} = 82.8\%$

10.1 (ii) $4/\sqrt{2} = 2.83$. Change x_2 by 1/3, and the ratio is $(3 + 1/3)/\sqrt{(1 + 1/9)} = 3.16$.

10.3 The biggest difference in standard deviations is between types A and D, and an approximate 99% confidence interval is:

$$(4.83 - 2.76) \pm 2.58\sqrt{(4.83)^2/20 + (2.76)^2/20))}$$

which becomes

$$2.07 \pm 3.21$$

None of the six such 99% confidence intervals excludes 0, and there is no strong evidence of a difference in standard deviations.

10.4 (i) 24.5, 28.5 and 25.25 for mines one, two and three respectively. The standard deviation of these three data is 2.126.
(iii) $\sqrt{[(2.126)^2 - (2.102)^2/4]} = 1.85$

10.6 A plot of grain size (y) against stirring rate shows a clear quadratic relationship. If x_1 is coded stirring rate, (rate in rpm $- 15$)/5, and indicator variables x_2 and x_3 correspond to furnace B and C

$$y = 82.1 + 0.620x_1 - 0.266x_1^2 + 6.20x_2 - 4.20x_3$$

The standard deviations of the four coefficients are 0.158, 0.027, 2.73 and 2.73, respectively. There is very strong evidence of a quadratic effect of stirring rate on grain size. Furnace B seems to be associated with higher y-values than furnace C and to a lesser extent A.

References

Abbess, C., Jarrett, D. and Wright, C. C. (1983) Bayesian methods applied to accident blackspot studies. *The Statistician*, **32**, 181–87.

Adamson, P. T. (1989) Robust and exploratory data analysis in arid and semi-arid hydrology. *Department of Water Affairs Technical Report TR 138*, Pretoria.

Anderson, V. L. and McLean, R. A. (1974) *Design of Experiments: a realistic approach*, Marcel Dekker, New York.

Archer, D. R. (1981) Seasonality of flood and the assessment of seasonal flood risk. *Proceedings of the Institution of Civil Engineers*, Part 2, **70**, 1023–35.

Atkinson, A. C. and Donev, A. N. (1992) *Optimum Experimental Designs*, Oxford Scientific Publications, Oxford.

Avon County Council (1984) Safety of Pedestrians and Pedal Cyclists at small and mini-roundabouts. County Engineer and Surveyor.

Bajpai, A. C., Calus, I. M. and Fairley, J. A. (1968) *Statistical Methods for Engineers and Scientists*, Wiley, London.

Barnett, V. (1974) *Elements of Sampling Theory*, English University Press, London.

Bayes, T. (1763) An essay towards solving a problem in the doctrine of chances, *Philosophical Transactions of the Royal Society*, **53**, 370–418. Reprinted with a biographical note by G. A. Barnard in *Biometrika*, 1958, **45**, 293–315.

Box, G. E. P. and Draper, N. R. (1969) *Evolutionary Operation: a statistical method for process improvement*, Wiley, New York.

Box, G. E. P., Hunter, W. G. and Hunter, J. S. (1978) *Statistics for Experimenters*, Wiley, New York.

Central Statistical Office (1987) *Social Trends 17*, HMSO, London.

Central Statistical Office (1992) *Social Trends 22*, HMSO, London.

Chatfield, C. (1983) *Statistics for Technology*, Chapman and Hall, London.

Cho, H. R. and Chan, D. S. T. (1987) Mesoscale atmospheric dynamics and modelling rainfall fields. *Journal of Geophysical Research*, **98**(D8), 9687–92.

County Surveyors' Society Standing Advisory Group on Accident Reduction (1987) Small and mini-roundabouts, Report No. 1/4.

Cowpertwait, P. S. P. (1991) Further developments of the Neyman–Scott clustered point process for modeling rainfall. *Water Resources Research*, **27**(7), 1431–38.

Cowpertwait, P. S. P., Metcalfe, A. V., O'Connell, P. E. Mawdsley, J. A. and Threlfall, J. L. (1991) Stochastic generation of rainfall time series. *Foundation for Water Research*, Report Number FR0217.

Cox, D. R. (1972) Regression models and life tables (with Discussion). *Journal of the Royal Statistical Society B*, **34**, 187–220.

Dalal, S. R., Fowlkes, E. B. and Hoadley, B. (1989) Risk analysis of the space shuttle:

pre-Challenger prediction of failure. *Journal of the American Statistical Association* **84**, 945–57.

Dally, J. W., Riley, W. F. and McConnell, K. G. (1984) *Instrumentation for Engineering Measurements*, Wiley, New York.

David, F. N. (1955) Dicing and gaming (a note on the history of probability). *Biometrika*, **42**(1), 1–15.

Daw, R. H. and Pearson, E. S. (1972) Studies in the History of Probability and Statistics. XXX. Abraham De Moivre's 1733 derivation of the normal curve: a bibliographical note. *Biometrika*, **59**(3), 677–80.

Devore, J. L. (1982) *Probability and Statistics for Engineering and the Sciences*, Brooks/Cole, Monterey, California.

Dodd, E. L. (1923) The greatest and the least variate under general laws of error. *Transactions of the American Mathematical Society*, **25**, 525–39.

DuMouchel, W. (1989) Bayesian Metaanalysis, in *Statistical Methodology in the Pharmaceutical Sciences* (ed. D. A. Berry), Marcel Dekker, New York, pp. 509–29.

Efthimiadu, I., Tham, M. T. and Willis, M. J. (1993) Engineering control and product quality assurance in the process industries, in *Proceedings of the First Newcastle International Conference on Quality and its Applications* (ed. J. F. L. Chan), Penshaw Press, Newcastle upon Tyne.

Ettrick, T. M., Mawdsley, J. A. and Metcalfe, A. V. (1987) The influence of catchment antecedent conditions on seasonal flood risk. *Water Resources Research*, **23**(3), 481–88.

Fisher, R. A. (1921) On the probable error of a coefficient of correlation deduced from a small sample. *Metron*, **1**(4), 1–32.

Fisher, R. A. and Tippett, L. H. C. (1928) Limiting forms of the frequency distribution of the largest or smallest member of a sample. *Proceedings of the Cambridge Philosophical Society*, **24**, 180–90.

Futter, M. R. (1990) Predicting short term flood risks. Ph.D. thesis, University of Newcastle upon Tyne.

Futter, M. R., Mawdsley, J. A. and Metcalfe, A. V. (1991) Short term flood risk prediction: a comparison of the Cox regression model and a conditional distribution model. *Water Resources Research*, **27**(7), 1649–56.

Gerald, C. F. and Wheatley, P. O. (1989) *Applied Numerical Analysis*, 4th edn, New York, Addison-Wesley.

Ghani, A. A. A. and Metcalfe, A. V. (1985) Spectral predictions of flood risk, in *Proceedings of 4th International Symposium on Multivariate Analysis of Hydrological Processes*, 15–17 July 1985, Colorado State University, Fort Collins, pp. 744–54.

Green, H. (1977) Accidents at off-side priority roundabouts with mini or small islands. TRRL Laboratory Report 774. Transport and Road Research Laboratory, Department of the Environment, Crowthorne.

Greenfield, A. A. (1974) Theoretical basis to the study of fatigue failure time in the rolling four ball test. Paper no. CDL/MT/6/74, Metals Technology Unit, Corporate Development Laboratory, British Steel Corporation, Sheffield.

Greenfield, A. A. (1993) Communicating statistics. *Journal of the Royal Statistical Society A*, **156**(2) 287–97.

Greenfield, A. A. and Siday, S. (1980) Statistical computing for business and industry. *The Statistician*, **29**(1), 33–55.

Hanson, S. J. and Wilson, C. D. L. (1979) The safety of small roundabouts in Newcastle upon Tyne. Unpublished internal report by Transport Operations Research Group,

University of Newcastle upon Tyne, in collaboration with the Tyne and Wear County Council.

Hauer, E. (1980) Bias-by-selection: overestimation of the effectiveness of safety counter-measures caused by the process of selection for treatment. *Accident Analysis and Prevention*, **12**, 113–17.

Hauer, E. (1983) An application of the likelihood/Bayes' approach to the estimation of safety countermeasure effectiveness. *Accident Analysis and Prevention*, **15**(4), 287–98.

Hauer, E. (1989a) Comparison Systems in Road Safety Studies: An Analysis. Transport Safety Studies Group, Department of Civil Engineering, University of Toronto.

Hauer, E. (1989b) Comparison Systems in Road Safety Studies: An Empirical Inquiry. Transport Safety Studies Group, Department of Civil Engineering, University of Toronto.

Hill, I. D. (1962) Sampling inspection and defence specification DEF-131. *Journal of the Royal Statistical Society A*, **125**, 31–87.

Hosking, R. M. (1981) Fractional differencing. *Biometrika*, **68**(1), 165–76.

Hosking, R. M. (1984) Modelling persistence in hydrology. *Water Resources Research*, **20**(12), 1898–1908.

Institution of Highways and Transportation (1991) *Urban Safety Management Guidelines*, Institute of Highways and Transportation in collaboration with the Department of Transport, London.

Jennison, C. and Turnbull, B. W. (1989) Interim analyses: the repeated confidence interval approach (with discussion). *Journal of the Royal Statistical Society B*, **51**(3), 305–61.

Kamarulzaman bin Ibrahim and Metcalfe, A. V. (1993) Bayesian overview for evolution of mini-roundabouts as a road safety measure. Accepted for publication in *The Statistician*.

Kattan, M. R. (1993) Statistical process control in ship production. *Quality Forum*, **19**(2), 88–92.

Kendall, M. and Plackett, R. L. (1977) *Studies in the History of Statistics and Probability Vol II*, Charles Griffin, London, pp. 26–29, 63–66.

Kotz, S. and Johnson, N.L. (eds) (1982) *Encyclopedia of Statistical Sciences*, Wiley, New York.

Lalani, N. (1975) Introduction of mini, small and large roundabouts at major/minor priority junctions – impact on accidents. Greater London Safety Unit.

Laurence, C. J. D. (1980) Roundabouts – evolution, revolution and the future. *The Highway Engineer*, **27**(5), 2–10.

Leadbetter, M. R., Lindgren, G. and Rootzén, H. (1982) *Extremes and Related Properties of Random Sequences and Processes*, Springer-Verlag, New York.

Lee, P. M. (1989) *Bayesian Statistics*, Edward Arnold, London.

Lindley, D. V. (1985) *Making Decisions*, 2nd edn, John Wiley, London.

Lindley, D. V. and Scott, W. F. (1984) *New Cambridge Elementary Statistical Tables*, Cambridge University Press, Cambridge.

Lindley, D. V. and Smith, A. F. M. (1972) Bayes' estimates for the linear model (with discussions). *Journal of the Royal Statistical Society B*, **34**, 1–41.

Lipschutz, S. (1968) *Linear Algebra*. Schaum's Outline Series. McGraw-Hill, New York.

Liptrot, D. (1991) A comparison of two oxy-fuel gas cutting processes at Redpath Engineering, Servies. M.Sc. dissertation, Department of Civil Engineering, University of Newcastle upon Tyne.

Lockyer, K. and Oakland, J. S. (1981) How to sample success. *Management Today*, July.

Metcalfe, A. V. (1991a) Towards TQM in the water industry: a sampling approach. *Total Quality Management*, **2**(1), 69–73.

Metcalfe, A. V. (1991b) Probabilistic modelling in the water industry. *Journal of the Institution of Water and Environment Management*, **5**(4), 439–49.

Metcalfe, A. V. (1992) The role of engineering control in quality management. *Total Quality Management* **3**(3), 331–34.

Montgomery, D. C. (1991) *Design and Analysis of Experiments*, Wiley, New York.

Moore, P. G. (1972) *Risk in Business Decisions*, Longman, London.

Moran, P. A. P. (1959) *The Theory of Storage*, Methuen, London.

Mousa, A. H. N. (1976) Optimisation of rope-range bleaching of cellulostic fabrics. *Textile Research Journal*, **46**, 493–6.

Nicholson, A. J. (1985) The variability of accident counts. *Accident Analysis and Prevention*, **17**(1), 47–56.

Norman, P. and Naveed, S. (1990) A comparison of expert system and human operator performance for cement kiln operation. *Journal of the Operational Research Society*, **41**(11), 1007–19.

O'Hagan, A., Glennie, E. B. and Beardsall, R. E. (1992) Subjective modelling and Bayes' linear estimation in the UK water industry. *Journal of Royal Statistical Society C*, **41**, 563–77.

Plackett, R. L. (1972) The discovery of the method of least squares. *Biometrika*, **59**(2), 239–51.

Press, W. H., Flannery, B. P., Reukolsky, S. A. and Vetterling, W. T. (1992) *Numerical Recipes*. Cambridge University Press, Cambridge.

Road Accidents in Great Britain (1970–1979) Department of the Environment Scottish Development Department and Welsh Office, London: Her Majesty's Stationary Office.

Rodriguez-Iturbe, I., Cox, D. R. and Isham, V. (1987) Some models for rainfall based on stochastic point processes. *Proc. R. Soc. London, A*, **410**, 269–88.

Rutherford, E. and Geiger, H. (1910) The probability variations in the distribution of α particles. With a note by H. Bateman. *Philosophical Magazine* 6, **20**, 698–707.

Salameh, A. and Metcalfe, A. V. (1992) Experimental design: luxury or necessity? *Total Quality Management*, **3**(3), 301–6.

Scott, P. P. (1986) Modelling time-series of British road accident data. *Accident Analysis and Prevention*, **18**(2), 109–17.

Shaw, R. (1982) W ave Energy: a design challenge, Ellis Horwood, Chichester.

Shewhart, W. A. (1931) *Economic Control of Quality of Manufactured Product*, Reinhold, Princeton, NJ.

Slove, M. L. (1964) Advantages of CE-HDP bleaching for high brightness kraft pulp production. *Tappi*, **47**(2), 170A–173A.

Tanner, J. C. (1958) A problem in the combination of accident frequencies. *Biometrika*, **45**, 331–42.

Tawn, J. A. (1988) An extreme-value theory model for dependent observations. *Journal of Hydrology*, **101**, 227–50.

Teasdale, M. A., Bell, M. G. H. and Silcock, D. T. (1991) The definition of 'Controls' for assessment of area-wide road safety schemes. PTRC 19th Summer Annual Meeting, University of Sussex, England.

Todorovic, P. (1978) Stochastic models for floods. *Water Resources Research*, **14**(2), 345–56.

TRRL (1975) Accidents at offside priority roundabouts. Transport and Road Research Laboratory, Department of the Environment, Crowthorne.

Tsang, W.-W. (1991) Estimation of short term flood risk. M.Sc. dissertation, Department of Civil Engineering, University of Newcastle upon Tyne.

Vogel, R. M., McMahon, T. A. and Chiew, F. H. S. (1993) Flood frequency model selection in Australia. *Water Resources Research*, **146**, 421–47.

Water Research Centre (1986) *Sewerage Rehabilitation Manual*. WRC Engineering, Frankland Road, Swindon, UK.

Wetherill, G. B. (1982) *Elementary Statistical Methods*, 3rd edn, Chapman and Hall, London.

Wetherill, G. B. and Brown, D. W. (1991) *Statistical Process Control*, Chapman and Hall, London.

Wright, C. C., Abbess, C. R. and Jarrett, D. F. (1988) Estimating the regression-to-mean effect associated with road accident blackspot treatment: towards a more realistic approach. *Accident Analysis and Prevention*, **20**(3), 199–214.

Software references

DEX

Tony Greenfield
Middle Cottage
Little Hucklow
Derbyshire SK17 8RT
UK
Telephone: 0298 872326
Fax: 0298 872559

LOTUS

Lotus Development Corporation
55 Cambridge Parkway
Cambridge
MA 02142
USA
Telephone: 617-577-8500
Fax: 617-693-0968

MINITAB

Minitab, Inc.
3081 Enterprise Drive
State College
PA 16801-3008
USA
Telephone: 814-238-3280
Fax: 814-238-4383

Index